W9-CRB-420

Understanding Humic Substances
Advanced Methods, Properties and Applications

Understanding Humic Substances
Advanced Methods, Properties and Applications

Edited by

Elham A. Ghabbour
Soil, Water and Environmental Research Institute, Alexandria, Egypt and Northeastern University, Boston, USA

Geoffrey Davies
Northeastern University, Boston, USA

ROYAL SOCIETY OF CHEMISTRY

The proceedings of the third Humic Substances Seminar held on 22–23 March 1999 at Northeastern University, Boston, Massachusetts

The front cover illustration is taken from the contribution by T.M. Miano and J.J. Alberts, p. 157

Special Publication No. 247

ISBN 0-85404-799-9

A catalogue record for this book is available from the British Library

© The Royal Society of Chemistry 1999

Published by The Royal Society of Chemistry,
Thomas Graham House, Science Park, Milton Road,
Cambridge CB4 0WF, UK

For further information see our web site at www.rsc.org

Printed and bound by MPG Books Ltd, Bodmin, Cornwall.

"...at moments of very intense creativity, chemistry, art and religion join."

Morris Schnitzer

Foreword

1 OVERVIEW

I recently had the opportunity to write an account of progress made during the second half of this century on the chemistry of humic substances. As I was writing the article I noticed that advances in this field were directly related to how well we could adapt newly developed methods and instruments. For example, in the late 1940s wet chemistry done with beakers, flasks, test tubes, burettes, pipettes and balances was predominant. The major instruments available at that time were pH meters powered by batteries and colorimeters requiring filters for changing wavelengths.

In the early 1950s, recording UV and visible spectrophotometers became available and, in the mid 1950s, IR spectrophotometers. The arrival of the latter held out high hopes that with the aid of IR spectrophotometry it would be possible to obtain significant new information on the chemical structure of humic substances. Unfortunately, these hopes were not fulfilled, as researchers began to realize that the IR beam was unable to penetrate the matrices of the humic acid, fulvic acid and humin molecules, and only registered the presence of OH and CO_2H groups on the outside surfaces of the samples.

The early 1960s saw the arrival of gas chromatographs, which allowed researchers to separate complex mixtures of humic acid and fulvic acid oxidation products, and also of organic soil extracts containing alkanes, alkenes, fatty acids and esters. In the mid-1960s mass spectrometers, which could be attached to gas chromatographs, came on the market. This made it possible to not only separate complex mixtures of organics but also to identify the separated compounds. At about the same time, electron-spin resonance (ESR) spectrometers became available, which enabled researchers to measure concentrations of free radicals in humic materials and to obtain information on the nature of the free radicals. ESR also made it possible to study the symmetry and coordination of complexes formed by humic substances with paramagnetic metals such as Fe^{3+}, Cu^{2+} and Mn^{2+}.

In the early 1980s first liquid-state and then solid-state ^{13}C NMR became available. This was an important addition to our arsenal of equipment. It showed, for the first time, that aliphatic C in humic substances was as important as aromatic C, and that the older theories that humic substances were almost completely aromatic were no longer valid. The mid-1980s saw the arrival of pyrolysis-soft ionization mass spectrometry, which could be applied to the *in situ* analysis of humic substances in soils.

Other types of mass spectrometry, such as Curie-point pyrolysis-gas chromatography–mass spectrometry, provided important structural information on humic substances. From the chemical, spectroscopic, and mass spectrometric data, two- and three-dimensional model structures were proposed for humic acid, soil organic matter, and a whole soil with the aid of computational chemistry. During the past 50 years, I was fortunate to have participated, along with other scientists, in the developments described in this paragraph. I was a witness to the evolution of the chemistry of humic substances from wet chemistry to computational chemistry.

2 NOMENCLATURE

The current nomenclature of humic substances (humic acid, fulvic acid and humin) originated over 200 years ago.[1] Early workers[2] believed that the three humic fractions were different substances, well separated by the 'classical' extraction procedure. The latter consists of extraction of the soil with dilute base, acidification of the alkaline extract to produce a coagulate that is humic acid, while the material remaining in solution is fulvic acid. Thus, humic acid is soluble in base but not in acid, while fulvic acid is soluble in both base and acid. The humic material that cannot be extracted from the soil by base is referred to as humin.

We know now from chemical, ^{13}C NMR and mass spectrometric analyses that the main structural features of the three humic fractions are similar so that the assumptions of earlier workers are incorrect. Thus, the terms humic acid, fulvic acid and humin do not stand for distinct chemical substances. Each of these fractions consists of hundreds of compounds, which appear to be associated at the molecular levels by mechanisms not yet well understood.

While it is true that the terms humic acid, fulvic acid and humin have no chemical meanings, these designations have been used over the past 200 years in thousands of scientific papers. What are we to do now? Change these terms, and start to express ourselves in the language of chemistry, but lose the connection with the voluminous literature and with current researchers?

For an answer to this very complex problem we may want to turn to philosophy. The medieval Jewish philosopher Maimonides, who lived in Spain, Morocco, and Egypt, was concerned with the relationship between man and God. He proposed that the first thing to do in this discussion was to define God. "How can we define God?" he asked. His answer was that we can define God by His characteristics. For example, we can say that God is merciful. He heals the sick, He feeds the hungry, He lifts up those who have fallen, *etc*. It seems to me that we can apply the same approach to the definition of humic substances. Each humic fraction can be defined by its ^{13}C NMR spectrum, its IR spectrum, its mass spectrum, its elemental analysis, and so on. In this manner we can retain the designations humic acid, fulvic acid and humin without having to substitute them by other terms, and so avoid turmoil and confusion in our field of research.

3 CHEMISTRY AS ART

In 1961 I was offered the opportunity by Agriculture Canada to do post-doctoral studies at an English University. I decided to spend a year in the Organic Chemistry Department of the Imperial College of Science and Technology in London, England. After some correspondence, I was invited by Sir Derek Barton, Nobel Laureate, who was at that time Professor of Organic Chemistry at Imperial College. While most of the students and researchers at the College worked on natural products (steroids, alkaloids, and so on), I arrived there equipped with a glass bottle containing 200 g of purified fulvic acid, which I had extracted in Ottawa from a Spodosol Bh horizon. I had decided to work on elucidating the chemical structure of fulvic acid because I thought at that time that fulvic acid had a simpler molecular structure than humic acid. The work that Sir Derek and I did is described in the literature[3] and I will not discuss it here. During my stay at Imperial College, Sir Derek and I had many discussions not only on my immediate objectives but also on more general topics such as science and philosophy. One morning, Sir Derek approached me saying: "Morris, chemistry is not a science, chemistry is an art." I was somewhat surprised by what he said. In the 1960s, science was the greatest endeavor one could pursue, it was pure and virginal, how could one compare it to art? Nonetheless, I never forgot what he said to me. In the 1970s and 1980s, my co-workers and I made some, at least what I thought important discoveries on the chemical structure of humic substances. Each time we made such discoveries I got very excited, euphoric at having caught a glimpse of something that had never been seen before. At such moments I felt that chemistry and art had joined, so that chemistry had become art as Sir Derek had told me many years earlier. I also had another feeling at such moments: I felt that I was closer to God. So, at moments of very intense creativity, chemistry, art and religion join. Most chemists refrain from writing or speaking about the spiritual aspects of chemistry. But as a chemist who has worked for 50 years on materials that originate from the outer crust of the earth and as a retired person, I feel that these experiences deserve to be discussed.

Advances in the chemistry of humic substances depend on developments in analytical, physical and organic chemistry and how effectively we can apply the newly developed methods and instruments to our own research. It is very likely that these trends will continue in the future. We can also look forward to increasing applications of computational chemistry to elucidate the arrangements of molecular constituents in humic substances and the nature of interactions of humic substances with metals, minerals, pesticides and herbicides. It is important that the computational chemistry be based on data generated by 'classical' analytical, organic and physical chemistry.

Ottawa   Morris Schnitzer
March, 1999

References

1.   F.K. Achard, *Crell's Chem. Ann.*, 1786, **2**, 391.
2.   C. Sprengel, *Kastner's Archiv. Ges. Naturlehre*, 1826, **8**, 145.
3.   D.H.R. Barton and M. Schnitzer, *Nature*, 1963, **190**, 217.

Preface

This book complements the volume *Humic Substances: Structures, Properties and Uses* published by the Royal Society of Chemistry in 1998. Both books describe front-line research on humic substances (HSs), the remarkable brown biomaterials in animals, coals, plants, sediments, soils and waters. HSs' functions include water retention, pH buffering, photochemistry, redox catalysis, solute sorption and metal binding. No other natural materials have so many functions in so many different places.

HSs are highly functionalized, carbon-rich molecules with a strong tendency to aggregate. This helps HSs to hide their identities as macromolecules or self-assembling systems. HSs are chameleons that can behave like lipids, polysaccharides or proteins depending on the circumstances. We need to know HSs' molecular structures in order to understand their properties.

The importance of HSs in the environment and human health is encouraging the best minds with the best tools in a concerted effort to solve one of nature's greatest mysteries. This book benefits from the wisdom of Drs. Morris Schnitzer and Cornelius Steelink, two of this century's greatest contributors to humic substances research. They remind us of the history and importance of research into HSs and ask us not to forget that many questions about HSs have been answered. Drs. Schnitzer and Steelink make it plain that advances in HS research have always followed the proper application of some analytical technique. Some of the most fundamental questions about HSs are being addressed with the sophisticated tools and methods described in this book. New experimental work and well-known data are stimulating sophisticated molecular modeling that keeps our sights set on molecular structures.

This book is derived from Humic Substances Seminar III, which was held at Northeastern University, Boston, Massachusetts, USA on March 22–23, 1999. We were honored by the presence of Drs. James Alberts (President), Michael Hayes (Immediate Past President), Nicola Senesi (Past President) and C. Edward Clapp (Treasurer) of the International Humic Substances Society, together with other eminent authors from 15 countries. It was a memorable meeting, as you will discover from reading the excellent papers that follow.

ACKNOWLEDGEMENTS

We thank the authors and reviewers for their contributions and unwavering co-operation. Financial support from Arctech, Inc. and our other sponsors is gratefully acknowledged. We thank Northeastern University for providing facilities for the Humic Substances Seminar series. Michael Feeney ably managed the Seminar III presentations. Beth Rushton and the staff of the Barnett Institute handled many of the organizational details with efficiency and charm. The talented undergraduates of the Humic Acid Group were excellent hosts and we thank them for honest effort gladly given. Last but not least, we thank Janet Freshwater and her staff at the Royal Society of Chemistry for ensuring timely publication of the authors' fine work.

Boston, Massachusetts   Elham A. Ghabbour
August, 1999   Geoffrey Davies
                                                        Editors

Contents

What is Humic Acid? A Perspective of the Past Forty Years   1
   Cornelius Steelink

Abiotic Catalytic Humification of Organic Matter in Olive Oil Mill Wastewaters   9
   N. Senesi, G. Brunetti, E. Loffredo and T.M. Miano

Use of ^{13}C NMR and FTIR for Elucidation of Degradation Pathways during
Senescence and Litter Decomposition of Aspen Leaves   19
   Robert L. Wershaw, Jerry A. Leenheer and Kay R. Kennedy

A Biogeopolymeric View of Humic Substances with Application to Paramagnetic
Metal Effects on ^{13}C NMR   31
   Robert L. Cook and Cooper H. Langford

Evaluation of Different Solid-state ^{13}C NMR Techniques for Characterizing
Humic Acids   49
   Baoshan Xing, Jingdong Mao, Wei-Guo Hu, Klaus Schmidt-Rohr,
   Geoffrey Davies and Elham A. Ghabbour

NMR Evidence for Crystalline Domains in Humic Substances   63
   Wei-Guo Hu, Jingdong Mao, Klaus Schmidt-Rohr and Baoshan Xing

Changes of Colloidal State in Aqueous Systems of Humic Acids   69
   E. Tombácz and J.A. Rice

Humic Acid Pseudomicelles in Dilute Aqueous Solution: Fluorescence and
Surface Tension Measurements   79
   R. von Wandruska, R.R. Engebretson and L.M. Yates III

Atomic Force Microscopy of pH, Ionic Strength and Cadmium Effects on Surface
Features of Humic Acid   87
   C. Liu and P.M. Huang

Molecular Size Distribution of Humic Substances: A Comparison between Size
Exclusion Chromatography and Capillary Electrophoresis   101
   Maria De Nobili, Gilberto Bragato and Antonella Mori

Characterization of Humic Acids by Capillary Zone Electrophoresis and Matrix
Assisted Laser Desorption/Ionization Time-of-flight Mass Spectrometry   107
  L. Pokorná, D. Gajdošová and J. Havel

MALDI-TOF-MS Analysis of Humic Substances – A New Approach to Obtain
Additional Structural Information?   121
  G. Haberhauer, W. Bednar, M.H. Gerzabek and E. Rosenberg

Laser Spectroscopy of Humic Substances   129
  C. Illenseer, H.-G. Löhmannsröben, Th. Skrivanek and U. Zimmermann

Comprehensive Study of UV Absorption and Fluorescence Spectra of Suwannee
River NOM Fractions   147
  Gregory V. Korshin, Jean-Philippe Croué, Chi-Wang Li and
  Mark M. Benjamin

Fluorescence Behaviour of Molecular Size Fractions of Suwannee River Water.
The Effect of Photo-oxidation   157
  T.M. Miano and J.J. Alberts

Changes in Chemical Composition, FTIR and Fluorescence Spectral
Characteristics of Humic Acids in Peat Profiles   169
  M. Takács and J.J. Alberts

Chromatographic Separation of Fluorescent Substances from Humic Acids   179
  M. Aoyama

A Study of Non-uniformity of Metal-binding Sites in Humic Substances by X-Ray
Absorption Spectroscopy   191
  Anatoly I. Frenkel and Gregory V. Korshin

Application of High Performance Size Exclusion Chromatography (HPSEC) with
Detection by Inductively Coupled Plasma-Mass Spectrometry (ICP-MS) for the
Study of Metal Complexation Properties of Soil Derived Humic Acid Molecular
Fractions   203
  Sandeep A. Bhandari, Dula Amarasiriwardena and Baoshan Xing

Interaction of Organic Chemicals (PAH, PCB, Triazines, Nitroaromatics and
Organotin Compounds) with Dissolved Humic Organic Matter   223
  Juergen Poerschmann, Frank-Dieter Kopinke, Joerg Plugge and
  Anett Georgi

Humasorb-CS™: A Humic Acid-based Adsorbent to Remove Organic and
Inorganic Contaminants   241
  H.G. Sanjay, A.K. Fataftah, D.S. Walia and K.C. Srivastava

Stimulation of Plant Growth by Humic Substances: Effects of Iron Availability 255
  Y. Chen, C.E. Clapp, H. Magen and V.W. Cline

Using Activated Humic Acids in the Detoxification of Soils Contaminated with
Polychlorinated Biphenyls. Tests Conducted in the City of Serpukhov, Russia   265
    Alexander Shulgin, Alexander Shapovalov, Yuriy Putsykin, Cicilia
    Bobovnikov, Galena Pleskachevski and Andrew J. Eckles, III

Subject Index   271

WHAT IS HUMIC ACID? A PERSPECTIVE OF THE PAST FORTY YEARS

Cornelius Steelink

Department of Chemistry, University of Arizona, Tucson, AZ 85721, USA

1 THE PERENNIAL QUESTION

"What is humic acid?" A college newspaper reporter asked me that question shortly after Humic Substances Seminar III in March 1999. Her question reminded me of the identical question posed by Jon Jakob Berzelius 190 years ago in one of his publications.

Do we have an answer to that question today? For science newspaper reporters there is a simple answer: 'Humic Acid is a complex polyelectrolyte of variable composition found in nearly every scoop of dirt'.[1] But for working scientists in the field there are many answers to that question. I have thought of all the conceptual and operational descriptions of humic substances in the past 40 years of my involvement in this field. I decided to put together an historical perspective of changes I have seen in humic substances (HSs) science over the past forty years. This will not be a formal history of the subject. It will be one man's thoughts on his favorite field of research.

Forty years ago I attended the First International Symposium on Humic Acid at University College in Dublin, Ireland.[2] It was a meeting dominated by soil scientists, agronomists and coal scientists. The majority of the talks focused on structure, biosynthesis, isolation and reactions. Everyone at the Symposium was awaiting the forthcoming book by the leading humic acid authority in the world, Maria Kononova, of the Dokuchaev Institute of Soil Science in Moscow. She did not disappoint us.[3]

Forty years later, in 1999, I listened to the papers at Seminar III at Northeastern University. Once again I was struck by the fact that certain questions in the HS field have never changed. The newcomers in the field today are asking the same fundamental questions as the newcomers did 40 years ago. Nevertheless, these questions continue to provide incentives for research in the field.

1.1 Molecular Structure

A number of investigators are still asking the same question today as Berzelius did 190 years ago: what is the chemical (or molecular) structure of soil humic acid? Berzelius' student Mulder published an empirical formula of

$$C_{40}H_{30}O_{15}$$

for soil HA in 1840.[4] Since that time, countless investigators have viewed HA as a monomolecular species with a specific structure. These investigators have proposed numerous empirical and structural formulas for HA in attempts to fit a bewildering array of experimental data to a universal model. I, too, have been guilty of publishing HA model structures.

On the other hand, a number of investigators in recent years have concluded that the search for a universal structural formula for HA is meaningless. In their view, humic substances must be viewed as macromolecular species with fractal, micellar or other supramolecular properties. They say HSs are complex systems whose properties are not fully explained by understanding its component parts.

Still, the desire for a classic structural formula remains strong among many of us. It is a continuing search by many HA scientists. Therefore, I was astounded to see Mulder's empirical formula reappear in a journal this year. It was proposed as a general formula for soil humic acids at Seminar II.[5] After normalizing the empirical formula to reflect a multiple of the C-9 lignin building block and adding nitrogen, Jansen et al. put forth the structural formulas shown in Figure 1.

Experimental evidence to support this model was derived from all sorts of advanced spectroscopic and analytical measurements that lend credence to the formula. Only three weeks after Seminar III I was further astounded to see a study of binding sites of humic acid. Those sites were identical to the structural features of this TNB formula and were being examined as scavengers for uranium by a University of California, Berkeley scientist Heino Nitsche.[1] Both newcomers to this field as well as old hands are still intrigued with the goal of finding a discrete chemical structure for these elusive substances.

For many investigators, the answer to the question: WHAT IS HUMIC ACID? still remains A DISCRETE MOLECULAR STRUCTURE, at least for soil humic acid.

1.2 Biosynthesis of Humic Acid

Humic acid has also been defined as a product of a specific biosynthetic pathway. One of the 1959 Dublin Symposium papers described the origins of humic acids from leachates of plant leaves. Analytical methodology was crude in those days: paper chromatography, ultraviolet absorption spectroscopy and titration were the common tools to identify natural products. Yet remarkable deductions were made from the analytical measurements of these leachates and their interaction with the soil. Forty years later I listened to a presentation on the spectra of humic substances in leaf leachates. Once again I was struck by the fact that certain questions in the HSs field have never changed. Where do humic substances come from? How are they biosynthesized? Today there are sophisticated analytical tools to illuminate these processes. We can monitor processes over periods of time with a variety of non-intrusive spectroscopic probes. Compelling evidence can now be presented for the origin of soil humic acid from leaf leachates.

Another proposal for the origin of soil humic acid is the Maillard reaction. Over the past 80 years this proposal has been buried and resurrected many times. When a reducing sugar reacts with amino compounds (amino acids, peptides, proteins), brown polymeric substances result.[6] These Maillard products appear to have the properties of humic acid. The biosynthetic pools of amino acids and sugars certainly are available in plant cells and soil. Once again, this proposal is being examined, because now we have spectroscopic

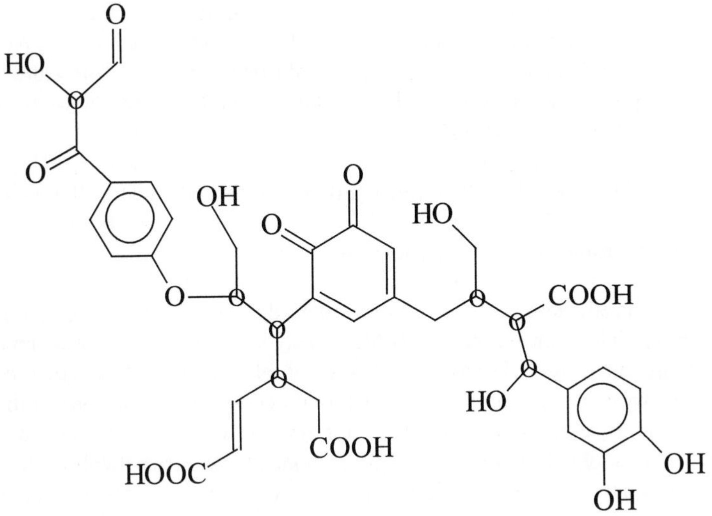

(a)

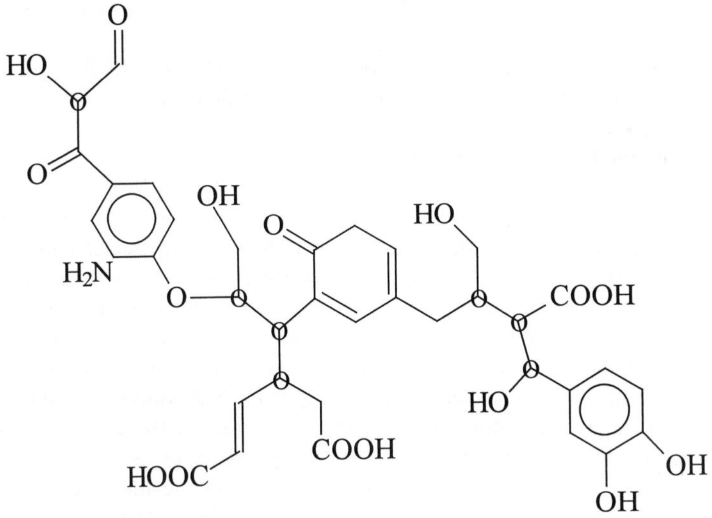

(b)

Figure 1 *Molecular structure of (a) Steelink and (b) TNB humic acid monomers showing chiral centers as open circles.*[5,10]

probes (^{15}N and ^{13}C NMR)[7] that can monitor the formation of these Maillard products.

Forty years ago,[8] a hypothesis was put forth that humic acid actually was an *in situ* plant cell product. It claimed that humic acid formation began in the leaf as a result of cell senescence. In the intervening years, not much could be found in the literature about this theory. This proposal has been revived, examined, tested and found to be alive and healthy.[9]

For some scientists, therefore, the answer to the question WHAT IS HUMIC ACID? would be: AN END PRODUCT OF A SPECIFIC BIOSYNTHETIC SEQUENCE.

1.3 Isolation, Fractionation, and Purification

For many years, humic acid has been operationally defined. That is, it is defined by the method of its removal from its original source. Forty years ago the most common source was soil. Humic acid was defined as a base-soluble, acid-insoluble fraction of soil leachate. There were many modifications of this procedure including pre-treatments and post-treatments, but the base extraction definition was the same in 1959 as it was in 1859.

Since 1959 sophisticated separation methods have dramatically changed this definition. The most fundamental change has been achieved with resin chromatography, including XAD, GPC and HPSEC. Nowadays, I am impressed with the different ways in which humic substances were characterized by their chromatographic treatment. It reminded me of my previous work with lignin chemistry in which a specific lignin substrate was identified by its extraction procedure. Thus, lignins were classified as Native lignin, Bjorkman lignin, Klason lignin, etc. depending on the solvent methodology used to remove them from wood.

Today, for some the answer to the question WHAT IS HUMIC ACID? would be THE FINAL FRACTION OF A SPECIFIC EXTRACTION PROCEDURE.

1.4 Supramolecular Concepts and Structures

In the past decade there has been a growing interest in non-traditional concepts of molecular architecture for humic substances. Instead of single monomeric species, humic substances are being described as micelles, colloids, aggregates, vesicles, fractals, clathrates and surfactants, just to name a few. Molecular surface areas are being measured and their morphological features described. Advanced computer modelling is predicting shapes, folding patterns and chelation sites, much as was done with proteins, polysaccharides, nucleic acids and lignins.[10] Humic Substances Seminar III gave maximum exposure to these developing concepts. Certainly, these new ideas will be thought provoking and may be considered heresy by some of us. But it will provide challenge and dialogue, as all new ideas should in a robust and healthy scientific gathering.

For some scientists, the answer to the question WHAT IS HUMIC ACID? would be A SUPRAMOLECULAR SPECIES DERIVED FROM TERRESTRIAL PLANTS.

1.5 Structure and Instrumentation

Most scientific research follows new advances in instrumentation. Since the 1959 Dublin Symposium on Humic Acid there have been dramatic changes in analytical and spectroscopic technology. With each change has come a new insight into the nature of HSs.

In 1959, destructive chemical degradation was the major method of probing the structure of humic acid. Oxidation, reduction and pyrolysis reactions were used to degrade the humic material. From the fragments in the reaction mixture we attempted to piece together the shape of the original molecule. Since that time whole ranges of non-destructive methods have been developed in the field of chemistry. Humic acid research has benefitted from each of these new methods. Here are a few examples of non-destructive methods that we have become familiar with.

Spectroscopy	*Other*
ESR	HPLC
^{13}C NMR solution	GPLC
^{13}C NMR solid state	Atomic Force Microscopy
^{1}H-NMR	XAFS
FT-IR	
X-ray Spectroscopy	
Mass Spectroscopy and all its variations	

It would take an entire monograph to describe each of these technologies as it applies to humic substances. So I will briefly describe my personal experience with electron spin resonance (ESR) to illustrate what unanticipated research horizons are opened by one of these techniques. In 1962 we published our first ESR studies of soil humic acid and proposed that the stable free radical was a quinone-hydroquinone moiety and/or a phenoxy radical. Since that time other investigators in Russia, England, Canada, Italy, U.S. etc. have expanded the scope of that work. For example, by examining the ESR spectra of metal-humate complexes, scientists have been able to deduce structures of active sites in HA molecules. Spin trapping ESR techniques are the latest developments employed to probe reactive sites.[5]

My ESR odyssey led to an entirely unanticipated field. Our research focus turned to lignin, a biosynthetic precursor to soil humic acid. Like many natural product chemists, we used retrobiosynthetic analysis to give us structural clues. The structure of lignin was well enough known to allow us to use model compounds as substrates for investigation, mainly phenylpropane derivatives. We were able to generate stable free radicals from these model compounds, even under physiological conditions (pH 6; aqueous solution; enzymatic oxidation). We could experimentally confirm the structure of the hypothetical free radical precursors, which had been proposed by many authors over the previous 30 years. In one of those rare exhilarating moments in a chemist's career we were able to synthesize a solid, stable, phenoxy radical from a model lignin dimer.[11] It was stable over a period of weeks. It could easily explain the nature of the deeply colored stable radical moiety in Kraft lignin. In addition, it was a perfect match for a section of the TNB building block of Davies et al.[5] and Jansen et al.[10] By traveling a retro-bioisynthetic pathway away from humic acid to lignin we were able to return to humic acid with important new ideas. Other investigators have done the same thing[10] and then tested their model structures with sophisticated instrumental analysis.

If we follow the new developments in chemical structure determination we can predict in what direction HS research will go. In the next year or so there should be some direct spectroscopic evidence for the nature of hydrogen bonds in humic biopolymers, thanks to very recent developments in NMR programs such as TROSY and J-coupling.

1.6 Structure and Symposia

Every gathering of scientists at a humic substances symposium has three features:
    (a) Progress reports on research projects. These are incremental additions to our knowledge of a specific area of humate research.
    (b) Presentation of new ideas or non-traditional concepts.
    (c) Synergistic interaction among symposium participants, with the generation of new, unanticipated ideas that dramatically alter the direction of research.

The majority of papers at any scientific gathering fall into category (a): that is, they are an extension of current investigations. New ideas or nontraditional ideas are rare. But when they do occur genuine excitement and debate occur. New proposals for the structure of humic acid have generated heated debates in International Humic Substances Society (IHSS) Symposia.

Category (c) is a rare occurrence indeed. I remember one in the past forty years. It was the Gordon Research Conference on ENVIRONMENTAL SCIENCE: WATER held on June 28-July 2, 1976 in Andover, New Hampshire. The conference theme was 'Organic Materials in Water.' Russell Christman of the University of North Carolina hosted it. The meeting included a very diverse group of scientists. There were representatives from water treatment engineering, environmental science and humic substance fields.

One of the most intriguing topics at the Gordon Conference was the origin of chloroform in treated municipal waters. John Rook of the Rotterdamwaterwerk in the Netherlands proposed that the reaction of humates with chlorine produced chloroform. Over coffee, Rook and I identified the probable structural moiety in humic acid that could generate chloroform. Barbara Baum of Harvard joined us to help with the mechanism of the haloform reaction, which was the basis of our joint proposal.

The meeting signalled a major shift in HS research towards aquatic humates and their environmental roles.[12] For many investigators, Federal environmental grant money became available for HS research. New academic research groups became involved in HS studies. Hydrologists, chemical engineers and other 'non-humic' scientists became involved in humic-related research. Many of the newcomers at the Gordon Conference became leaders of IHSS in the next decade.

A few years later (September 11, 1981) a group of HS scientists held another pivotal meeting to organize the IHSS. One important goal of the new Society was to establish a set of HS reference standards for soils and waters. Once defined, isolated and purified by agreed-upon procedures, these samples would be made available to the HS community. I consider this a major step in the field of HS studies. It has finally eliminated uncertainty and doubt that many experimentalists felt when they were comparing data on humic acid samples with literature values for other humic acid samples. Thanks to this IHSS decision and its implementation, HS researchers can now exchange data with one another, based on commonly accepted reference standards.

1.7 Structure and Commercial Applications

More and more attention is being devoted to the commercial applications of humic acids at seminars and symposia. The traditional markets for humic substances (boiler scale inhibitors, plant growth promoters, soil conditioners, drilling mud additives, etc.) have always been displayed at scientific meetings by industry representatives. Recently, however, humic acid products are being marketed as detoxifiers of soil and water.

At HS Seminar III, a number of papers were presented that demonstrated how humic acids could immobilize toxic heavy metals, dechlorinate toxic organohalides and selectively bind radioisotopes. For example, the United States Department of Defense is interested in removing TNT and its toxic degradation products from the soil around hundreds of ammunition dumps around this nation. Research has shown that humic acid irreversibly binds to these products and converts them to amine fertilizers.

This capacity to bind metals and organic compounds is well known to IHSS members. But the molecular architecture of the binding sites and the mechanism of selective binding remain an open field of investigation. A number of investigators are now involved in synthesizing model compounds or developing isotopically labeled humates in order to identify mechanisms and structure-activity relationships.

So structural studies and commercial applications are intimately linked. Here, the driving force is the demand to clean up the environment. This, in turn, stimulates entrepreneurs to meet this demand. And this, in turn, requires researchers to examine the mechanism of detoxification and maximize this process and/or the chemical nature of humic acid.

2 FINAL THOUGHTS

As we continue to publish in a broader range of international journals, humic substances become more visible in the science community. For this reason the field of humic substances is attracting more and more investigators from "non-traditional" disciplines. Even archeologists are becoming involved. In recent weeks I have read articles about archeologists studying ancient metal-humate complexes in the peat soils of Spain and Scotland.[13] Anthropologists at the University of Arizona are trying to date very ancient fulvic acid (Is there a Jurassic fulvic acid?). Interdisciplinary scholarship has always been a great benefit and stimulus to my own work, and I am sure it has been to many other HS scientists.

3 CONCLUSION

As aficionados of humic substances, we live in an exciting era of novel experimental and theoretical developments. The HS Seminars expose us to these new ideas and stimulate our creative thought processes. What new ideas will emerge next year?

At the end of the Seminar I asked one of the speakers: "Tell me, what is Humic Acid?" He thought for a while and then answered: 'Twenty years ago, I thought I was approaching the answer. Today I am not so sure. I am rethinking the whole concept. See me next year and I'll let you know.'

References

1.   H. Nitsche, *Chemical and Engineering News*, 1999, April 12, **44**.
2.   "Symposium on Humic Acid", *Scientific Proceedings of the Royal Dublin Society, Series A*. 1960, **1**, 1-191.
3.   M. M. Kononova, 'Soil Organic Matter', Pergamon Press, New York, 1961.

4.   G. J. Mulder, *J. prakt. Chem.*, 1840, **21**, 203, 321.
5.   M. D. Paciolla, S. Kolla, L. T. Sein, Jr., J. M. Varnum, D. L. Malfara, G. Davies, E. A. Ghabbour and S. A. Jansen, in 'Humic Substances: Structures, Properties and Uses', G. Davies and E. A. Ghabbour, (eds.), Royal Society of Chemistry, Cambridge, 1998, p. 203.
6.   T. P. Labuza, G. A. Reineccius, V. M. Mounier, J. O'Brien and J. W. Baynes, (eds.), 'Maillard Reactions in Chemistry, Food and Health', Royal Society of Chemistry, Cambridge, 1994.
7.   C. Steelink, in 'Humic Substances in the Global Environment and Implications on Human Health', N. Senesi and T. M. Miano, (eds.), Elsevier, Amsterdam, 1994.
8.   C. Steelink, *J. Chem. Educ.*, 1963, **40**, 379.
9.   A. Radwan, G. Davies, A. Fataftah, E. A. Ghabbour, S. A. Jansen and R. J. Willey, *J. Appl. Phycol.*, 1997, **8**, 553.
10.   L. T. Sein, Jr., J. M.Varnum and S. A. Jansen, *Environ. Sci. Technol.*, 1999, **33**, 546.
11.   C. Steelink and R. E. Hansen, *Tet. Lett.*, 1966, **1**, 105.
12.   C. Steelink, *J. Chem. Educ.* 1977, **54**, 599.
13.   "Digging in the Mire for the Air of Ancient Times", *Science*, 1999, **284**, 900.

ABIOTIC CATALYTIC HUMIFICATION OF ORGANIC MATTER IN OLIVE OIL MILL WASTEWATERS

N. Senesi, G. Brunetti, E. Loffredo and T. M. Miano

Istituto di Chimica Agraria, University of Bari, Bari, Italy

1 INTRODUCTION

The disposal of wastewaters from olive-oil extraction processes is a major and unresolved economic and environmental problem for olive-oil producing countries such as those in the Mediterranean area.

Olive-oil mill wastewaters (OWs) contain a relatively high amount of organic matter (up to 16 %) mostly in the form of sugars, pectins, organic acids, phenols, lipids, amides and other nitrogenated compounds. Since these wastewaters derive from a mechanical extraction process of a natural product, the olive, they are free of any inorganic and organic contaminant. Thus, they can be considered environmentally safe and can be recycled in agriculture as a liquid organic amendment with enormous advantages for most soils of the Mediterranean area, which typically are deficient in organic matter.

Similar to other organic waste materials, the direct application of "fresh" organic matter contained in these wastewaters to soil may often result in more adverse than beneficial effects on global soil status and fertility.[1] Because of this, organic waste materials of various origin and nature are generally properly processed before soil application to give at least partial stabilization and humification of organic matter with production of humic-like materials.[2]

Catalytic treatment with a polyenzymatic mixture has been attempted to induce humification processes of organic matter contained in OW.[3] The results were only partly successful because of intrinsic difficulties of promoting and maintaining efficient enzymatic activity in the adverse biological conditions in these wastewaters.

Soil mineral components are well known to play an important role in the abiotic formation of humic substances (HSs) from decomposition products of plant and animal residues and microbial metabolites, such as phenolic and N-containing compounds. Wang et al.[4] have reviewed results obtained on the catalytic effects of soil clays, primary minerals, and the soil oxides and oxyhydroxides of Fe, Al, Si and Mn in promoting the oxidative polymerization of various phenolic and similar compounds to give HSs. Manganese oxides have been shown to be powerful catalysts in these reactions, which give humic acids (HAs) with a relatively high degree of humification, and compositional, structural and functional properties similar to those of natural soil HAs.[5,6]

The objective of this work was to test the catalytic efficiency of a number of mineral compounds that are similar to those naturally occurring in soil in inducing and controlling the humification of organic matter contained in OW.

2 MATERIALS AND METHODS

2.1 Catalytic Treatment of Wastewater

An average OW sample was collected from an olive oil mill in Southern Italy. Aliquots (2.5 L) of the OW sample were treated separately with 25 g of bentonite, iron(III) oxide (FeOx) and manganese(IV) oxide (MnOx), all of commercial origin. Two parallel experiments were conducted for 3 and 7 weeks at a room temperature ranging from 13 to 21°C, in cylindrical glass containers provided with continuous mechanical stirring and air bubbling from the bottom. During the experiment no correction was made for pH changes and distilled H_2O was added daily to maintain the initial volume of the OW sample.

Before starting the experiment and after 3 and 7 weeks of treatment, the pH and redox potential of OW samples were measured by conventional methods.

2.2 Isolation of Humic Acids

The HAs were isolated from the untreated OW sample and OW after catalytic treatments for 3 and 7 weeks with a conventional procedure[7] adapted to materials in water solution and suspension.

Briefly, each OW sample was treated with HCl until a pH of ~ 1.0 was reached. The suspension was centrifuged to separate the precipitated material, which was then treated with NaOH to reach a pH of ~ 12.0 in order to dissolve the humic material. After centrifugation, the supernatant was treated with HCl to reach a pH ~ 1.0. The precipitated HA was separated by centrifugation, redissolved in 0.1 M NaOH and reprecipitated by addition of HCl to pH ~ 1.0. This procedure was repeated two times. The centrifuged HA was washed twice with distilled H_2O, then finely suspended in distilled H_2O, dialysed with a Spectrapor membrane (size exclusion limit 8 kDa) to remove salts and finally freeze-dried.

2.3 Analyses of Humic Acids

Carbon, H, N and S contents of the isolated HAs were measured with a Fisons Model EA 1108 Elemental Analyzer. Oxygen concentration was calculated by difference. Experimental data were corrected for ash and moisture contents.

Total acidity and carboxylic group concentrations were determined by conventional titration methods.[7] Phenolic hydroxyl concentrations were obtained by difference.

The E_4/E_6 ratios were determined by dissolving 1.0 mg of each HA in 5 mL of $NaHCO_3$ and the pH was adjusted to 8.3 with NaOH. The adsorbances at 465 and 665 nm were measured on a Perkin Elmer Lambda 15 UV/vis spectrophotometer. The ratio of these adsorbances gave the E_4/E_6 ratio.

FTIR spectra of HAs were recorded in KBr pellets in the 4000 to 400 cm^{-1} wavenumber range with a Nicolet 5 PC FTIR spectrophotometer. The KBr pellets were obtained by pressing a mixture of 1 mg HA and 400 mg spectrometry grade KBr under vacuum.

Fluorescence spectra in the emission, excitation and synchronous-scan modes were obtained on aqueous solutions of HAs at a concentration of 100 mg L^{-1} after overnight equilibration at room temperature and adjustment to pH 8 with 0.05 M NaOH. Spectra were recorded with a Perkin Elmer LS-5 luminescence spectrophotometer equipped with a

Perkin Elmer Data Station 3600 for data generation and processing with PECLS software. Emission and excitation slits were set at 5-nm band width and a scan speed of 120 nm min^{-1} was selected for both monochromators. Emission spectra were recorded over the range 380-550 nm at a constant excitation wavelength of 360 nm. Excitation spectra were obtained over a scan range of 300 to 500 nm by measuring the emission radiation at a fixed wavelength of 520 nm. Synchronous-scan excitation spectra were measured by scanning simultaneously for both the excitation (varied from 300 to 550 nm) and emission wavelengths while maintaining a constant, optimised wavelength difference $\Delta\lambda = \lambda_{exc} - \lambda_{em}$ = 18 nm.[8]

The ESR spectra were obtained at 293 ± 2K with a Bruker ER-200D SRC spectrometer operating at X-band frequency with 100 KHz magnetic field modulation. Solid HA samples were packed in quartz ESR tubes (4 mm o.d., 3 mm i.d.). Field ranges scanned were 500 and 10 mT using modulation amplitudes of 5 and 0.63 mT, respectively, a microwave frequency of 9.52 GHz and a microwave attenuation of 13 dB (corresponding to a microwave power of about 10 mW). The absolute free radical concentration expressed in spins g^{-1}, line widths expressed in mT, and spectroscopic splitting factors (g-values) were calculated using standard procedures and equations.[9]

3 RESULTS AND DISCUSSION

3.1 Wastewaters

The initial pH value of the untreated OW was acidic and remained almost constant for 3 weeks, whereas it increased by about 1 unit after 7 weeks of any treatment (Table 1). After 3 and especially 7 weeks of any treatment, the redox potential of OW increased from an initial reductive value of - 316 mV for untreated OW to positive values, which reached 110 mV for the OW treated with MnOx (Table 1). The above trends suggest that the three catalytic treatments, and especially that with MnOx, were effective in oxidizing the organic matter contained in OW.

Table 1 *pH and redox potential (mV) values of the OWs*

Samples	pH	Redox potential
Untreated OW	5.2	-316
OW treated with bentonite		
3 weeks	5.2	37
7 weeks	6.0	48
OW treated with FeOx		
3 weeks	4.9	38
7 weeks	6.2	75
OW treated with MnOx		
3 weeks	5.2	-185
7 weeks	6.1	110

3.2 Humic Acids

The elemental composition and atomic ratios of HAs isolated from variously treated OWs are shown in Table 2, together with the data for two native HAs from Terra Rossa soils typical of the Mediterranean area.[10] Generally, the contents of C and H, and the C/N ratio of HAs tended to decrease, whereas the contents of N and especially of O, and C/H and O/C ratios tended to increase as a result of the treatment of OW with either MnOx or FeOx. The effects were apparent after 7 weeks of treatment with FeOx and after 3 weeks of treatment with MnOx. The elemental composition of the HAs obtained after 7 weeks of treatment with MnOx approached sensibly that of natural soil HAs. This result shows the particular efficiency of MnOx as a catalyst in inducing the humification of organic matter in OW.

Table 2 *Elemental composition (g kg^{-1}) and atomic ratios of HAs isolated from OWs and of two native HAs from Terra Rossa soils of the Mediterranean area*

Origin of HAs	C	H	N	S	O	C/N	C/H	O/C
Untreated OW	675	102	13	10	200	61.7	0.6	0.2
OW treated with bentonite								
3 weeks	677	97	15	4	207	52.8	0.6	0.2
7 weeks	691	97	15	6	191	53.4	0.6	0.2
OW treated with FeOx								
3 weeks	715	102	16	4	163	51.7	0.6	0.2
7 weeks	615	69	26	12	278	27.9	0.7	0.3
OW treated with MnOx								
3 weeks	580	62	20	10	328	34.4	0.8	0.4
7 weeks	607	69	26	13	285	27.0	0.7	0.4
Terra Rossa soil-I	528	47	45	3	337	13.7	0.9	0.5
Terra Rossa soil-II	557	55	50	13	325	13.0	1.8	0.4

The acidic functional group composition, the E_4/E_6 values and the ash content of HAs from OWs are shown in Table 3, in comparison to the corresponding values of two typical Terra Rossa soil HAs.[10] The COOH group and the phenolic OH group contents increased markedly with time with any of the three treatments. After 7 weeks of any treatment the

Table 3 *Functional group composition (mol kg^{-1}), E_4/E_6 ratio and ash content (%, 550°C) of the HAs isolated from OWs and of two native HAs from Terra Rossa soils of the Mediterranean area*

Origin of HAs	COOH	Phenolic OH	Total acidity	E_4/E_6	Ash Content
Untreated OW	1.21	0.51	1.74	2.0	0.8
OW treated with bentonite					
3 weeks	1.65	1.45	3.10	2.3	0.7
7 weeks	2.49	5.20	7.69	6.4	1.2
OW treated with FeOx					
3 weeks	1.60	1.42	3.01	2.1	0.2
7 weeks	2.61	4.38	6.98	4.3	2.3
OW treated with MnOx					
3 weeks	1.89	2.90	4.78	2.5	0.2
7 weeks	3.00	5.27	8.27	6.6	1.8
Terra Rossa soil-I	3.1	3.0	6.1	4.6	1.5
Terra Rossa soil-II	4.1	4.4	8.5	4.9	4.0

acidic groups in HAs reached values more than two times (for COOH) and about one order of magnitude (for phenolic OH) higher than those in the HA from untreated OW. The final content of COOH groups was a little lower and that of phenolic OH groups and total acidity were a little higher than the corresponding values of native soil HAs. After 7 weeks of any treatment, the E_4/E_6 ratio also increased to values a little higher than those of native soil HAs. These effects resulted from any of the three treatments but were more evident for the MnOx treatment, thus confirming the efficiency of this catalyst in humification.

The FTIR spectra of the HAs obtained from untreated OW and from treated OW with MnOx for 3 and 7 weeks, together with that of a typical Terra Rossa soil HA are shown in Figure 1. The FTIR spectrum of the HA from the untreated OW was markedly different from that of soil HA, whereas a substantial modification of the spectrum of the HA occurred, approaching that of soil HA, after 3, and especially 7 weeks of treatment with MnOx. In particular, from comparison of the HA from untreated OW with the HA from 3-week treated OW and the HA from 7-week treated OW, the following main modifications could be observed: (a) the aliphatic character (aliphatic CH peaks at 2924 and 2853 cm^{-1} and at 1460-1440 cm^{-1}) of HA decreased; (b) the aromatic character (aromatic C=C vibrations in the range 1650-1630 cm^{-1}) increased; (c) the relative intensity of the broad OH-stretching band at 3420-3400 cm^{-1} increased; (d) the intensity of the band of carbonyl groups (aldehydic, ketonic, ester) at 1745 cm^{-1} decreased; and (e) the intensity of the C=O stretching band at 1710-1719 cm^{-1} and the C-O stretching and OH deformation band at about 1230 cm^{-1} of COOH groups increased.

In conclusion, the FTIR spectrum of HA from OW treated with MnOx for 7 weeks appeared much more similar to that of a native soil HA than the FTIR spectrum of HA from untreated OW. This result confirms the catalytic effect of MnOx on humification of organic matter contained in OW. The FTIR spectra (not shown) of HAs obtained from treatments with bentonite and FeOx showed a similar but not so evident trend, thus confirming the limited catalytic efficiency of these compounds in the humification of OW.

Fluorescence spectra in the emission, excitation and synchronous-scan modes of the HAs obtained from untreated OW and from OW treated with MnOx for 7 weeks, together with the corresponding fluorescence spectra of a Terra Rossa soil HA are shown in Figure 2. Similar to FTIR spectra, the shapes of fluorescence spectra of HA from untreated OW were very different and those of HA from MnOx-treated OW were more similar to the corresponding spectra of native soil HAs. In particular, as a result of MnOx treatment it could be observed that (a) the emission maximum shifted to a longer wavelength (from 458 to 468 nm), approaching the emission maximum (486 nm) of the soil HA (Figure 2a); (b) the intensity of the excitation peak at intermediate wavelength (392-393 nm) decreased markedly relatively to that at long wavelength (469 nm), approaching the intensity ratio of the corresponding excitation peaks of a soil HA (Figure 2b); and (c) the synchronous-scan peak at 332 nm disappeared, the intensity of the peak at intermediate wavelength (396-406 nm) decreased markedly relative to that at long wavelength (477 and 507 nm), and the shape of the synchronous-scan spectrum approached that of a soil HA. These effects can be ascribed to an increase of condensed aromatic ring systems, of molecular weight and molecular complexity, and of the degree of humification of HA resulting from the MnOx treatment.[8] The fluorescence spectra (not shown) of HAs resulting from both bentonite and FeOx treatments of OW showed similar effects, but to much less extent.

The ESR spectra recorded over a large scan range (500 mT) (not shown) of the HAs studied showed marked but difficult to explain modifications as a function of the type and time of treatment of OW with the various catalysts. The values the ESR parameters

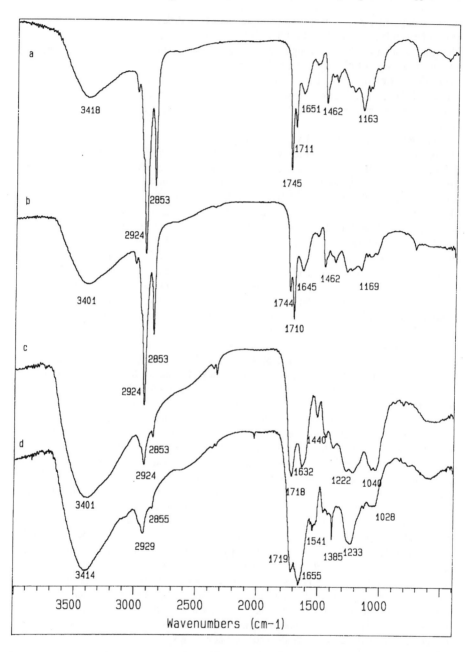

Figure 1 *FTIR spectra of HAs isolated from (a) untreated OW, (b) OW treated with MnOx for 3 weeks and (c) 7 weeks, and (d) a Terra Rossa soil HA*

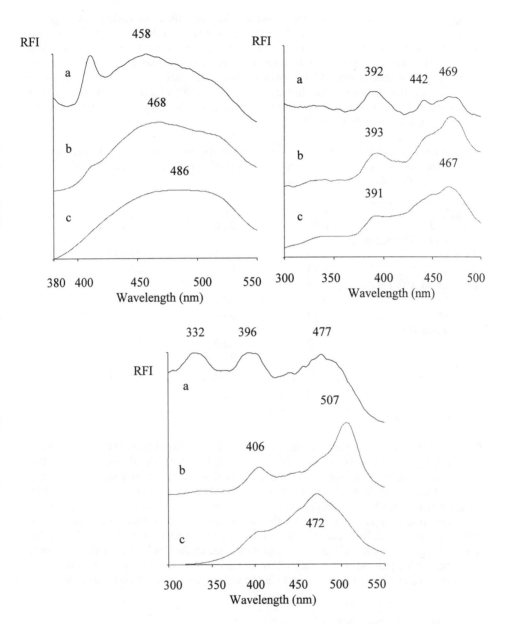

Figure 2 *Fluorescence emission (top, left), excitation (top, right) and synchronous-scan (bottom) spectra of HAs isolated from(a) untreated OW, (b) OW treated with MnOx for 7 weeks and (c) a Terra Rossa soil. RFI = relative fluorescence intensity*

(concentration, linewidth and g-factor) of organic free radicals in the studied HAs together with those of a Terra Rossa soil HA are presented in Table 4. The free radical concentration increased with time for any treatment and approached the value typical of a soil HA. This result suggests the formation of an increasingly extended aromatic network in oxidative polymerization and humification processes that allow the stabilization of newly formed free radicals.[9] The similarity of the linewidth values and g-factors of HAs from untreated and treated OWs to those of a soil HA indicates a similar, semiquinonic nature of organic free radicals in all HAs.[9]

Table 4 *Concentration (C, spin g^{-1} x 10^{17}), linewidth (lw, mT) and g-factor of organic free radicals in the HAs isolated from OWs and of a native HA from a Terra Rossa soil of the Mediterranean area*

Origin of HAs	C	lw	g-factor
Untreated OW	0.59	0.80	2.0044
OW treated with bentonite			
3 weeks	0.57	0.77	2.0034
7 weeks	1.99	0.87	2.0033
OW treated with FeOx			
3 weeks	0.85	0.83	2.0032
7 weeks	2.39	0.88	2.0032
OW treated with MnOx			
3 weeks	0.72	0.80	2.0041
7 weeks	2.46	0.80	2.0039
Terra Rossa soil-I	5.60	0.80	2.0032

4 CONCLUSIONS

Results of chemical and spectroscopic analyses of HAs isolated from OWs treated for 3 and 7 weeks with three different mineral compounds, bentonite, Fe(III) oxide and Mn(IV) oxide, similar to those encountered in soil, show that the first two compounds have a limited efficiency in promoting and controlling humification of organic matter present in these wastewaters, whereas Mn(IV) oxide can act as an efficient catalyst of humification.

These results appear promising for the development of a large-scale catalytic process aiming to convert, at least partially, the organic matter in OWs and other similar waste materials from the agro-food industry into humic materials similar to native soil humic substances. This process would thus allow an environmentally safe and agronomically-efficient recycling of organic waste materials in agriculture as soil organic amendments.

ACKNOWLEDGEMENTS

This research was supported in part by the Consiglio Nazionale delle Ricerche (CNR), Italy, Project "Reflui Oleari" and in part by the Ministero dell'Università e della Ricerca Scientifica e Tecnologica (MURST), Italy.

References

1. N. Senesi, *Sci Total Environ.,* 1989, **81/82**, 521.
2. N. Senesi, T. M. Miano and G. Brunetti, in 'Humic Substances in Terrestrial Ecosystems', A. Piccolo, (ed.), Elsevier, Amsterdam, 1996, p. 531.
3. G. Brunetti, N. Senesi, T. M. Miano and G. Benedetti, *Scienza e Tecnica Agraria,* 1995, **1-3**, 3.
4. T. S. C. Wang and P. M. Huang, in 'Interactions of Soil Minerals with Natural Organics and Microbes', P. M. Huang and M. Schnitzer, (eds.), Soil Science Society of America, Madison, 1986, p. 251.
5. H. Shindo and P. M. Huang, *Soil Sci. Soc. Am. J.*, 1984, **48**, 927.
6. T. S. C. Wang, M. C. Wang, Y. L. Ferng and P. M. Huang, *Soil Sci.*, 1983, **135**, 350.
7. M. Schnitzer, in 'Methods of Soil Analysis, Part 2, Chemical and Microbiological Properties', B. L. Page, R. H. Miller and R. D. Keeney, (eds.), 2nd edn., Agronomy Monograph No. 9, Soil Science Society of America, Madison, 1982, p. 581.
8. N. Senesi, T. M. Miano, M. R. Provenzano and G. Brunetti, *Soil Sci.*, 1991, **152**, 259.
9. N. Senesi, *Adv. Soil Sci.*, 1990, **14**, 77.
10. N. Senesi, G. Brunetti and T. M. Miano, in 'Tecnologie e Impianti per il Trattamento dei Reflui di Frantoi Oleari', P. Amirante, G. C. Di Renzo and C. Bruno, (eds.), Conte Editore, Lecce, 1993, p. 58.

USE OF ^{13}C NMR AND FTIR FOR ELUCIDATION OF DEGRADATION PATHWAYS DURING SENESCENCE AND LITTER DECOMPOSITION OF ASPEN LEAVES

Robert L. Wershaw, Jerry A. Leenheer and Kay R. Kennedy

U.S. Geological Survey, Denver Federal Center, Denver, CO 80225

1 INTRODUCTION

1.1 Leaves as a Source of Humic Substances

Decomposing leaves are an important input for the formation of natural dissolved organic carbon compounds (DOC) and soil organic matter (humus). Fallen leaves and other dead plant material accumulate in litter layers on soil surfaces. As degradation of the plant tissue proceeds, the degradation products are leached from the litter layer by rainwater and transported into the underlying soil and into nearby streams. The chemical reactions that transform the organic components of plant cells into soil humus and DOC are not well understood. In order to provide insight into these reactions, Wershaw et al.[1-3] used ^{13}C NMR spectrometry to elucidate the degradation pathways of the major components of leaf tissue during senescence and aqueous leaching of the leaves. The resulting degradation products can undergo polymerization reactions. Evidence for polymerization and its possible pathways will be the subject of future studies.

1.2 Leaf Senescence

The degradation of plant cells is initiated by senescence. During leaf senescence of deciduous plants in the Fall, polymeric species such as polysaccharides and proteins are hydrolyzed to soluble monosaccharides and amino acids that are transported to storage organs for subsequent use when the plants reawaken in the spring.[4-6] Leshem[4] pointed out that oxygen-containing free radical species such as superoxide, hydroxyl, peroxyl, alkoxyl, and polyunsaturated fatty acid and semiquinone free radicals are active during plant senescence. It is these free radicals that cause the catabolic breakdown of plant tissue components.[7,8] In addition to degradation by active-oxygen species, peptide hydrolases depolymerize proteins to amino acids during senescence.[9] The catabolic breakdown of plant components that takes place during senescence is the first step in humification, which leads to the formation of humus in natural soil systems and compost in man-made systems.

Striking evidence of these reactions is provided by the dramatic color changes of senescing leaves. The green chlorophyll pigments in the chloroplasts of the leaves undergo photo-oxidation and enzymatic hydrolysis; evidence also exists for enzymatically-catalyzed oxidation.[10] Loss of the chlorophylls unmasks the accessory pigments of the chloroplasts such as the yellow to red-orange carotenoids. The carotenoids pigments degrade more slowly than the chlorophylls and therefore the yellow color persists after bleaching of the green color.[11]

In some trees such as Oaks (*Quersus*) the senescent leaves also develop bright reddish colors that have been attributed to anthocyanin pigments.[11] The changing of the anthocyanin colors often seen in the Fall provides us with further evidence of the profound changes that are taking place in the leaves. The anthocyanins may be accessory pigments in the chloroplasts that are more slowly degraded than the chlorophylls, or they may reside in other organelles in the leaves. In either case, the colors of these pigments are dependent on pH and their association with other phenolic compounds.[12] Any alteration of the environment of the anthocyanins brought about by cell-membrane degradation and the mixing of fluids from one type of organelle with that of another can lead to a color change. The destruction of cell membranes is caused by the increased concentration of free radicals during senescence.[13,14]

1.3 NMR Characterization

Wershaw et al.[2] demonstrated that the ^{13}C NMR spectra of leaves of different species are distinctive. The spectra of the nonsenescent leaves are composed of bands that represent the functional groups of the major chemical components of the leaves. The spectrum of each chemical component consists of characteristic groups of bands that occur in well-defined spectral regions. Of particular interest are the bands in the aromatic region of the spectra because major differences were observed in the spectra of the leaves of different species in this region. For a given species one of three ^{13}C NMR aromatic-carbon spectral patterns characteristic of hydrolyzable tannins, nonhydrolyzable tannins or lignins generally was predominant.

Evidence of β-oxidation of lipids, ether demethylation, preferential solution of noncellulosic carbohydrates, and oxidative degradation of phenolic compounds was found by comparison of the ^{13}C NMR spectral patterns of the unfractionated leachates and leachate fractions with those of the senescent leaves from which they were derived.[3] The leachate fractions were isolated from the water extracts with a fractionation procedure developed for the separation of groups of compounds of different polarities. The spectra of the fractions were generally better resolved than the unfractionated leachate.

The oxidative degradation of phenolic compounds observed in the present study was similar to that described by Wershaw,[1] where the oxidative degradation of the lignin components of leaves follows the sequence of O-demethylation, hydroxylation followed by ring-fission, chain-shortening and oxidative removal of substituents. Oxidative ring-fission leads to the formation of carboxylic acid groups on the cleaved ends of the rings and in the process transforms phenolic groups into aliphatic alcoholic groups. The carbohydrate components are broken down into aliphatic hydroxy acids and aliphatic alcohols.[1,3]

1.4 Choice of Aspens

To extend this work to the elucidation of the reactions that take place after the leaves are leached in a natural system, we have initiated a characterization of the DOC derived from Aspen (*Populus tremuloides*) leaves. Aspen leaves were chosen because many Aspen groves are composed of a single Aspen species with few other deciduous plants present.[15]

Therefore, in such groves the leaf leachate in the Fall will be derived largely from Aspen leaves. In this paper we demonstrate that the Aspen leachate can be separated into fractions that have distinctive ^{13}C NMR spectra. The information obtained from these spectra will be used to follow the transformation of Aspen leaf-derived natural organic compounds (NOM) to dissolved organic carbon (DOC) in natural water systems. The results presented here are the first part of a study of the changes that Aspen NOM undergoes as it moves from the litter layer to soil water, groundwater and surface water.

2 MATERIALS AND METHODS

2.1 Leaf Collection

Aspen leaves were collected from two sampling sites in the foothills of the Rocky Mountains, Gilpin County, Colorado and at the Denver Botanic Gardens. On March 11, 1997, leaves were collected from the soil surface and beneath the snow near the side of Robinson Hill Road, Gilpin County, Colorado at about 9000 feet elevation. The time of the sampling was after a significant melting of the snow; some of the collected leaves were resting directly on the frozen soil surface and others were on a layer of frozen snow melt an inch or two under the snow surface. Frozen snow melt was also collected. The stand was composed almost entirely of Aspen trees 4 inches to 1 foot in diameter at breast height. In addition to the single collection at site 1, Aspen leaves were collected throughout the year from a single tree located at another Gilpin County site at about 8200 feet elevation and from a small grove located at the Denver Botanic Gardens.

2.2 Leaching of Leaves

The leaching and fractionation of the leachates were performed as previously described.[1] Briefly, leachates of senescent leaves were prepared by soaking 50 g of leaves for 24 hr in 1 L of distilled water. After separation from the leaves, each leachate sample was poured through a plug of glass wool in a funnel, then pressure filtered through a prewashed Gelman Supor* 0.45 µ pore size polysulfone membrane filter and freeze-dried. The final pH values of the leachates were between 5.3 and 5.7.

In order to simulate the multiple leaching that leaves undergo in nature, some leaves were subjected to two leachings. Unleached leaves were first soaked for 24 hr in distilled water, then the leached leaves were soaked a second time for 72 hr in distilled water.

2.3 Fractionation of Leachates

The leachates from the various leaching experiments were all separated into fractions of different polarities by sequential adsorption on XAD macroreticular resins.[1] Briefly, each leachate (approximately 1 L) was first pumped onto a 150 mL XAD-8 column without pH adjustment. The adsorbed material (fraction 1) was eluted with 100 mL of 75% (v/v) acetonitrile/water solution. The solvents were removed in a rotary vacuum evaporator and the DOC was freeze dried. The pH of the leachate that passed through the XAD-8 column in the first step of the procedure was lowered to 2. The leachate was pumped through the XAD-8 column again. The adsorbed fraction 2 was eluted with 100 mL of 75% (v/v)

---

*The use of tradenames in this report is for identification purposes only and does not constitute endorsement by the U.S. Geological Survey.

acetonitrile/water solution and freeze dried. The pH of the eluted solution from the second step was adjusted to 5, its volume was reduced to 500 mL by rotary evaporation, the pH was adjusted to 2 and the leachate then was pumped onto the XAD-8 column for a third time. The adsorbed fraction 3 was eluted with 100 mL of 75% (v/v) acetonitrile-water and freeze dried.

The leachate that passed through the XAD-8 column was adjusted to pH 4 and dried on a rotary evaporator. The residue was dissolved in 100 mL of 0.1 N HCl, filtered through a glass wool plug to remove silica and pumped through a 100 mL XAD-4 column. The column was eluted with 100 mL of 75% (v/v) acetonitrile/water and vacuum evaporated to 20 mL. Residual HCl and water were removed by repeated additions and evaporations of anhydrous acetonitrile. The dried sample (fraction 4) was taken up in 50 mL of water and freeze dried. Fraction 5 was not isolated (previous studies have shown it to be a very minor fraction).[1] However, the original numbering sequence for the fractions was retained in order to allow comparison of the fractions isolated in this study with previous results.

The leachate that passed through the XAD-4 resin was deionized on 80 mL columns of MSC-1H cation exchange resin and Duolite A-7 anion exchange resin in series. The volume of effluent was reduced by rotary evaporation and the hydrophilic neutral material remaining in the effluent was freeze dried (fraction 6). The organic acids sorbed on the A-7 resin (fraction 7) were not isolated because previous work has shown that this fraction is very small.

The fractions from the second leachate (leaves leached for 96 hr total, section 2.2) are identified by a "B" appended to each of the fraction numbers.

2.4 NMR Measurements

Solid-state, cross-polarization magic-angle-spinning (CPMAS) ^{13}C spectra of the freeze dried leachate samples and the crushed leaves were measured on a 200 MHz Chemagnetics CMX spectrometer with a 7.5 mm-diameter probe. The spinning rate was 5 kHz. The acquisition parameters for the solid samples were contact time of 1 msec, pulse delay of 1 sec and a pulse width of 4.5 μsec for the 90° pulse. Interrupted decoupling (dipolar dephasing) CP experiments were performed as described previously.[2] A coupling time (τ) of 40 μsec provided satisfactory results.

Direct polarization magic angle spinning (DPMAS) measurements were also made with the Chemagnetics CMX spectrometer using the same probe and spinning rate as in the CPMAS experiments. An excitation pulse of 2.00 μsec duration (corresponding to a 40° tip angle) was applied to the samples. Pulse delays of 8 to 48 sec were used to determine when complete relaxation was attained. It was found that the receiver delay had to be increased from 18 to 20 μsec to obtain flat baselines in the DPMAS spectra.

The liquid-state ^{13}C NMR spectrum of fraction 6, which dried as a syrup, was measured in a solution of approximately 200 mg mL^{-1} of the sample dissolved in D_2O in a 10-mm diameter tube with a Varian XL 300 spectrometer at 75.429 MHz. Quantitative spectra were obtained using inverse gated-decoupling in which the proton decoupler was on only during the acquisition of the free induction decay (FID) curve. An 8 sec delay time and a 45° tip angle were used. The sweep width was 30 kHz.

2.5 IR Measurements

Pellets for infrared analysis were prepared by grinding together approximately 5 mg of sample with 250 mg KBr in a mortar and pressing the mixture in a die. The spectra of the pellets were measured with a Perkin-Elmer 2000 fourier transform spectrometer. Attenuated total reflectance (ATR) measurements of leaf surfaces were made on dried leaves with a Herrick Scientific Corporation SplitPea™ apparatus.

3 RESULTS AND DISCUSSION

3.1 NMR Spectra of Leaves

CPMAS and DPMAS spectra of unleached and leached Aspen leaves are shown in Figure 1. Comparison of CPMAS spectra of the unleached and leached leaves with the corresponding DPMAS spectra clearly demonstrates that the CP experiment underestimates carbon atoms with no attached protons such as carbonyl carbon atoms in carboxylic acids, esters and amides in the region between 160 and 180 ppm. Aromatic carbons in the region between 110 and 160 ppm are also suppressed in the CPMAS spectra. These results indicate that the aromatic rings are highly substituted.[16] Another difference between the CPMAS and the DPMAS spectra occurs at 30 ppm. A sharp, well-resolved band at 30 ppm is present in each of the DP spectra but is absent in the CP spectra. This band represents methylene groups in aliphatic chains. The absence of the 30 ppm band in the CP spectra indicates that the methylene groups are in liquid fats or waxes because molecular motion in liquids inhibits cross polarization from taking place.

3.2 IR Spectra of Leaves

The ATR IR spectrum of the surface of a dried Aspen leaf clearly demonstrates the presence of waxes on the surface (Figure 2). The most prominent bands in this spectrum are at 2918 and 2850 cm^{-1}. These bands are characteristic of the stretching modes of methylene groups. The bands at 1473 and 1463 cm^{-1} are characteristic of the methylene CH deformation modes and those at 730 and 720 cm^{-1} of CH_2 skeletal vibrations. The band at 1733 cm^{-1} probably represents the carbonyl groups of esters and the 1715 cm^{-1} band may due to $\alpha\beta$ unsaturated groups adjacent to the ester carbonyl group or to aryl esters. The carbonyls of free aliphatic acids would also occur in this region, however the absence of an OH stretching band at about 3500 cm^{-1} mitigates against this possibility. The band at 1179 cm^{-1} may represent a C-O- stretching mode.[17,18]

3.3 NMR Spectra of Leachate

The CPMAS spectra of an unfractionated leaf leachate and the leaves from which the leachate was derived are shown in Figure 3. The vertical scales have been adjusted to render the 73 ppm bands of the two spectra equal. The leachate spectrum has bands in the same regions as the leaf spectrum. However, a few of the bands are shifted or absent, and the relative intensities of the bands are different in the two spectra. The aliphatic bands between 0 and 50 ppm are much weaker and less well-resolved in the leachate spectrum than in the leaf spectrum, probably indicating that these bands mainly represent aliphatic carbon atoms in insoluble waxes and cutin in the leaf tissue.[2] The methyl ether band near 55 ppm that normally is attributed to unaltered lignin is absent from the leachate spectrum. The carbohydrate bands of the leaves are characteristic of crystalline cellulose fibrils, whereas those in the leachate indicate noncrystalline carbohydrate species.[2] The relative

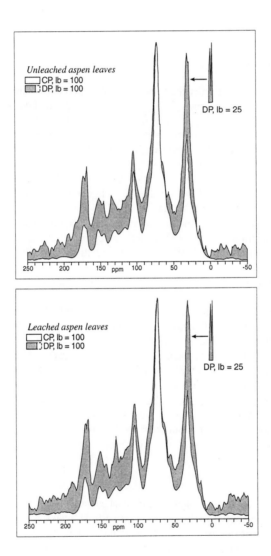

Figure 1 *CPMAS and DPMAS spectra of unleached and leached Aspen leaves. Spectra are displayed with line broadening (lb) of 100 Hz; the aliphatic bands between 20 and 30 ppm are also displayed with lb=25 Hz to show detail*

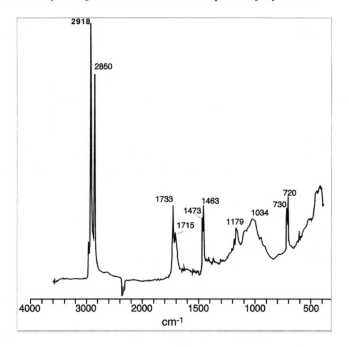

Figure 2 *Attenuated total reflectance infrared (ATR IR) spectrum of dried Aspen leaf*

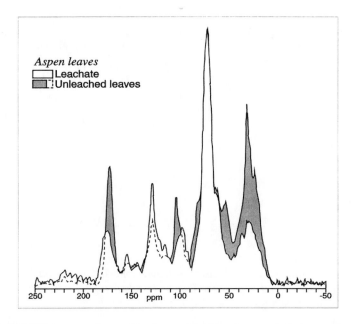

Figure 3 *CPMAS spectra of an unfractionated leaf leachate and the leaves from which the leachate was derived*

intensities of the aromatic carbon bands between 110 and 160 ppm are generally greater in the leachate spectrum than in the leaf spectrum. The carbonyl band is broader and diminished in intensity in the leachate spectrum.

Further insight into the degradation processes leading to the formation of DOC from senescent leaves may be obtained by fractionating leaf leachates into fractions of increasing polarity and measuring their ^{13}C NMR spectra.[2] Spectra of fractions of leachates extracted from leaves collected in the Fall and from leaves collected at the end of the Winter are shown in Figure 4. Leachate fractions from two consecutive leachings for each batch of leaves are given in Figure 4. The spectra have been plotted with a line broadening of 25 Hz rather than 100 Hz that normally is used to provide information on the intrinsic line widths of the bands. No significant change in line width was noted when line broadenings of less than 20 Hz were applied to the data. The possible structural component assignments for the bands of the spectra of leaf leachate fractions have been described previously.[1,3] The aliphatic bands of the Aspen leachate fractions are very similar to those in the spectra of mixed hardwood leaf compost leachate and *Acer campestre* leachate fractions. The pattern of aromatic bands, however, is distinctive and appears to be derived mainly from lignin structural units.

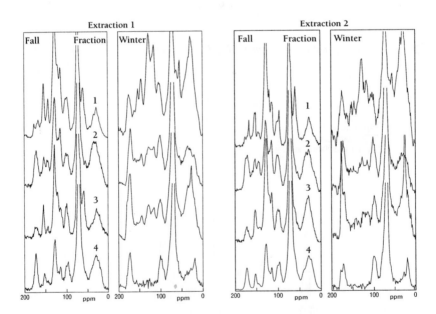

Figure 4 *Spectra of fractions of leachates from leaves collected in the Fall and from leaves collected at the end of the Winter. Leachate fractions from two consecutive leachings for each batch of leaves are shown*

The spectrum of fraction 1 of the first extraction of the Fall leaves is a good starting point for discussion of the spectra. All of the bands for p-hydroxycinnamyl units are present in this spectrum in approximately the right ratios (Figure 5). In addition to the aromatic bands, a shoulder at about 56 ppm, which probably represents methoxyl groups,

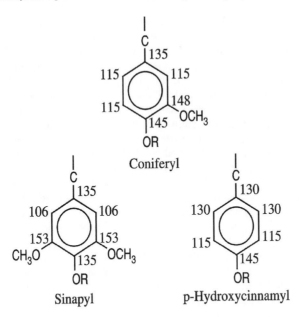

Coniferyl

Sinapyl

p-Hydroxycinnamyl

Figure 5 *Chemical shifts of lignin monomeric units*

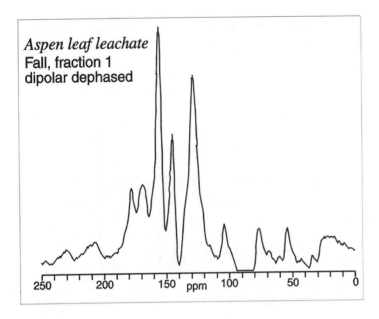

Figure 6 *Interrupted decoupled (dipolar dephased) spectrum of fraction 1 of leachate of Fall leaves*

is also present. The 56 ppm band is more clearly seen in the interrupted decoupled spectrum (Figure 6). The methoxyl band in conjunction with the aromatic bands is considered diagnostic of lignin units. Additional aromatic bands at 167, 156, and 123 ppm are present in the spectrum. These bands may represent hydrolyzable tannin units.[2] The presence of carbohydrate bands at 63, 75, and 101 ppm indicates that at least some of the aromatic units are probably conjugated to carbohydrates, because one would not expect free carbohydrates to be hydrophobic enough to be retained in fraction 1.

The major aromatic bands of fraction 2 are in the same positions as those of fraction 1. However, the spectrum of the carbohydrate groups is different in fraction 2. The C-6 band at 63 ppm is absent in fraction 2 and a new band at about 173 ppm appears in the spectrum. These changes indicate that some of the plant carbohydrate components that are present in fraction 1 have been oxidized to uronic acids or that fraction 2 has higher concentrations of naturally occurring pectins than fraction 1. A carboxylic acid carbonyl stretching band in the IR spectrum of fraction 2 at 1734 cm^{-1} that is absent from the IR spectrum of fraction 1 provides further evidence for uronic acids. The 63 ppm band reappears in the spectrum of fraction 3. There is also a weak 173 ppm band in this fraction 3. The 63 ppm band is absent in fraction 4 and there is a strong 173 ppm band.

The relative intensities of the bands in the NMR spectra of the fractions indicate that the aromaticity of the fractions decreases from fraction 1 to fraction 4. This reduction in aromaticity apparently causes a progressive reduction in hydrophobicity, which in turn leads to the order of elution of the fractions.

The spectra of the fractions in the second extraction are very similar to the corresponding fractions in the first extraction. However, bands of fraction 1B are generally sharper and better resolved than those of fraction 1. This difference probably indicates that structural units in fraction 1B have undergone less alteration than those in fraction 1.

Quantitative liquid-state spectra of Fall fraction 6 samples from each of the extractions are also given in Figure 7. These spectra are very similar to those of fraction 6 samples measured in previous studies.[1,3] Comparison of the integrated areas under the

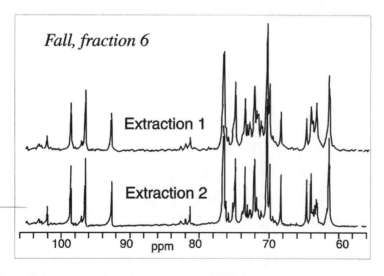

Figure 7 *Quantitative liquid-state spectra of Fall fraction 6 samples from extractions 1 and 2*

anomeric bands of the spectra with the areas under the other carbohydrate bands in the spectra in those studies indicated that about half of each sample consisted of saccharides and half of sugar alcohols or inositols.

The carboxylic acid bands of the fractions of the leachate from leaves that were collected from the ground at the end of the Winter are generally stronger than those of the corresponding fractions from the Fall leaf leachate (Figure 4). The bands of some of the Winter fractions are also broader than corresponding bands in the Fall fractions, indicating that there is more heterogeneity in the constituent structural units that make up the fractions. As expected, these data show that the degradation continued through the Winter without any major changes in reaction mechanisms but with an increased heterogeneity of the products. The continued degradation through the Winter has resulted in fraction 4 constituting a significantly larger percentage of the total weight of the Winter leachate than the Fall leachate (Tables 1 and 2), and fraction 1 is a much lower percentage of the total weight of the Winter leachate than the Fall leachate.

Table 1 *Weights of leachate fractions from 100 g of Fall leaves (collected 10/5/97)*

Fraction	First extract (g)	Second extract (g)	Total (g)	Percent
1	0.5015	0.8377	1.3392	48.3
2	0.1324	0.2053	0.3377	12.2
3	0.1100	0.1413	0.2513	9.1
4	0.3800	0.4626	0.8426	30.4
Total			2.7708	100.0

Table 2 *Weights of leachate fractions from 100 g of Winter leaves (collected 5/10/98)*

Fraction	First extract (g)	Second extract (g)	Total (g)	Percent
1	0.2640	0.0884	0.3524	21.2
2	0.2178	0.1816	0.3994	24.1
3	0.0480	0.0210	0.0690	4.2
4	0.4280	0.4104	0.8384	50.5
Total			1.6592	100.0

Wershaw et al.[1] have proposed that fractions isolated from a compost leachate using a similar fractionation scheme to the one used here represent different stages in the oxidation of lignin and that fraction 1 contains the least degraded lignin fragments and fraction 4 the most degraded fragments. The changes in abundance of fractions 1 and 4 in the Fall and Winter samples noted above provide further evidence that fraction 1 is composed of the least degraded material and fraction 4 the most degraded.

4 CONCLUSIONS

Comparison of the integrated areas of bands in the DPMAS spectra of Aspen leaves with the integrated areas of the bands in the CPMAS spectra indicate that the concentrations of carbon atoms with no attached protons, such as substituted carbons in aromatic rings and carbonyl carbons, are underestimated in the CPMAS spectra. In addition, bands representing liquid materials, such as cuticle waxes, are present in the DPMAS spectra but absent in the CPMAS spectra. The presence of aliphatic waxes on the leaf surfaces was confirmed by ATR IR measurements.

The CPMAS NMR spectrum of the unfractionated leachate has bands in the same regions as the leaf spectrum. However, some of bands are shifted and the relative intensities of the bands are different. The bands in the spectra of the leachate fractions are better resolved than the bands in the unfractionated leachate. This increased resolution provides additional structural information. The aliphatic bands of the Aspen leachate fractions are similar to those that have been measured previously for leachate from mixed hardwood leaf compost and from *Acer campestre* leaves. The pattern of the aromatic bands, however, is distinctive in the Aspen leachate, and appears to be derived mainly from lignin structural units; p-hydroxycinnamyl units predominate. The aromatic units appear to be conjugated to carbohydrates. Fraction 1 of the leachate contains the least degraded material that is closest in chemical structure to the unaltered leaf components. Fraction 4 consists of the most degraded material.

References

1. R. L. Wershaw, J. A. Leenheer, K. R. Kennedy and T. I. Noyes, *Soil Sci.*, 1996, **161**, 667.
2. R. L. Wershaw, K. R. Kennedy and J. E. Henrich, in 'Humic Substances: Structures, Properties and Uses', G. Davies and E. A. Ghabbour, (eds.), Royal Society of Chemistry, Cambridge, 1998, p. 29.
3. R. L. Wershaw, J. A. Leenheer and K. R. Kennedy, in 'Humic Substances: Structures, Properties and Uses', G. Davies and E. A. Ghabbour, (eds.), Royal Society of Chemistry, Cambridge, 1998, p. 47.
4. Y. Y. Leshem, *Free Radical Biology & Medicine*, 1988, **5**, 39.
5. R. Aerts, *J. Ecology*, 1996, **84**, 597.
6. K. T. Killingbeck, *Ecology*, 1996, **77**, 1716.
7. S. Philosoph-Hadas, S. Meir, B. Akiri and J. Kanner, *J. Agric. Food Chem.*, 1994, **42**, 2376.
8. S. Strother, *Gerontology*, 1988, **34**, 151.
9. M. B. Peoples and M. J. Dalling, in 'Senescence and Aging in Plants', L. D. Noodén and A. C. Leopold, (eds.), Academic Press, San Diego, 1988, p. 181.
10. S. Gepstein, in 'Senescence and Aging in Plants', L. D. Noodén and A. C. Leopold, (eds.), Academic Press, San Diego, 1988, p. 85.
11. J. E. Sanger, *Ecology*, 1971, **52**, 1075.
12. T. Goto and T. Kondo, *Angew. Chem. Int. Ed. Engl.*, 1991, **30**, 17.
13. M. N. Merzlyak, G. A. F. Hendry, N. M. Atherton, T. V. Zhigalova, V. K. Pavlov and O. V. Zhiteneva, *Biokhimiya*, 1993, **58**, 240.
14. J. E. Thompson, in 'Senescence and Aging in Plants', L. D. Noodén and A. C. Leopold, (eds.), Academic Press, San Diego, 1988, p. 51.
15. W. F. Mueggler, 'Aspen Community Types of the Intermountain Region', U.S. Department of Agriculture Forest Service General Technical Report INT-250, 1988.
16. R. L. Wershaw, G. R. Aiken, J. A. Leenheer and J. R. Tregellas, in preparation.
17. L. J. Bellamy, 'The Infra-red Spectra of Complex Molecules', Methuen, London, 1958.
18. N. B. Colthup, L. H. Daly and S. E. Wiberley, 'Introduction to Infrared and Raman Spectroscopy', Academic Press, San Diego, 1990.

A BIOGEOPOLYMERIC VIEW OF HUMIC SUBSTANCES WITH APPLICATION TO PARAMAGNETIC METAL EFFECTS ON ^{13}C NMR

Robert L. Cook[1] and Cooper H. Langford[2]

[1] Nova Research and Technology, Nova Chemicals, Calgary, Alberta, Canada T2E 7K7
[2] Department of Chemistry, University of Calgary, Calgary, Alberta, Canada T2N 1N4

1 INTRODUCTION

Physical scientists most often use reductionist methods. They analyze fundamental components at molecular and atomic levels and determine their function. Then explanations are constructed with the rules that govern fundamental behavior. An analogy is to describe chess with the rules for moving the pieces. The complexity of the results of the rules prevent even a master (and the best chess computer programs) from calculating all the possible moves and responses. Rather, chess players base strategy on the use of patterns at the higher level of complexity. This other strategy studies emergent patterns at the level of complex behavior. Like clever chess players, students of humic substances (HSs) do more than predict all complexities resulting from the molecular constituents. They also search for patterns of behavior of the HSs mixture itself.

We believe there are patterns at the complex level. Recognizable regularities exist in HSs behavior despite their different environments. HSs are formed by transformation processes driven by large energy flows, and HSs structures are far from thermodynamic equilibrium (that is, CO_2, C(s), H_2O and inorganic ions). Thus, we can take the regularities to be those of a complex, self-organizing system. The organizing features connect to what can be explained at the molecular level. The emergent features of HSs depend on the many conformational and aggregation processes available and resemble the tertiary and quaternary structures of proteins.

The wise chess player analyzes all possibilities the rules allow for a modest number of moves ahead. But he or she also uses strategic patterns that would be hard to find from the move rules. Likewise, the wise HSs researcher exploits both reductionist and emergent strategies.

1.1 Biogeopolymers

The definition of a polymer is based on a large molecule constructed from similar parts. This definition fits common industrial polymers with a small number of distinct "similar parts" (monomer units). In the biochemical domain we must accept a larger variety of similar units, for example twenty or so distinct amino acids in proteins. Biopolymers have many distinct monomer units. However, biopolymers do not have ambiguous primary structures. There are well-defined covalent links connecting the monomers. At the same time, biopolymer chemistry emphasizes secondary, tertiary and quaternary structures that depend on weaker interactions.

We propose to extend the term polymer to include the class biogeopolymers, which include HSs and other naturally occurring organic materials. Biogeopolymers are usually described as "mixtures" because physical separation techniques can overcome the weaker interactions governing tertiary and quaternary structure. Nevertheless, these interactions can confer many of the same functional regularities (emergent properties) that characterize industrial and biopolymers. These include solution viscosity behavior, light scattering, glass transition, partial crystallinity, polymer fluorescence and some other features that will be described below. We propose extension of the concept of polymers to systems with characteristic polymer behavior (emergent properties) even if their conventional primary structure is unknown. Thus, we define a biogeopolymer as an operationally definable substance composed from a set of "similar parts" that exhibits characteristic polymer behaviour including properties dependent upon tertiary and quaternary structure.

1.2 Some Accomplishments of Reductionist Stategies

However we regard the behavior of humic biogeopolymers, it is true that they have many more "monomer units" than biopolymers such as proteins and nucleic acids. One classic study will illustrate the point.

Schnitzer's laboratory pioneered the definition of constituents of HSs identified by mild oxidative degradation.[1] Table 1 shows aromatic structures observed from some fractions of Armadale fulvic acid (FA). There are more than 30 structures in this small part of the overall organic chemistry. More recent pyrolysis mass spectrometry has increased the number of fragments.[2] The variety of similar units is large but 'similarity' does remain at the level of monomer classes.

The power of physical separation methods to resolve biogeopolymers is well illustrated by the separations effected with XAD resin adsorption.[3,4] A sample of Suwannee River FA is first resolved into seven fractions (Table 2) and then in further stages into at least 31. This again illustrates the biogeopolymer complexity. Preliminary fraction separation is reminiscent of adsorption used to identify tacticity classes (atactic, syndiotactic and isotactic) in industrial polymers. The fractionation helped to elucidate the behavior of carboxylate groups that mostly define the acidity of humic acids (HAs).[3,4] NMR and titrimetric studies of the fractions defined the structural environments of 68-90% of the total carboxylate content.

Much has been learned about HSs from fractionation and degradation studies. We argue below that this approach and experimental procedures that address the complex system directly are complementary.

1.3 Evidence of Polymeric Behavior from Glass Transition, Ordering of Phases and Non-Linear Sorption Isotherms

Industrial polymers and biopolymers have glass transition temperatures. For a one-component system, a glass transition temperature is a sharp transition point monitored by differential scanning calorimetry (DSC). The glass transition depends on the degree of ordering (see X-ray and NMR evidence below) and macromolecular mobility. Less mobile systems have higher and sharper glass transitions. More mobile, e.g., solvent swelled systems have lower and less sharp glass transitions. Glass transitions may overlap in a many-component system. A heterogeneous system with structural and molecular weight heterogeneities will produce only a broad, weak glass transition. The more components in

Table 1 *Chemical structures of compounds identified by mild oxidative degradation of Armadale FA[1]*

$CH_3(CH_2)_xCO_2CH_3$
  6 $n = 12$
  10 $n = 13$
  14 $n = 14$
  19 $n = 16$

$CH-CO_2CH_3$
$\parallel$
$CH-CO_2CH_3$
  11

$CH_2-CO_2-C_8H_{17}$
$|$
$(CH_2)_2$
$|$
$CH_2-CO_2-C_8H_{17}$
  26

1 $R_1 = CO_2CH_3$; $R_4 = OCH_3$; $R_2 = R_3 = R_5 = R_6 = H$
2 $R_1 = R_2 = CO_2CH_3$; $R_3 = R_4 = R_5 = R_6 = H$
3 $R_1 = R_3 = CO_2CH_3$; $R_2 = R_4 = R_5 = R_6 = H$
4 $R_1 = R_4 = CO_2CH_3$; $R_2 = R_3 = R_5 = R_6 = H$
5 $R_1 = CO_2CH_3$; $R_3 = R_5 = OCH_3$; $R_2 = R_4 = R_6 = H$
7 $R_1 = CO_2CH_3$; $R_3 = R_4 = OCH_3$; $R_2 = R_5 = R_6 = H$
8 $R_1 = CO_2CH_3$; $R_2 = R_3 = OCH_3$; $R_4 = R_5 = R_6 = H$
9 $R_1 = CH_2COCH_3$; $R_2 = CO_2CH_3$; $R_3 = R_4 = R_5 = R_6 = H$
12 $R_1 = CH_2CO_2CH_3$; $R_3 = CO_2CH_3$; $R_2 = R_4 = R_5 = R_6 = H$
13 $R_1 = R_3 = CO_2CH_3$; $R_4 = OCH_3$; $R_2 = R_5 = R_6 = H$
15 $R_1 = CH_2CO_2CH_3$; $R_2 = CO_2CH_3$; $R_4 = OCH_3$; $R_3 = R_5 = R_6 = H$
16 $R_1 = R_2 = R_3 = CO_2CH_3$; $R_4 = R_5 = R_6 = H$
17 $R_1 = R_2 = R_4 = CO_2CH_3$; $R_3 = R_5 = R_6 = H$
18 $R_1 = R_3 = R_5 = CO_2CH_3$; $R_2 = R_4 = R_6 = H$
20 $R_1 = CH_2CO_2CH_3$; $R_3 = R_4 = CO_2CH_3$; $R_2 = R_5 = R_6 = H$
21 $R_1 = R_2 = R_4 = CO_2CH_3$; $R_3 = OCH_3$; $R_5 = R_6 = H$
22 $R_1 = R_3 = R_5 = CO_2CH_3$; $R_2 = OCH_3$; $R_4 = R_6 = H$
23 $R_1 = R_2 = R_3 = R_4 = CO_2CH_3$; $R_5 = R_6 = H$
24 $R_1 = R_2 = R_4 = R_5 = CO_2CH_3$; $R_3 = R_6 = H$
25 $R_1 = R_2 = R_3 = R_5 = CO_2CH_3$; $R_4 = R_6 = H$
27 $R_1 = R_2 = R_3 = R_4 = CO_2CH_3$; $R_5 = OCH_3$; $R_6 = H$
28 $R_1 = R_3 = R_4 = R_5 = CO_2CH_3$; $R_2 = OCH_3$; $R_6 = H$
29 $R_1 = R_2 = R_3 = R_4 = R_5 = CO_2CH_3$; $R_6 = H$
30 $R_1 = R_2 = R_3 = R_4 = R_5 = CO_2CH_3$; $R_6 = OCH_3$
31 $R_1 = R_2 = R_3 = R_4 = R_5 = R_6 = CO_2CH_3$
32 $R_1 = CH_2CO_2CH_3$; $R_2 = R_3 = R_4 = R_5 = R_6 = CO_2CH_3$

Table 2 *Summary of the determined carboxyl group structures in Suwannee River FA[3]*

Acid group structure	Content, mmol/g	Total carboxyl groups, %
Keto acids (pK$_a$ < 3.0)	0.20	3.0
Aromatic and olefinic acids (pK$_a$ < 3.0)	0.46	7.7
Aromatic and olefinic acids (pK$_a$ > 3.0)	0.86	14.3
Alpha ether and ester acids (pK$_a$ < 3.0)	1.28	21.3
Additional alpha ether and ester acids (1 additional acid assumed)	1.28	21.3
Additional alpha ether and ester acids (2 additional acids assumed)	2.56	42.6
Total accounted acids	4.1-5.4	68-90

a system, the broader is the glass transition. Thus, glass transition temperature measurements on a humic material would seem difficult. Nevertheless, two coal derived HAs have been reported to show glass transitions.[5,6] They may be more homogeneous than soil derived HAs, but they are still complex heterogeneous systems. It has been shown that both these humic glass transitions are weakened and broadened when the system is solvent swelled.[5,6]

No glass transition temperature has been detected for soil derived humic materials. This suggests that soil HAs have more monomer types then coal derived HAs, or it may be because of low instrument sensitivity. Most DSC units are designed for industrial polymers and biopolymers with sharper and more intense glass transitions. It would be useful to run DSCs on HSs fractions

The lack of an observable glass transition for soil derived humic materials does not exhaust the methods of searching for order. Debye-Scherrer X-ray diffraction (XRD) shows evidence of increased ordering with aging and mineralization of humic biogeopolymers.[7,8] The more ordered phases seemed to be mainly aromatic. One would expect these carbons to be in a more rigid environment than other carbons in the system.

Recent proton spin-lattice relaxation time in the rotating frame $(T_{1\rho}(^1H))$ measurements confirm this for a humic biogeopolymer.[9] The aromatic carbons in a fulvic biogeopolymer have longer $T_{1\rho}(^1H)$ and thus appear to be in a rigid environment. This may also indicate phase ordering in fulvic biogeopolymeric systems. It is worth noting that NMR can detect much shorter range order than either XRD or DSC techniques. Extremely small rigid domains and very small zones of condensed matter can be detected and monitored by NMR. 'Order' in the alkyl region of some ^{13}C NMR spectra of HSs with signals analogous to polyethylene has been observed; these zones probably arise from cross-linked fatty acids.[10]

A dual mode kinetic model has been used to evaluate HSs sorption data.[11-14] A condensed/non-condensed polymer model has been used to explain the experimental data. These polymer models explain non-linear isotherms, competitive effects in multi-solute systems and concentration dependent enthalpies of sorption. None of these effects can be explained with a simple phase partition model. Sorption studies on whole soils indicate that the humic components behave like polymers.

Evidence from any one of these three approaches would be a weak basis for adopting the biogeopolymer viewpoint without knowing HSs primary structures. However, taken together they argue strongly for the polymer view of HSs behavior and establish a unified definition of a term such as "humic acid."

2 MATERIALS AND METHODS

2.1 Materials

Laurentian fulvic acid (LFA) was extracted from a forest podzol in an area controlled by Laval University (Quebec) and purified as described previously.[15,16] LFA has been the subject of elemental analyses, acid-base and metal titration curves, fluorescence, FTIR, ^1H NMR, magnetic circular dichroism (MCD),[17,20] and recently 1D and 2D ramped amplitude CP-MAS ^{13}C NMR. The composition of LFA is 45.1% C, 4.1% H, 1.1% N, 49.7% O, <1ppm Fe and <1% ash. This purified protonated form of LFA has a "bidentate" complexing capacity of 5.8 mmol/g.[21] We used cupric nitrate (Fisher Lot 874505) and 18 $M\Omega$ water produced by a Barnstead Nanopure system.

2.2 Instruments

All NMR spectra where obtained on a Bruker AMX2-300 spectrometer with a Bruker BL4 probe. The rotors used were 4 mm O.D. by 18 mm long zirconia rotors with Kel-F caps. All calibration such as magic angle setting and Hartman-Hahn matching (HH) was done with ^{13}C labeled glycine using the ketonic signal. Spectra were obtained at 75.469 MHz, a contact time of 3 msec, a 1 sec recycle time, a spinning rate of 8 kHz and with ramp-CP. It has been shown that these conditions give a quantitative spectrum of LFA.[22] A standard experiment where the contact time is varied revealed that $T_{1\rho}{}^1H \gg T_{CH}$, and thus the CP experiment works on this sample. A 1 sec recycle time allowed complete relaxation of the system (the spectra obtained at longer delays were identical to the spectra obtained with a delay of 1 sec).[23]

The ramped CP experiment has been discussed elsewhere.[9,22,24-26] The ramp was applied to the proton channel because its amplifier has higher linearity than the X (^{13}C) channel amplifier. The ramp was centered on the -1 sideband because (i) arcing would have occurred at the probe if the ramp had been set on the +1 sideband; (ii) it has been shown that the signal buildup at the -1 sideband is much faster than that at the centerband.[27] The ramp (Hz) was set at the spinning rate.

160,000 scans spectra were collected except for the "zero metal" loaded sample (14,400 scans). Paramagnetic ions make the field inhomogeneous across the whole sample volume and cause a large signal loss (which is a reason that these experiments are not quantitative). This requires a great increase in the numbers of scans.

300 mg of LFA was dissolved in the appropriate volume of water e.g. 100 mL for the 0.05 metal ion to bidentate binding sites (M:BS) ratio sample, 150 mL for the 0.1 M:BS ratio sample, and 200 mL for the 0.15 M:BS ratio sample. The volume was chosen so that the LFA in solution would just start to precipitate after the desired stoichiometeric amount of Cu(II) had been added. Once the LFA had fully dissolved the desired amount of Cu(II) was added, and stirred for another hour. Each solution was then allowed to air dry in the dark for 35 days, after which it was transferred to a sample vial. All solid samples were ground in a mortar before being packed into the rotor.

3 RESULTS AND DISCUSSION

3.1 A Meso-Structural Model of a Fulvic and a Humic Biogeopolymer Derived from ^{13}C NMR

The use of NMR to study biopolymers has been fruitful. A similar strategy should be fruitful in the study of biogeopolymers. NMR interrogates a HS sample as a whole and without bias. Description of functional groups distribution in HSs now depends mainly on NMR evidence. NMR has a special place in humic and fulvic biogeopolymer studies. NMR also is a powerful tool for dynamic properties analysis.

Due to the large number of distinct carbons, their slow relaxation and low receptivity, ^{13}C NMR of HSs is extremely time consuming. ^{15}N and ^{33}S spectral measurements also are time consuming since parameters of the experiment are similar to those for C and the concentrations in HSs are low. ^1H NMR is less valuable because of poor resolution caused by the inherent small chemical shift range of protons. An NMR technique known as cross polarization magic angle spinning (CPMAS) overcomes the slow relaxation and low receptivity by transferring the proton's polarization to carbon via a Hartman-Hahn (HH) condition in the rotating frame.

CPMAS NMR is inherently not quantitative because a stronger signal will arise in a CPMAS experiment where more protons are attached to the carbon observed. The opposite is true for a carbon more isolated from protons. It also has been shown that spinning speeds of 8 kHz or higher must be used for some humic samples in order to get the most quantitative results.[24] However, the faster one spins a sample the more exact the HH condition becomes. Because of the large variety of carbons in a humic sample, there will be many HH that only overlap if they are broad. Thus, no single HH will work quantitatively for the whole sample, especially when one needs to spin the sample at high spinning rates (in excess of 5 kHz, as needed for high field instruments, e.g., 7 Tesla).

Ramp CP-MAS can overcome the quantitative problems with the CPMAS technique if the contact time is optimized. Figure 1 shows a classical CPMAS pulse sequence and a ramp-CPMAS pulse sequence. The reason that ramp-CPMAS can give quantitative results is that it allows for multiple HH conditions to be covered, as shown in Figure 2. Ramp-CPMAS has been discussed[28-30] as has its application to HSs.[9,22,24-26] All the NMR data presented below were obtained with ramp-CPMAS under conditions that are independently confirmed to give quantitative results.

Laurentian FA and HA have been characterized by 1D, 2D and dynamic ^{13}C NMR. The results are reported in Tables 3 and 4.[9] Because of the dynamics of the systems, there

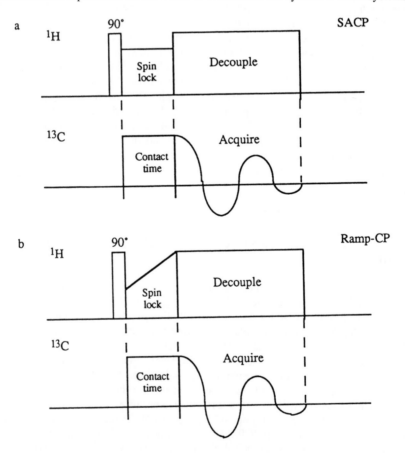

Figure 1 *a) Single amplitude CP and b) ramp-CP pulse sequences*

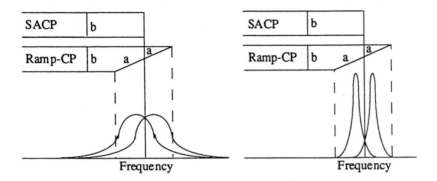

Figure 2 *A comparison of single amplitude and ramp-CP for a sample in which there are two subsystems, each having its own HH. A) shows the situation at slow sample spinning speeds (broad HH), while B) shows the situation at high sample spinning speed (narrow HH)*

Table 3 *Ramp-CPMAS ^{13}C NMR characterisation of LFA*

Chemical shift assignments (ppm)	Total observable carbon, %	$T_{1\rho}{}^{1}H$, msec	$T_2{}^{13}C$, msec
Ketonic (190-220)	8.8	7.3	11.9
Carboxyl (162-190)	33.8	4.9	13.1
Phenolic (145-162)	2.2	6.0	22.7
Aromatic (145-108)	12.0	5.5	11.4
O-C-O (96-108)	3.6	3.6	10.4
Carbohydrate (50-96)	17.8	2.6	12.3
Aliphatic (0-50)	21.8	2.8	8.0

Table 4 *Ramp-CPMAS ^{13}C NMR characterisation of LHA*

Chemical shift assignments (ppm)	Total observable carbon, %	$T_{1\rho}{}^{1}H$, msec	$T_2{}^{13}C$, msec
Ketonic (190-220)	6.5	2.4	7.8
Carboxyl (162-190)	15.6	3.2	8.3
Phenolic (145-162)	2.8	4.6	9.1
Aromatic (145-108)	18.7	4.1	7.8
O-C-O (96-108)	1.0	3.6	5.1
Carbohydrate 1 (60-96)	11.9	2.3	6.0
Carbohydrate 2 (50-60)	9.1	2.8	5.1
Aliphatic (0-50)	34.3	3.0	6.9

is not enough aromatic carbon to account for all the carboxylate functionality of the FA (see column 4 of Tables 3 and 4). However, there are more than enough aromatic carbons to account for the carboxylate functionality of the HA.

The relaxation time T_2 mainly is governed by the freedom of movement of the nucleus; the more mobile the nucleus, the longer the relaxation time. T_2 (^{13}C) data show that the carboxylate functionality is associated with the carbohydrate moieties of the FA and with the aromatic moieties of the HA because the relaxation times are most closely related in these cases. The $T_{1\rho}$ data are subject to the same interpretation.[9]

2-D dipolar dephasing (DD) data are shown in Figures 3 and 4. The diagnostic point is the spread on the Y-axis. The more a proton influences the signal the greater the dipolar dephasing and the less spread on the Y-axis. Methyl groups have very little DD effect because their high rate of rotation averages the influence of the three protons to nearly zero. The DD data show that the aromatic moieties in the FA are not highly functionalized since many protons produce a large dipolar dephasing effect. In contrast, the aromatic moieties of the HA appear to be highly functionalized (i.e., few protons), as indicated by the very weak DD effect. The T_2 (^{13}C) also shows the aliphatic moieties of the FA are the least mobile and the carbohydrate moieties the most mobile. By contrast, for the HA the T_2 (^{13}C) data show that the aromatic moieties are the most mobile.

The above data and discussion lead to a dynamic "meso-structural" model of the two Laurentian HSs. The FA meso-structure consists of: (i) large relatively immobile regions that are mainly aliphatic and largely unfunctionalized; (ii) relatively unfunctionalized

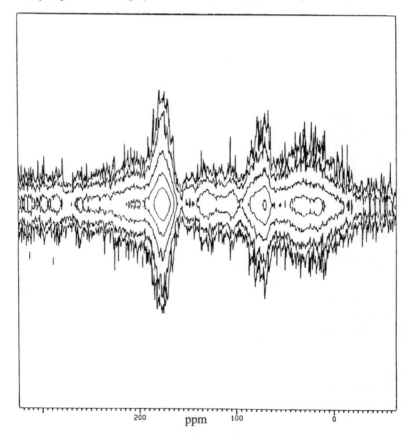

Figure 3 *The two dimensional DD spectrum of LFA. The vertical axis has arbitrary units*

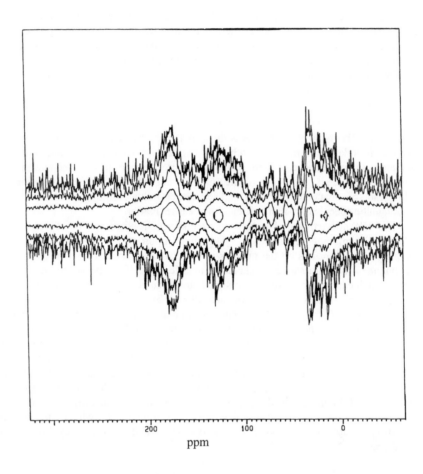

Figure 4 *The two dimensional DD spectrum of LHA. The vertical axis has arbitrary units*

more mobile regions that are mainly aromatic; and (iii) more mobile functionalized regions that are mainly carbohydrate. By contrast, the HA meso-structure consists of one major relatively immobile region associated with slightly more mobile functionalized units that are mainly aromatic.

3.2 A ^{13}C NMR Study of Metal Binding

Next we present a study that uses ^{13}C ramp-CPMAS NMR to study metal binding via paramagnetic relaxation. Paramagnetic species have long been nuisances in the NMR study of biogeopolymers. In fact, a pre-requisite to obtaining high quality quantitative and qualitative NMR spectra of biopolymers is the removal of paramagnetic species. A great deal of effort has been directed towards the removal of paramagnetic species (mainly iron) from biogeopolymers. However, once one has a fully protonated biogeopolymer, paramagnetic ions can be added to probe metal binding sites via NMR. The following discussion gives the underlying principles behind this method. The main point is that the moieties involved in paramagnetic probe binding and the nearest neighbors disappear from the spectrum. This is due to line broadening induced by the paramagnetic probe.

3.2.1 Paramagnetic Relaxation, Line Broadening and Spectral Simplification. The inverse relationship between relaxation in the spin-spin mode (or transverse relaxation, T_2) and line width is given by eq 1,[31]

$$T_2 = 1/\pi\Delta\nu_{1/2} \tag{1}$$

where $\Delta\nu_{1/2}$ (Hz) is the spectral line width at half-peak height. Paramagnetic ions cause an increase in relaxation rates of both spin-lattice and spin-spin modes. The magnetic moments of unpaired electrons are ca. 10^3 times greater than those of the nuclear magnetic moments. This causes them to generate much greater local fields. The fluctuations in these fields lead to enhanced relaxation and the larger fields resulting from the presence of paramagnetic species give more efficient nuclear relaxation.

The relaxation times of nuclei near paramagnetic sites can be represented by the Solomon-Bloembergen equations.[32,33] The most important factor in how much effect the paramagnetic centre has on the nucleus is the distance between the two, with a distance dependence $1/r^6$. Paramagnetic relaxation manifests itself in the spectra via line broadening, which can easily cause certain signals to "disappear" into the baseline. The regions affected by paramagnetic relaxation are altered by loss of apparent signal intensity. Because of the $1/r^6$ distance dependence of dipolar coupling, paramagnetic relaxation is highly diagnostic of the location of the paramagnetic ion. Figure 5 illustrates the effect on a hypothetical five component spectrum. In Figure 5a no paramagnetic ion is present, while in Figure 5b a paramagnetic ion is bound at a site. The bound paramagnetic ion causes this site's spectral component to broaden and it also can cause a smaller, but significant, spectral broadening of a neighboring site. Note that the overall envelope is simplified and the unperturbed components emerge for easier identification. The region of the spectrum that corresponds to the moiety or moieties to which the paramagnetic relaxation agent is bound and their near neighbors will be altered by loss of apparent signal intensity.

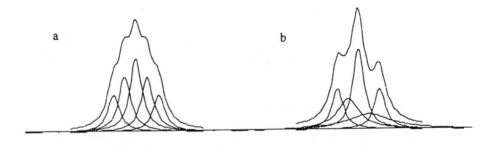

Figure 5 *The effect of a paramagnetic center on line broadening in NMR: a) is a hypothetical spectrum with no bound paramagnetic ion, b) is the same but with a bound paramagnetic ion*

Paramagnetic relaxation affects the CP process in CPMAS by reducing the $T_{1\rho}{}^{1}H$ of the protons. The $T_{1\rho}{}^{1}H$ can be so severely reduced that the protons relax before CP can take place. In a CP ^{13}C NMR experiment the carbon connected to these protons would be invisible. This leads to further simplification of the spectrum. The signals lose intensity from failure of CP.

3. 2. 2 Cu(II) Line Broadening of CPMAS ^{13}C NMR Spectra. Figure 6 shows the ^{13}C NMR spectra of LFA with the following Cu(II) loadings: 0.00, 0.05, 0.10 and 0.15 M:BS The spectra track the earlier parts of the titration curve and focus on the stronger metal ion binding sites in the LFA complex. The 0.00 M:BS spectrum is interpretable in both qualitative and quantitative fashion. As expected, all regions of the spectra lose signal due to the paramagnetic ions that increase the overall field inhomogeneity. We can not compare the absolute spectral intensities because paramagnetic effects vary. However, we can compare the spectra with each other in terms of relative intensities since the effects of paramagnetic loading influence the spectral regions differently.

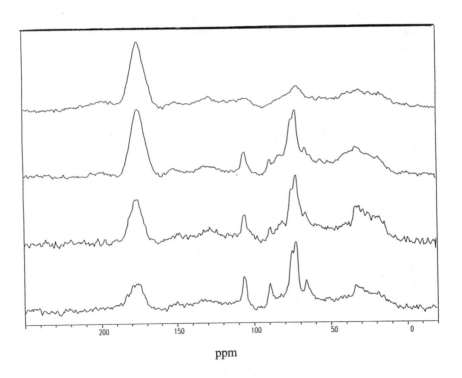

ppm

Figure 6 *Laurentian FA at different Cu(II) loadings. The top spectrum is with no Cu(II), followed by 0.05, 0.10 and 0.15 M:BS Cu(II) loaded spectra, respectively*

It often is assumed that the major and strongest metal binding sites in HA are phenol-carboxylate or phthalic type functionalities with carboxyl and hydroxyl groups adjacent to each other on an aromatic ring. A large number of studies of the binding of metal ions by organic matter in soils and fresh water have relied on aromatic model compounds including salicylic acid and phthalic acid.[21,34-39] This is perhaps a bias arising from degradation and fragmentation studies. Figure 6 exposes the limitation of this concept.

After noting the expected changes in the carboxylate region, the next most striking feature of Figure 6 is the alteration in the carbohydrate region (50-92 ppm) induced by metal loading. Although it appears that there is line narrowing rather than line broadening in the carbohydrate region, this is not the case. As shown in Figure 5 and discussed above, line broadening induced by paramagnetic ion binding reveals itself by the loss of resolution of the broadened signals and their weakening by loss of CP. No mechanism of paramagnetic line narrowing exists. This means that the envelope of carbohydrate signals is simplified by leaving narrow spectra of only those components that do not interact with the metal. This observation confirms the view that humic materials have broad lines because there are so many overlapping components. Perturbation is very pronounced even at 0.05 M:BS ratio, and it increases with metal loading. In contrast, it is difficult to detect any perturbation of the phenol (145-162 ppm) and aromatic (108-162 ppm) signals up to a 0.10 M:BS ratio. As expected, the other major region of perturbation is the carboxylate region (162-195 ppm). It appears that the strongest binding sites are associated with carboxylate groups and quite possibly acidic -OH groups on carbohydrate structures. These almost certainly are multidentate sites. There also is evidence of cooperativity.

It comes as no surprise that -COOH and -OH groups are the major functional groups involved in metal binding. As discussed above, the majority of the -COOH and -OH groups in FAs are on the carbohydrate moieties and in close proximity to one another. They are not on aromatic moieties in LFA, as 1D NMR, 2D NMR and NMR relaxation measurements indicate.[9] Other NMR results also support this conclusion.[3,4,40] The carboxylate containing carbohydrate moieties seem to be the strongest Cu(II) binders. Copper(II) in humic materials is in a highly distorted environment. If the humic ligand is polydentate, it may be easier for carbohydrate moieties to accommodate such a distorted environment than it is for aromatic moieties, which are more rigid and bulky. The flexible carbohydrate moieties could easily allow pseudo-chelation (see below) through cooperative interactions of ligand side chains.[41]

The NMR results are especially important for natural water chemistry, where heavy metal concentrations are small compared to humic carboxylate concentrations. Carbohydrate moieties are expected to be the major moieties involved in metal binding and metal speciation in this geochemically important limit.

On reaching 0.15 M:BS loading we see clear evidence that the phenolic and aromatic regions of the spectra are being specifically perturbed by the paramagnetic ion. These spectra do suggest that the phenolic groups participate in metal ion complexation at higher metal ion loading. It should also be noted that the aliphatic region (0-50 ppm) experiences significant and specific paramagnetic perturbation throughout. Either aliphatic side chains are neighbors to complexation sites or metal ion complexation induces a conformational change that brings the complexation site and aliphatic moieties closer together. This may indicate cooperative "pseudo-chelation." DD results (e.g., CH coupling strength) strongly indicate that the aromatic moieties are highly unsubstituted and poor in aliphatic sidechains.[9] Consequently, effects on the aliphatic region also tend to de-emphasize the role of aromatic moieties in the structure of the strongest metal binding sites and the importance of the carbohydrate moieties.

It is interesting to note that the NMR signature in the region between 50-108 ppm that emerges at higher loading is that of cellulose. Cellulose should not be a strong metal ion binding moiety. Our results suggest that cellulose may be a significant component of LFA.

3.2.3 A Model for Metal Binding. Metal binding moieties can be developed from the data presented above. The 1D and 2D ^{13}C NMR data indicate that the carbohydrate carbons are proton poor and highly functionalised as mainly carboxylic and -OH groups. The relaxation data also suggest that these functionalities are grouped together and co-located. The metal binding sites are modeled as polydentate O donor complexes from carboxylate and other oxygens in the carbohydrate region. The fact that the aliphatic moieties are also strongly affected by Cu(II) binding (even at metal binding as low as 0.10 M:BS ratio) emphasizes the notion of a polydentate organic ligand. This ligand may result from conformational change and "aggregation" of fulvic fractions around the metal by hydrogen bonding. Also conceivable are hydrophobic interactions that produce "pseudo-chelation," a cooperative binding of the metal and the interaction of ligand groups with each other.[41] Evidence from light scattering titrations suggests "cooperative" effects in the Cu(II) binding curve.[42]

This model would explain the large variations in the kinetics of metal uptake and release by FAs. Metal binding to simple oxygen functions takes place in the nsec to msec range for divalent ions such as Ni(II), Fe(II) and Cu(II). However, it can take an unfractionated FA hours to take up and release metal ions. With the model presented this variation in metal binding kinetics can be explained by conformational and aggregation effects. The slower the metal ion is taken up, the more change in the tertiary and quaternary structure of the FA system required to find the preferred binding conformation. This interpretation of metal binding explains the effects of pH, metal to humic material ratio, and electrolyte concentration on the rate of metal ion dissociation.[43-48]

3.2.4 Comparison with Studies using Fractionation. This study and that of Leenheer et al. have used very different approaches to study the moities involved in FA metal binding.[49] Leenheer et al.[49] used advanced fractionation methods to isolate the major metal binding fraction of the FA. Their[49] method constrains possible structures and simplifies the choices needed to construct models. Our method interrogates the whole sample. The emergent properties of the whole FA mixture are taken into consideration. Both studies found that the strongest metal binding sites are on carbohydrate moieties and involve carboxylic and -OH functionality. This was also suggested for two FAs extracted from soils from the flood plain of the Altamaha River in southern Georgia.[40]

The FA used by Leenheer et al.[49] is the well known Suwannee River aquatic FA isolated from the Suwannee River, Georgia. The FA used in our study was isolated from a boreal forest soil in Quebec, Canada. Wilson's samples are from a third type of environment. The conclusions drawn in the three studies are fairly general. These FAs have similar low sulfur and nitrogen contents.

Our study underlines the value of studying the whole FA sample rather than its separated fractions. The whole FA contributes collectively to metal binding through conformational and aggregational effects. Emergent properties associated with aggregation and conformation may be dominant factors in the kinetics of metal ion uptake and release by FA. Copper(II) perturbation of protons is an effective way of investigating these possibilities.

3.3 Metal Binding as Monitored by Light Scattering

Light scattering measurements also are informative. The strength of light scattering measurements is that photons are only an elastic probe of the size and shape of a particle. However, well-designed light scattering can give accurate conformational data at low concentrations. Cu(II) binding to the Mossy Point (Armadale) fulvic biogeopolymer as monitored by light scattering is shown in Figure 7. These data show that the aggregation curve is very steep past a break point.[42] This shape is not allowed by the algebra of one to one metal binding. This curve shape can only be explained by cooperative binding, e.g., the "pseudo-chelation" idea offered above. In contrast to the overall increase of Rayleigh scattering, the disymmetry ratio implied that the average particle size of the larger fractions was decreasing. This can be understood by assuming that smaller particles from the fulvic biogeopolymer co-operatively bind to Cu(II) to move those particles into the size range above $\lambda/20$ that contribute to disymmetry.

We see extensive evidence for the role of emergent cooperative properties such as conformational change, aggregation and "pseudo-chelation" in metal binding by biogeopolymers. The metal binding data also are consistent with the meso-structure model for a fulvic biogeopolymer presented above.

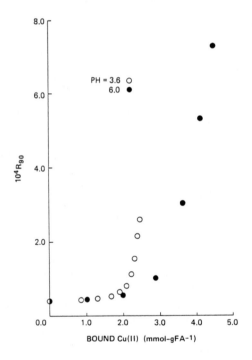

Figure 7 *Cu(II) Titration of Armadale FA monitored by Rayleigh scattering. No background electrolyte was added.[42]*

3.4 Scattering by HSs and Fractal Dimensionality

Light scattering studies are not limited to the study of metal binding effects. A compact hydrogen bonded aggregate system is expected to both dissociate and unfold as pH rises. Several studies have been interpreted in these terms. Recent information from scattering measurements employs fractional dimensionality in the interpretation of turbidity and small angle X-ray scattering.

Fractal dimensionality, a relatively new concept, is a very powerful idea for the analysis of biogeopolymers. The basic concept of fractal dimensionality can be illustrated as follows. A piece of string has one dimension, length, when fully extended and two dimensions when fully folded to completely fill a plane. Intermediate degrees of folding are described by dimensionality between 1 and 2. This concept easily can be extended into three dimensions. When a spring is fully drawn out it is 1-D, however when it is coiled it occupies space in 3-D. As the spring example shows, one can go from 1-D space without having to go through 2-D space. However when the spring is coiled it does not fully fill 3-D space rather it fills "2 plus" space. Thus fractal dimensionality allows for non-integer dimensionality.

We can expect a biogeopolymer to have a greater fractal dimensionality as a solid than as a liquid. This has been shown with small angle X-ray scattering. An aquatic humus fraction had a fractal dimensionality of 2.5 in the solid state but only 1.6 when it was dissolved.[50] In contrast, similar fractal dimensionality was found for a fulvic and humic biogeopolymer from the same source, 2.2 and 2.3, respectively.

The light scattering data suggest that increasing pH decreases the fractal dimensionality of the system due to the repulsion of deprotonated functional groups. A recent study based on turbidity using three humic biogeopolymers from three different sources shows this trend.[51] The largest change in fractal dimensionality was from 2.77 ± 0.10 (at pH 3) to 0.97 ± 0.13 (at pH 7). Turbidity based fractal dimensionality results were confirmed by an SEM analysis of one of the samples.

The fractal dimensionality is dynamic, Figure 8. For most cases the fractal dimensionality decreased with time. The fact that no equilibrium is established after 24 hours shows that the aggregational and conformational changes experienced by these biogeopolymers are complex and large in number. Some samples exhibited increases in fractal dimensionality, as shown in Figure 8 from the 16 hour data point to the 24 hour data point. Thus, the conformational changes of these biogeopolymers are so complex that approach to equilibrium may be non-monotonic.

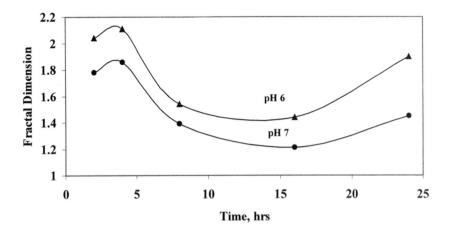

Figure 8 *A graphical representation of selected fractal dimensionality data as a function of time and pH.[51]*

A consistent message from NMR, XRD and light scattering is that aggregation and conformation are critical to the behavior of unfractionated humic materials. The controlling forces underlying the interactions can include van der Waals, H-bonding and charge transfer complexation. The number of options for the interplay of these weak forces in a humic mixture are enormous. The weak forces in biopolymers produce tertiary and quaternary structure. It is our view that "self-organizing" tendencies within this complexity account for many of the similarities of HSs from different environments.

3.5 Fulvic and Humic Biogeopolymer Binding and Folding

From the above discussion we see that the tertiary and quaternary structures of HSs systems must be considered. When a biogeopolymer "folds" itself into a low energy conformation one would not expect a smooth energy profile or a well-defined minimum nor a well-defined path to the minimum. The reason for this is that the folding of a biogeopolymer will be complex due to the complicated make-up of the system (recall the large number of monomers). The major driving forces of molecular conformational searches will be charge balancing, hydrogen bonding and hydrophobic interactions, which are weak enough to be interdependent in ways enhancing the complexity of the system. A hypothetical energy of biogeopolymeric folding is presented in Figure 9, a rough sided and

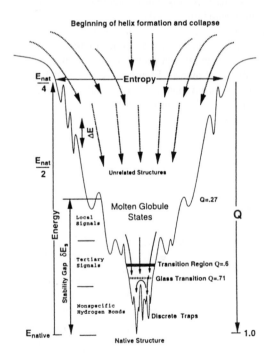

Figure 9 *A biogeopolymer folding landscape based on a viable protein folding landscape. Here, Q is the fraction of native-like contacts and the order parameters are represented by E. The depth of the funnel is on an energy scale, while the width represents entropy. For biogeopolymers there is no single native state. Also, the bottom of the funnel is expected to be less rough and with a higher energy.*[53]

bottomed teacup.[52] The wall of the funnel is rough due to the large number of local minima conformations one expects for this type of system. The floor of the energy profile is very rough since there can be many conformations with very similar energies but separated by significant energy barriers. Thus, we would not expect a single final conformation for a single biogeopolymer system, since the energy barrier between a conformation of only slightly higher to one of only slightly lower energy is high enough that change would not occur at room temperature. In other words, a single native conformation for a biogeopolymer system is not expected since there is not the regularity imposed by genetic encoding that exists in proteins. A preliminary molecular mechanics simulation of the qualitative plausibility of this model was carried out by Belliveau and Langford.[53] They took a model polymer of MW $\approx$ 1 kDa representing a typical FA composition and conducted a Monte-Carlo study of folding on cooling from 500°C using Biosym. The results were several distinct conformations of similar energy.[53]

ACKNOWLEDGEMENTS

We thank the Natural Science and Engineering Research Council of Canada for financial support. We also express our gratitude to Mrs. Qiao Wu and Dr. R. Yamdagni for technical assistance and to Mr. A. Bruccoleri for fruitful discussions. R. L. C. thanks Nova Chemicals for creating an environment in which extracurricular work is appreciated.

References

1. J. A. Neyroud and M. Schnitzer, *Can. J. Chem.*, 1974, **52**, 4123.
2. H. R. Schulten and M. Schnitzer, *Sci. Total Environ.*, 1992, **200**, 151.
3. J. A. Leenheer, R. L. Wershaw and M. M. Reddy, *Environ. Sci. Technol.*, 1995, **29**, 393.
4. J. A. Leenheer, R. L. Wershaw and M. M. Reddy, *Environ. Sci. Technol.*, 1995, **29**, 405.
5. E. J. Laboeuf and W. J. Weber, *Environ. Sci. Technol.*, 1997, **31**, 1697.
6. E. J. Laboeuf and W. J. Weber, *EOS Trans. Am. Geophys. Union*, 1998, **S92**, H221.
7. M. Schnitzer, H. Kodama and J. A. Ripmeester, *Soil Sci. Soc. Am. J.*, 1991, **55**, 745.
8. B. Xing and Z. Chen, *Soil Sci.*, 1999, **164**, 40.
9. R. L. Cook and C. H. Langford, *Environ. Sci. Technol.*, 1998, **32**, 719.
10. W. –G. Hu, J. Mao, K. Schmidt-Rohr and B. Xing, submitted to *Environ. Sci. Technol.*
11. W. Huang and W. J. Weber, *Environ. Sci. Technol.*, 1997, **31**, 2562.
12. B. Xing and J. J. Pignatello, *Environ. Sci. Technol.*, 1998, **32**, 614.
13. J. Lee, C. H. Langford and D. S. Gamble, *J. Agric. Food Chem.*, 1996, **44**, 3672.
14. J. Lee, C. H. Langford and D. S. Gamble, *J. Agric. Food Chem.*, 1996, **44**, 3680.
15. S. M. Griffith and M. Schnitzer, *Soil Sci.*, 1975, **120**, 126.
16. M. Schnitzer and S. I. M. Skinner, *Soil Sci.*, 1968, **105**, 392.
17. A. Bruccoleri, B. C. Pant, D. K. Sharma and C. H. Langford, *Environ. Sci. Technol.*, 1993, **27**, 889.
18. Z. D. Wang, B. C. Pant and C. H. Langford, *Anal. Chim. Acta*, 1990, **232**, 43.

19. Z. D. Wang, D. S. Gamble and C. H. Langford, *Environ. Sci. Technol.*, 1992, 26, 560.
20. Z. D. Wang, Ph.D. Thesis, Concordia University, 1989.
21. F. J. Stevenson, *Soil Sci.*, 1977, **123**, 10.
22. R. L. Cook and C. H. Langford, in preparation.
23. M. Schnitzer and C. M. Preston, *Soil Sci. Soc. Am. J.,* 1986, **50**, 326.
24. R. L. Cook, C. H. Langford, R. Yamdagni and C. M. Preston, *Anal. Chem.*, 1996, **68**, 3979.
25. R. L. Cook and C. H. Langford, *Polymer News*, 1999, **24**, 6.
26. R. L. Cook, Ph.D. Thesis, University of Calgary, 1997.
27. E. O. Stejskal, J. Schaefer and J. S. Waugh, *J. Magn. Reson.*, 1997, **28**, 10.
28. G. Metz, X. Wu and S. O. Smith, *J. Mag. Reson. Ser. A.,* 1994, **110**, 219.
29. G. Metz, M. Ziliox and S. O. Smith, *Solid State Nucl. Magn. Reson.*, 1996, **7**, 155.
30. D. Marks and S. Vega, *J. Mag. Reson. Ser. A.*, 1996, **118**, 157.
31. R. K. Harris 'Nuclear Magnetic Resonance Spectroscopy; A Physicochemical View', Longman, Essex, 1986, Ch. 3.
32. I. Solomon, *Phys. Rev.*, 1955, **99**, 559.
33. N. J. Bloembergen, *J. Chem. Phys.*, 1957, **27**, 572.
34. D. S. Gamble, M. Schnitzer and I. Hoffman, *Can. J. Chem.*, 1970, **48**, 3197.
35. R. D. Guy and C. L. Chakrabarti, *Can. J. Chem.*, 1976, **54**, 2600.
36. D. E. Wilson and P. Kinney, *Limnol. Oceanorg.*, 1977, **22**, 281.
37. J. A. Davis and J. O. Leckie, *Environ. Sci. Technol.*, 1978, **12**, 1309.
38. D. S. Gamble, A. W. Underdown and C. H. Langford, *Anal. Chem.*, 1980, **52**, 1901.
39. D. S. Gamble, C. H. Langford and A. W. Underdown, *Org. Geochem.,* 1985, **8**, 35.
40. M. A. Wilson, E. M. Perdue, J. H. Reuter and A. M. Vassallo, *Anal. Chem.*, 1987, **59**, 551.
41. C. H. Langford, in 'Coordination Chemistry; A Century of Progress', G. B. Kauffmann, (ed.), ACS Symposium Series 565, 1994, Ch. 33.
42. A. W. Underdown, C. H. Langford and D. S. Gamble, *Environ. Sci. Technol.*, 1985, **19**, 132.
43. J. A. Lavinge, C. H. Langford and M. K. S. Mak, *Anal. Chem.*, 1983, **59**, 2616.
44. J. G. Hering and F. M. M. Morel, *Environ. Sci. Technol.*, 1990, **24**, 242.
45. S. E. Cabaniss, *Environ. Sci. Technol.*, 1990, **24**, 583.
46. A. W. Rate, R. G. McLaren and R. S. Swift, *Environ. Sci. Technol.*, 1993, **27**, 1408.
47. C. H. Langford and R. L. Cook, *Analyst*, 1995, **120**, 591.
48. M. Bonifazi, B. C. Pant and C. H. Langford, *Environ. Technol.,* 1996, **17**, 885.
49. J. A. Leenheer, G. K. Brown, P. MacCarthy and S. E. Cabaniss, *Environ. Sci. Technol.*, 1998, **32**, 2410.
50. J. A. Rice and J. S. Lin, *Environ. Sci. Technol.*, 1993, **27**, 413.
51. N. Senesi, F. R. Rizzi, P. Dellino and P. Acquafredda, *Soil Sci. Soc. Am. J.*, 1996, **60**, 1773.
52. P. G. Wolynes, J. N. Onuchic and D. Thirumalai, *Science*, 1995, **267**, 1619.
53. S. Belliveu and C. H. Langford, ACS Environ. Div. Symp. On Computational Chemistry of Humics, 1996.

EVALUATION OF DIFFERENT SOLID-STATE ^{13}C NMR TECHNIQUES FOR CHARACTERIZING HUMIC ACIDS

Baoshan Xing,[1] Jingdong Mao,[1] Wei-Guo Hu,[2] Klaus Schmidt-Rohr,[2] Geoffrey Davies[3] and Elham A. Ghabbour[3]

[1] Department of Plant and Soil Sciences and [2] Department of Polymer Science and Engineering, University of Massachusetts, Amherst, MA 01003, USA
[3] The Barnett Institute and Chemistry Department, Northeastern University, Boston, MA 02115, USA

1 INTRODUCTION

Structural information is critical to understanding the reactions of humic acids (HAs) with organic and inorganic contaminants.[1-8] However, complex HA structures pose many analytical difficulties for their elucidation. Classical methods are based on elemental compositions.[9] Elemental data only show the average of molecular agglomeration, not a precise HA structure. Chemical degradation techniques provide valuable information on possible chemical constituents and building blocks of HAs,[10] but the cleaved HA components may be much different from the molecules that actually compose HA. Many spectroscopic methods have been employed to investigate HA compositions and structures. Nuclear magnetic resonance spectroscopy (NMR) has proven to be one of the most powerful tools.[11,12]

Solid-state ^{13}C direct polarization magic angle spinning (DPMAS) is one of the oldest NMR techniques for studying organic molecules. This technique has not been widely applied because in order to obtain a quantitative spectrum the recycle delay between scans must be five times longer than the longest T_1^C (^{13}C spin-lattice relaxation time) in a sample. In most glassy or crystalline materials, the longest T_1^Cs can be of the order of 5 sec to over 10,000 sec. Thus, forbiddingly long recycle delays have to be used to obtain a quantitative spectrum. The longest T_1^Cs in HAs can be tens of sec or even longer.[13] Therefore, it is often impractical to use only DPMAS to analyze HAs with long T_1^Cs.

The most popular solid-state ^{13}C NMR technique for studying HAs structures is cross polarization magic angle spinning (CPMAS).[14] The transfer of abundant ^{1}H spin to dilute ^{13}C spin by CP results in substantial sensitivity improvement.[15] However, there are several drawbacks that can lead to non-quantitative spectra.[13,14,16-19] The first is the reduced CP efficiency for unprotonated carbons, mobile components or regions having short $T_{1\rho}^H$ (proton spin-lattice relaxation time in the rotating frame). The second is spinning sidebands. Spinning sidebands reduce the intensity of centerbands, resulting in the loss of intensity and distortion of peak areas. The third is baseline distortions due to a dead time associated with each pulse.[20]

The total sideband suppression (TOSS) pulse sequence can be used to remove spinning sidebands before detection.[21,22] However, TOSS spectra are not quantitative at low spinning speeds because the intensity of the suppressed sidebands is not fully added to the centerband. Aromatic and C=O groups are underestimated because their sidebands are strong due to their large chemical-shift anisotropies.[23] Furthermore, ^{13}C spin-spin (T_2^C) relaxation that occurs during TOSS can lead to a differential decrease in signal intensities.[17]

High speed magic angle spinning (MAS) can reduce the sidebands but it also interferes with CP efficiency. Thus, CPMAS experiments are often run at a spinning speed below 5 kHz.[8] However, significant sidebands can occur at this speed. In order to overcome the dilemma, ramp-CPMAS was developed to establish a relatively efficient Hartmann-Hahn matching (HH) condition at a relatively high MAS speed.[24-26] Ramp-CPMAS was recently applied in studies of humic substances (HSs).[27,28] The applicability of this technique still requires further investigation because the long contact times used may render some carbon species with short $T_{1\rho}^H$ in HSs invisible.

DPMAS combined with T_1^C correction obtained from CP/T_1-TOSS spectra was first used to quantify polymer crystallinity.[29] We have applied this technique to HAs characterization. With this technique, sidebands are reduced to an insignificant proportion by high speed spinning. If HAs have long T_1^Cs, CP/T_1-TOSS spectra are used for correction.[30]

This study evaluated the above five solid-state ^{13}C NMR techniques and discusses their limitations and advantages. We hope that this paper will provide a better understanding of solid-state ^{13}C NMR techniques and their application in characterizing HSs and other organic materials.

2 MATERIALS AND METHODS

2.1 Origins and Preparation of Humic Acid Samples

Two humic acids (NHA and NYHA) were extracted from New Hampshire and New York soils and a third originating from a Leonardite sediment was purchased from International Humic Substances Society (IHSS). Details of extraction and purification were described elsewhere.[2,10] The locations and analytical data for these HAs are shown in Table 1.

Table 1 *Origins, locations, elemental compostions and ash contents of HAs[a]*

HA	Origins	%C	%H	%O	%N	% ash	%Fe
NHA	a bog soil in Rumney, New Hampshire, USA	52.9	5.40	39.7	2.00	0.25	0.024
NYHA	an alluvial farm soil, New York, USA	53.8	5.08	37.3	3.89	1.2	0.3
IHSS	a reference HA (Leonardite) from IHSS	59.2	4.08	35.6	1.12	2.3	NA

[a]Oxygen percentage was calculated assuming that the sum of C, H, O and N was 100%; NA = not available.

2.2 NMR Spectroscopy

For DPMAS, each humic acid sample was packed in a 4-mm-diameter zirconia rotor with a Kel-F cap. The samples were run at a ^{13}C frequency of 75 MHz in a Bruker DSX-300

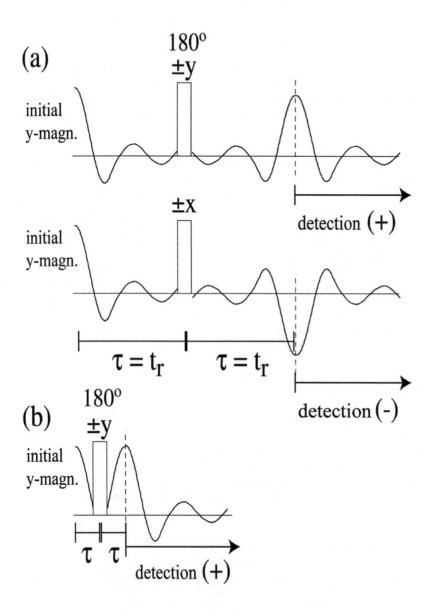

Figure 2 *Pulse sequence for the Hahn echo, which permits dead-time free detection. Limitations to the delay τ under MAS are discussed in the text*

short enough that T_2 relaxation during $2\tau = 2t_r$ is negligible enough. We have found that the T_2^Cs of solid HAs samples are usually on the order of a millisec or more, permitting echo periods of $2\tau = 150$ μsec or longer to be used without spectral distortions.

Alternatively, the echo time 2τ can be chosen to be very short. In this case the difference in the magnitude of the precession angle (phase) covered by a given isochromat in either of the two τ periods will be small. At low spinning speeds a quasi-static approximation can be used, where it is considered that the rotor, and thus the segments and interaction tensors, change their orientation insignificantly during 2τ. This is the approach taken in the CPMAS experiments described below. In the case of fast spinning, the 2τ periods (< 20 μsec) must be very short. Then it can be argued that the precession phases remain proportionally small and the MAS effects are of second order. In practice, spectra with $2\tau < 20$ μsec and with $2\tau = 150$ μsec are indistinguishable.

The optimum phase for the 180° pulse is different for the two approaches described. In the short-2τ case, the pulse phase should be along the direction of the initial magnetization. This minimizes errors due to deviations from the exact 180°-pulse length. Consider that the pulse length does not matter at all in the case of the pulse being applied along the magnetization. On the other hand, for $2\tau = 2t_r$ the EXORCYCLE approach[34] should be used, where the pulse phase is rotated through all quadrature directions (x, y, -x, -y) with respect to the magnetization direction.

The exact timing of the start of detection should be determined using a sample with very good sensitivity, such as a ^{13}C-labeled model compound that exhibits sharp peaks over the full spectral range ($\sim$ 30 - 180 ppm). For the 13kHz DPMAS spectra, timings that are off by as little as 2 μsec can result in significant spectral distortions. Once the timings are set on the model compound, spectra of HAs with excellent baselines are obtained using only constant phase correction (without linear/first-order correction).

3 RESULTS AND DISCUSSION

3.1 DPMAS and Corrected DPMAS

The elemental compositions calculated from DPMAS and corrected DPMAS spectra were consistent with those from an elemental analyzer (Table 2). These results indicate that both DPMAS and corrected DPMAS are reliable techniques for quantitative characterization of HAs. The procedure for calculating elemental HAs compositions from NMR spectra was provided by Mao et al.[30]

Table 2 *Elemental compositions from NMR calculation and chemical analysis*

HA	Methods	% C	% (O + N)	% H
NHA	Chemical analysis	52.9	41.7	5.40
	Corrected DPMAS	53.3	41.6	5.18
NYHA	Chemical analysis	53.8	41.2	5.08
	Corrected DPMAS	54.7	39.9	5.41
IHSS	Chemical analysis	59.2	36.7	4.08
	DPMAS	54.3	41.8	3.87

The advantage of DPMAS is that this technique uses direct polarization and the problems associated with CP can be avoided. For example, the reduced CP efficiency for unprotonated carbons, mobile components or regions having short $T_{1\rho}{}^H$ (proton rotating-frame spin-lattice relaxation time) is eliminated. Furthermore, DPMAS permits high spinning speeds, which reduce sidebands to an insignificant percentage. In addition, the residual sidebands are placed outside the region of the centerbands so that we can easily integrate and add them to the centerbands. At a spinning speed of 13kHz the total sidebands of the aromatic group were reduced to less than 8% of the centerband in our experiments. Moreover, with the introduction of a rotation-synchronized Hahn echo, the baseline distortion can be substantially minimized.

However, the sensitivity of DP experiments without CP is greatly reduced. Many more scans are needed. In our experiments it took just 4096 to 8192 NS for CP, but 8192 to over 30000 scans for DP to obtain a spectrum with a good signal-to-noise ratio. Furthermore, for a quantitative DPMAS spectrum, the relaxation delay should be five times longer than the longest $T_1{}^C$ in a sample. The $T_1{}^C$s in HAs samples vary widely. NYHA and IHSS HA had a relatively short $T_1{}^C < 3$ sec (Figure 3B and C). The CP/T_1-TOSS spectrum shows that IHSS HA was basically fully relaxed within 5 sec. Thus, with a recycle delay of only 5 sec an essentially quantitative DPMAS spectrum is obtained and many scans can be averaged to yield a good signal-to-noise ratio. These fully relaxed DPMAS spectra are then reliable references for the other intrinsically non-quantitative techniques.

It would be daunting to use only DPMAS for HAs samples with long $T_1{}^C$. Thus, CP/T_1-TOSS correction needs to be used for quantitative analysis of HAs. With CP/T_1-TOSS correction,[30] the recycle delay can be less that five times the longest $T_1{}^C$. Because of a relatively long $T_1{}^C$ (Figure 3A), CP/T_1-TOSS correction was used for characterizing NHA (Tables 2 and 3). This procedure has been applied to other HAs.[30] However, unreliable results due to large correction errors can occur if the second filter time (τ) is not long enough to allow the first CP/T_1-TOSS spectrum ($\tau = 500$ μsec) to relax by more than 50%.

3.2 Comparison of Different Methods

Assignment of HAs functional groups based on different chemical shift ranges is as follows:[10,28] 0-50 ppm, aliphatic groups; 50-60 ppm, methoxy groups; 60-96 ppm, carbohydrate groups; 96-108 ppm, C-O-C groups; 108-145 ppm, aromatic groups; 145-162, phenolic groups; 162-190 ppm, carboxylic groups; 190-220 ppm, carbonyl groups. Integration was performed according to these ranges (Table 3). A CP/T_1-TOSS-based correction can be performed for HAs with non-fully-relaxed DPMAS spectra. In the following discussion, the results of DPMAS or corrected DPMAS (also referred to as DPMAS later) will be used as the standards for comparison because both techniques generated quantitative results when compared to elemental analysis (Table 2).

The CPMAS technique gave higher percentages for carbons between 0-108 and 190-220 ppm and lower percentages between 108-190 ppm (Table 3). The carbon between 0-108 ppm is sp^3-C and between 108-220 ppm it is sp^2-C. Compared with DPMAS, CP/TOSS gave a higher percentage of functional groups between 0-96 ppm and a lower percentage of functional groups within the range 108-190 ppm. For anomeric (96-108 ppm) and carbonyl (190-220 ppm) groups, CP/TOSS produced larger proportions in some HAs but lower proportions in others. Nevertheless, the differences between data for these two groups from CP/TOSS and DPMAS measurements were small.

Compared to CPMAS, CP/TOSS had lower percentages of functional groups within the ranges 0-108 ppm and 190-220 ppm, and a higher percentage of functional groups within the range 108-190 ppm. These percentages were closer to the integration results of DPMAS. There is no clear trend for ramp-CPMAS for the HAs tested when compared to DPMAS (Table 3).

Table 3 *Integration results of DPMAS, CPMAS, CP/TOSS and ramp-CPMAS Spectra*

HA	Methods	0-50 ppm	50-60 ppm	60-96 ppm	96-108 ppm	108-145 ppm	145-162 ppm	162-190 ppm	190-220 ppm
NHA	DPMAS	17.0	7.11	18.0	6.48	29.3	9.73	11.2	1.21
	CPMAS	22.9	8.33	24.4	5.57	18.3	6.16	9.26	5.17
	CP/TOSS	21.9	7.46	20.6	4.54	22.7	8.40	11.0	3.36
	ramp-CP	25.5	7.68	15.1	4.01	27.2	10.1	10.1	0.33
NYHA	DPMAS	22.4	4.78	11.0	3.04	29.7	8.58	17.9	2.58
	CPMAS	31.0	7.64	20.2	5.68	15.4	4.52	10.9	4.71
	CP/TOSS	28.3	6.99	17.0	3.79	19.8	5.87	14.7	3.62
	ramp-CP	45.5	2.49	14.1	1.16	20.3	4.25	11.3	0.87
IHSS	DPMAS	15.2	2.13	4.68	3.70	43.0	13.3	15.0	2.98
	CPMAS	31.2	5.46	11.9	4.65	24.2	5.70	9.36	7.44
	CP/TOSS	27.0	3.11	6.61	3.79	38.1	9.58	9.94	1.91

The spectra from DPMAS, CPMAS, CP/TOSS and ramp-CPMAS of NHA, NYHA and and IHSS HA are shown in Figure 4. DPMAS and CP/TOSS displayed very similar spectral patterns for all HAs, with an almost identical number of peaks for a given HA. However, more peaks were exhibited in CPMAS than in CP/TOSS or DPMAS because of sidebands. Thus, it is difficult to conclude how many peaks are really exhibited by a HA sample. Again, ramp-CPMAS spectra were variable: the spectrum of NHA was similar to that from DPMAS while that of NYHA was different, particularly as regards the shape and size of the peaks (Figure 4).

The most frequently used solid-state ^{13}C technique is CPMAS because the CP enhancement via the abundant ^{1}H $(I = 1/2)$ spins can provide up to four times the signal from DPMAS. Furthermore, in CPMAS experiments recycle delays between each scan are determined by the longest $T_1{}^H$s of a sample, not the longest $T_1{}^C$s as for DPMAS. Generally, $T_1{}^H$ is significantly shorter than $T_1{}^C$. Therefore, in a given amount of time many more CPMAS experiments can be carried out. One drawback of CPMAS is that CP enhancement varies from one species of carbon to another, and even the same species could vary between HAs samples. The other drawback of CPMAS is the existence of spinning sidebands (Figure 4).

CPMAS overestimated the sp^3-C and underestimated the sp^2-C except for carbonyl-C (Table 3). The sp^2-C has a larger chemical shift anisotropy, and thus has larger sidebands than sp^3-C. As a result, the centerband area of sp^2-C is reduced more than that of sp^3-C. At spinning rates below 11 kHz, some of the sidebands of aromatic carbons occur in the sp^3 region, increasing the relative sp^3-C percentage. This is one important reason why CPMAS overestimates sp^3-C. The higher percentage of carbonyl-C integrated from CPMAS is a combination of carbonyl-C and the sidebands of aromatic-C. The detailed reasons for the

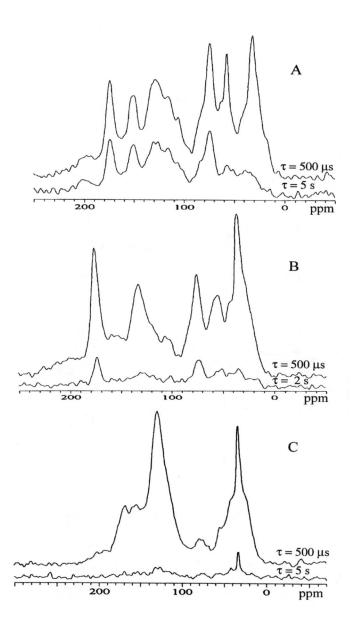

Figure 3 *CP/T$_1$-TOSS spectra showing ^{13}C relaxation of HAs with different T$_1{}^C$. A = partially relaxed spectrum of NHA at τ = 5 sec; B = nearly fully relaxed spectrum of NYHA at τ = 2 sec; C = fully relaxed spectrum of IHSS at τ = 5 sec*

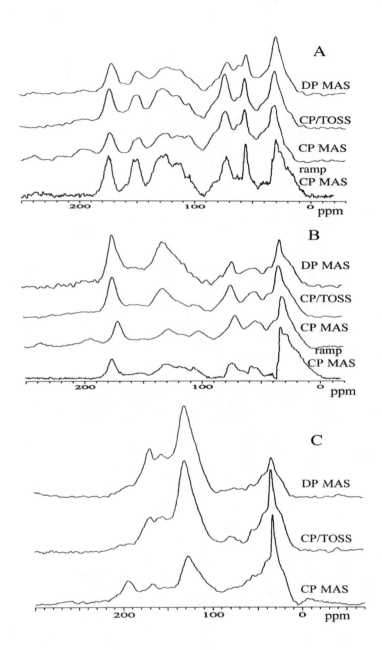

Figure 4 *Comparisons of DPMAS, CP/TOSS, CPMAS and ramp-CPMAS NMR spectra.*
A= NHA (partially relaxed DPMAS spectrum); B = NYHA (nearly fully relaxed
DPMAS spectrum); C = IHSS HA (fully relaxed DPMAS spectrum)

non-quantitative features of CPMAS have been discussed elsewhere.[13,14,16-19]

TOSS was developed to eliminate spinning sidebands from MAS spectra.[21,22] There are two advantages of CP/TOSS over simple CPMAS. The first is that a good TOSS can eliminate all the sidebands so that the spectrum shows only the true peaks for a HA sample. The second is that implementation of CP/TOSS can avoid baseline distortion arising from the dead time. However, because of their larger chemical shift anisotropy, sp^2-C are less completely refocused than sp^3-C during the TOSS sequence. Thus, CP/TOSS underestimates sp^2-C as compared to DPMAS (Table 3). On the other hand, the T_2 relaxation during the TOSS sequence with its duration of two rotation periods or more tends to decrease the signal of protonated aliphatic carbons more strongly.

The comparison of CP/TOSS or CPMAS with DPMAS (Table 3, Figure 4) indicates that CP/TOSS was consistently better than CPMAS for both quantification and qualification. Spinning sidebands of CPMAS distort the pattern and area of NMR spectra, particularly in the sp^2 region. Though it is hard to obtain absolutely quantitative spectra using CP/TOSS, comparisons of different CP/TOSS spectra for samples from similar sources can be made by assuming the same CP efficiencies for the similar samples. It is recommended that CP/TOSS be used instead of CPMAS for measurements in a $\geq$300-MHz spectrometer. We are currently examining the spectra and results obtained with lower-field spectrometers.

Ramp-CPMAS was developed in an attempt to solve the CP efficiency problem in the conventional CPMAS technique. In CPMAS experiments, proton and carbon field strength are adjusted to meet the HH condition. Broad matching profiles centered around the HH condition exist at a low spinning speed. However, if the spinning speed is increased, the HH matching profiles split into a series of narrow matching bands separated by the rotor frequency. Compared to the matching sidebands, the CP rate at the matching centerband corresponding to the exact HH condition is lower. To establish an efficient matching condition under a high speed MAS condition, Metz et al.[24,25] introduced an amplitude ramp on either of the radiofrequency channels during the contact time. It substantially improved the performance of the CP experiments at high spinning speeds. It was found that compared with conventional CP, a linear ramp of amplitudes centered at the sideband and covering its entire width greatly increased the cross-polarization rate and final signal intensity. Furthermore, the broadened matching profile permitted some deviation from an exact HH condition. Cook et al.[27,28] applied this technique in the study of a fulvic acid (FA) and a HA and reported much improved results over CPMAS for the HSs used.

However, it should be noted that the transfer rate in the ramped CP is smaller than in regular CP at lower spinning speeds. The reason is that the matching condition is fulfilled only during part of the contact time. This is no problem for model substances with a long $T_{1\rho}^{H}$ relaxation time during CP. However, for HAs the short $T_{1\rho}^{H}$ relaxation times of the order of a few millisec result in significant and often differential signal loss during long ramped CP periods. It may even render some carbon species with short $T_{1\rho}^{H}$ invisible.[25]

These effects make ramped CP non-quantitative and Cook et al.[27] had to determine the appropriate contact time empirically for each sample, using the solution ^{13}C NMR spectrum of the same HA as a standard. Acquisition of the solution NMR spectrum took about 4 days. Several ramp-CPMAS spectra need to be run to obtain the right contact time.[27] Hence, the contact time determination procedure can be time-consuming and arbitrary. Solution ^{13}C NMR spectra of HAs samples cannot be assured as quantitative spectra. Solution-state NMR has the following limitations for characterizing HSs: 1) samples may not completely dissolve, 2) solvents can modify HSs structurally, and 3) humin cannot be dissolved at all in an aqueous phase. As one can see, if the determination of the contact time for a ramp CP experiment is based on solution NMR spectra, the ramp CP spectra can be only as good as those from solution NMR. As pointed out previously, the results from this study show the inconsistency of the ramp technique (contact time = 2.5 msec) as compared to DPMAS.

4 CONCLUSIONS

Several commonly used NMR techniques have been compared to quantitative DPMAS and corrected DPMAS. Although it provides high sensitivity, CPMAS using a $\geq$300-MHz spectrometer is not reliable for quantitative or even qualitative characterization of HAs. In this situation, CP/TOSS is highly recommended because it not only provides clear qualitative information but also yields better semi-quantitative results than CPMAS. With suitable empirically determined contact times, ramp-CPMAS produces approximately quantitative spectra for some HAs but non-quantitative spectra for others. Thus, this technique needs further investigations and improvements. The selection of solid-state ^{13}C NMR techniques should be made with care. The choice depends on whether good quantitation is required, which will usually be time consuming using high-speed DPMAS, or whether only qualitative or semi-quantitative information is to be obtained, which can be achieved most efficiently using CP/TOSS, possibly with ramped CP, at intermediate spinning speeds.

ACKNOWLEDGMENTS

We thank Dr. Robert Cook for providing us with ramp-CPMAS spectra. This work was in part supported by the U.S. Department of Agriculture, National Research Initiative Competitive Grants Program (97-35102-4201 and 98-35107-6319) and the Federal Hatch Program (Project No. MAS00773).

References

1.   D. F. Cameron and M. L. Sohn, *Sci. Total Environ.*, 1992, **113**, 121.
2.   M. H. B. Hayes, in 'Humic Substances in Soils, Peats and Waters: Health and Environmental Aspects', M. H. B. Hayes and W. S. Wilson, (eds.), The Royal Society of Chemistry, Cambridge, 1997, p. 3.
3.   B. Xing, W. B. McGill and M. J. Dudas, *Environ. Sci. Technol.*, 1994, **28**, 1929.
4.   B. Xing and J. J. Pignatello, *Environ. Sci. Technol.*, 1997, **31**, 792.
5.   B. Xing and J. J. Pignatello, *Environ. Sci. Technol.*, 1998, **32**, 614.
6.   B. Xing, *Chemosphere*, 1997, **35**, 633.
7.   B. Xing, *J. Environ. Sci. Health*, 1998, **B33**, 293.
8.   B. Xing and Z. Chen, *Soil Sci.*, 1999, **164**, 40.
9.   D. S. Orlov, 'Humus Acids of Soils', Amerind Publishing, New Delhi, India, 1985.
10.   F. J. Stevenson, 'Humus Chemistry: Genesis, Composition, Reactions', 2nd edn. Wiley, New York, 1994.
11.   M. A. Nanny, R. A. Minear and J. A. Leenheer, 'Nuclear Magnetic Resonance

Spectroscopy in Environmental Chemistry', Oxford University Press, New York, 1997.

12. C. M. Preston, *Soil Sci.*, 1996, **161**, 144.
13. P. Kinchesh, D. S. Powlson and E. W. Randall, *Eur. J. Soil Sci.*, 1995, **46**, 125.
14. M. A. Wilson, 'NMR Techniques and Applications in Geochemistry and Soil Chemistry', Pergamon Press, Oxford, 1987.
15. E. O. Stejskal, J. Schaefer and J. S. Waugh, *J. Magn. Reson.*, 1977, **28**, 105.
16. R. Fründ and H. -D. Lüdemann, *Sci. Total Environ.*, 1989, **81/82**, 157.
17. P. Kinchesh, D. S. Powlson and E. W. Randall, *Eur. J. Soil Sci.*, 1995, **46**, 139.
18. C. M. Preston, J. A. Trofymow, G. G. Sayer and J. Niu, *Can. J. Bot.*, 1997, **75**, 1601.
19. R. L. Wershaw and M. A. Mikita, 'NMR of Humic Substances and Coal: Techniques, Problems and Solutions', Lewis Publishers, Chelsea, MI, 1987.
20. K. Schmidt-Rohr and H. W. Spiess, 'Multidimensional Solid-State NMR and Polymers', Academic Press, San Diego, 1994.
21. W. T. Dixon, *J. Chem. Phys.*, 1982, **77**, 1800.
22. W. T., Dixon, J. Schaefer, M. D. Sefcik, E. O. Stejskal and R. A. McKay, *J. Magn. Reson.*, 1982, **49**, 341.
23. D. E. Axelson, 'Solid State Nuclear Magnetic Resonance of Fossil Fuels', Multiscience Publications, Canadian Government Publishing Center, Supply and Services Canada, 1985.
24. G. Metz, X. Wu and S. O. Smith, *J. Magn. Reson. Ser. A*, 1994, **110**, 219.
25. G. Metz, M. Ziliox and S. O. Smith, *Nucl. Magn. Reson.*, 1996, **7**, 155.
26. O. B. Peersen, X. Wu and S. O. Smith, *J. Magn. Reson. Ser. A*, 1993, **106**, 127.
27. R. L. Cook, C. H. Langford, R. Yamdagni and C. M. Preston, *Anal. Chem.*, 1996, **68**, 3979.
28. R. L. Cook and C. H. Langford, *Environ. Sci. Technol.*, 1998, **32**, 719.
29. W. Hu and K. Schmidt-Rohr, *Polymer* (submitted).
30. J. Mao, W. Hu, K. Schmidt-Rohr, G. Davies, E. A. Ghabbour and B. Xing, in 'Humic Substances: Structures, Properties and Uses,' G. Davies and E.A. Ghabbour, (eds.), Royal Society of Chemistry, Cambridge, 1998, p. 79.
31. D. A. Torchia, *J. Magn. Reson.*, 1978, **30**, 613.
32. E. L. Hahn, *Phys. Rev.*, 1950, **80**, 580.
33. A. C. Kolbert, D. P. Raleigh and R. G. Griffin, *J. Magn. Reson.*, 1989, **82**, 438.
34. G. Bodenhausen, R. Freeman and D. L. Turner, *J. Magn. Reson.*, 1977, **27**, 511.

NMR EVIDENCE FOR CRYSTALLINE DOMAINS IN HUMIC SUBSTANCES

Wei-Guo Hu,[1] Jingdong Mao,[2] Klaus Schmidt-Rohr[1] and Baoshan Xing[2]

[1]Department of Polymer Science and Engineering, University of Massachusetts, Amherst, MA 01003, USA
[2]Department of Plant and Soil Sciences, Stockbridge Hall, University of Massachusetts, Amherst, MA 01003, USA

1 INTRODUCTION

Soil organic matter (SOM) represents a major component of the world surface carbon reserves. Humic substances (HSs) represent up to 90% or higher of SOM. Even though HSs contents in mineral soils are generally in the range of 1-5%, they significantly affect soil properties and performance and contribute to plant growth through their effects on the physical, chemical and biological properties of soils. HSs are also of environmental importance.[1] They are involved in the transportation and subsequent concentration of minerals and metals and responsible for the enrichment of uranium and other metals in various bioliths such as coals. Furthermore, aquatic HSs are carriers of organic xenobiotics and can reduce the toxicity of certain heavy metals to aquatic organisms. Depending on environmental conditions, HSs can serve as oxidizers or reducing agents and they affect photochemical processes in natural waters, including the photoalteration of xenobiotics. Moreover, HSs influence the sorptive behavior of soils for many organic and inorganic contaminants.[2-6]

The structures of individual SOM fractions regulate their reactivity, properties and functions but they are poorly understood. A better understanding of SOM structures and particularly of HSs would help to determine their origin, genesis, reactivity and roles in environmental processes.

Significant fractions of SOM are poorly soluble or insoluble and contain macromolecules with molecular weights of >1 kDa. Therefore, solid-state nuclear magnetic resonance (NMR) is the method of choice for investigating their chemical and physicochemical structures. Modern solid-state NMR can provide information on composition, segmental dynamics, domain sizes and local ordering of macromolecules.[7,8] Because of the highly diverse and irregular chemical structure of SOM, the application of solid-state NMR to SOM has been limited mainly to compositional characterization.[9-13]

HSs are considered to be amorphous. In this paper we report the identification and characterization of poly(methylene) crystallites in various HSs with cross polarization (CP) and direct polarization (DP) NMR experiments. Implications of these crystallites for soil properties are briefly discussed.

2 MATERIALS AND METHODS

Two samples were studied: (1) ARC humic acid (called ARC HA in the later text) and (2) German peat humic acid (German HA). ARC HA is a product of Arctech, Inc., Chantilly, Virginia and German HA was extracted from a peat sample obtained near Bad Pyrmont, Germany. The NMR experiments were performed on a Bruker DSX-300 spectrometer at a ^{1}H frequency of 300 MHz and a ^{13}C frequency of 75 MHz. Two kinds of experiments were conducted: (1) cross-polarization (CP) from ^{1}H to ^{13}C and (2) direct-polarization (DP) by a ^{13}C 90° pulse. All the experiments were performed under magic angle spinning (MAS) conditions with proton decoupling during detection. The spinning speed was 4 kHz for CP experiments and 13 kHz for DP experiments. At these spinning speeds the sideband intensity of $(CH_2)_n$ carbons is less than 5% of the centerband and sidebands of other major components are much less than the $(CH_2)_n$ signal in the vicinity of 30 ppm. In the CP experiments, the CP time was 0.8 msec and the recycle delay was 1 sec. In the DP experiments the recycle delays were 4 to 5 sec. The ^{1}H decoupling power was 70 kHz. The number of scans (NS) for the CP spectra was 1,000 to 4,000 and for the DP spectra was 16,000.

3 RESULTS AND DISCUSSION

3.1 NMR Identification of Poly(methylene) Crystallites

Figure 1a shows the CP and DP ^{13}C NMR spectra of ARC HA. The signal extends over a wide range (10 - 210 ppm), and can be assigned to aliphatic, aromatic, carbonyl and other chemical groups.[14] There is a strikingly sharp peak at 32.9 ppm. In the DP spectrum, the 32.9 ppm peak is relatively weaker and the 31 ppm peak becomes prominent. The 32.9 ppm and 31 ppm peaks are assigned to crystalline and amorphous $(CH_2)_n$ units ("poly(m)ethylene-like" chains), respectively. Even though the amorphous and crystalline regions have the same chemical structure, they exhibit slightly different chemical shifts due to their different conformations: in the crystalline state $(CH_2)_n$ chains are all-trans, while in the amorphous state the chains have both trans and gauche conformations. Due to the γ-gauche effect,[15] the crystalline and amorphous signals appear at different chemical shifts. The chemical shift of 32.9 ppm corresponds to orthorhombic $(CH_2)_n$ crystals.[16]

Compared to the CP spectrum in Figure 1a, the DP spectrum of the crystalline $(CH_2)_n$ signal is suppressed while the amorphous $(CH_2)_n$ signal is better represented. This is because the crystalline $(CH_2)_n$ chains have less mobility than the amorphous chains, which produces two effects: (1) The ^{13}C T_1 relaxation time in the crystalline region is much longer than in amorphous regions. Because the recycle delay in the DP experiments was not long enough for the crystalline magnetization to fully relax, the crystalline signal is underrepresented, while the amorphous signal is shown much better since its ^{13}C T_1 is much shorter than 5 s as a result of the chain mobility; (2) The CP efficiency in crystals is much higher due to the lack of chain mobility. Therefore, in addition to the chemical shift of 32.9 ppm, comparison of the relative intensity of 32.9 ppm and 31 ppm signals in the CP and DP spectra confirms the crystalline nature of the 32.9 ppm signal. The crystalline signal is underrepresented due to the incomplete relaxation.

The CP and DP spectra of German HA are shown in Figure 1b. As in Figure 1a, the crystalline signal at 32.9 ppm is suppressed in the DP spectrum of German HA. It is clear that the fraction of the crystalline component in German HA is smaller than in ARC HA.

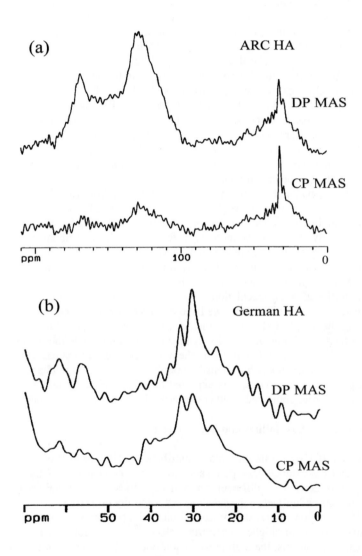

Figure 1 *Cross-polarization (CP) and direct-polarization (DP) ^{13}C NMR spectra of (a) ARC HA and (b) German HA. In both samples the crystalline signal at 32.9 ppm and amorphous signal at 31 ppm are clearly seen*

Signal intensity at 30-33 ppm in various SOM samples and its assignment to $(CH_2)_n$ groups has been documented in the literature.[12,17] It also has been reported that part of the $(CH_2)_n$ spectral contribution has a short ^1H-^1H dipolar dephasing time and a long ^{13}C T_1 (see ref. 17, 18). However, the characteristic two-peak crystalline-amorphous structure was never resolved in the spectra due to line broadening introduced by weak decoupling, short acquisition times and data smoothing. Therefore, the semicrystalline nature of this component was not recognized.

The $(CH_2)_n$ segments persist to a late stage in decomposition of soil and aquatic organic matter[12,19-21] and this is likely the mechanism of kerogen formation.[22] It is well known that synthetic high density polyethylene (HDPE) is extremely difficult to degrade in the natural environment.[23] All these facts suggest that the PE-like component is long-lasting in soil. We propose that this is at least partly due to the semicrystalline nature of the poly(methylene) component. Being inaccessible to chemicals[24] or enzymes, crystalline domains in polymers are much harder to biodegrade than their amorphous counterparts. For example, in poly(ε-caprolactone, $[-(CH_2)_5-COO-]_n$, PCL), the amorphous regions are degraded prior to the degradation of the crystalline regions. Therefore, we believe that in addition to the chemical stability of the $(CH_2)_n$ molecules, the highly compact structure of the $(CH_2)_n$ crystal contributes to making the residence time of the $(CH_2)_n$ component longer than that of some other components in HSs.

Investigating forest soil organic matter by solid-state NMR, Kögel-Knabner et al.[17] found that the $(CH_2)_n$ signal in the neighborhood of 30 ppm can be decomposed into more rigid and more mobile parts and that the rigid fraction increases with soil profile depth. Since the most likely reason for the mobility difference of $(CH_2)_n$ segments is a difference in packing order (crystalline vs. amorphous), in view of our results it must be expected that the more rigid part is crystalline;[17] because it is more resistant to permeation by small molecules, it has a longer residence time than the amorphous chains.

Contrary to the common speculation that rigid chains would be easier to crystallize in humic substances, the crystallites that we have observed are composed of highly flexible chains. Having the simplest chemical structure and the widest application, polyethylene (PE) is by far the most thoroughly studied synthetic polymer. Despite the flexibility of the polyethylene chains, the crystals formed by them are very rigid as manifested by the fact that highly crystalline PE fibers have tensile moduli higher than for steel and are used to make bullet-proof vests.[22,25] Thus, it is expected that the $(CH_2)_n$ crystallites will exhibit very different properties compared to other non-crystalline aliphatic components in SOM.

3.2 Relation between Crystallites and Soil Properties

The presence of $(CH_2)_n$ crystallites may contribute specific properties to soil. Since crystalline regions are much less permeable to small molecules than the amorphous regions, they must have very different sorption capacities. The crystallites may act as physical cross-links and influence the swelling and solubility properties.

The $(CH_2)_n$ component may derive from aliphatic biopolymers in algal cell walls and the protective layers of higher terrestrial plants.[20,27,28] Despite their initial low concentrations in organisms, the high preservation potential of such structures will result in their selective enrichment during diagenesis.[22] This process was found to be related to the formation of kerogen.[22]

The formation of humic substances is not well understood. This is largely due to the diversity of their chemical structures. Because of the chemical regularity of the polyethylene-like component, further investigations to isolate and thoroughly characterize it regarding chain length, end groups and its reactions since the beginning of the decomposition of plant residue may be more straightforward than for other components. Due to its chemical stability, the crystalline component remains unchanged over long times. Therefore, it could serve as an "internal standard" of the evolution of SOM. In short, the semicrystalline PE-like component could bring a new perspective to the understanding of soil organic matter.

ACKNOWLEDGEMENTS

We sincerely thank Dr. Elham A. Ghabbour and Dr. Geoffrey Davies (Barnett Institute, Northeastern University) for isolating, purifying and chemically analyzing the ARC and German HAs used in this study. This work was in part supported by the U.S. Department of Agriculture, National Research Initiative Competitive Grants Program (97-35102-4201 and 98-35107-6319), the Federal Hatch Program (Project No. MAS00773), and a Faculty Research Grant from the University of Massachusetts at Amherst.

References

1. F. J. Steveson, 'Humus Chemistry: Genesis, Composition, Reactions', 2nd edn., Wiley, New York, 1994.
2. G. Davies, A. Fataftah, A. Cherkasskiy, E. A. Ghabbour, A. Radwan, S. A. Jansen, S. Kolla, M. D. Paciolla, L. T. Sein, Jr., W. Buermann, M. Balasubramanian, J. Budnick and B. Xing, *J. Chem. Soc., Dalton Trans.*, 1997, 4047.
3. B. Xing and J. J. Pignatello, *Environ. Sci. Technol.*, 1996, **30**, 2432.
4. B. Xing and J. J. Pignatello, *Environ. Toxicol. Chem.*, 1996, **15**, 1282.
5. B. Xing and J. J. Pignatello, *Environ. Sci. Technol.*, 1997, **31**, 792.
6. B. Xing and J. J. Pignatello, *Environ. Sci. Technol.*, 1998, **32**, 614
7. K. Schmidt-Rohr and H. W. Spiess, 'Multidimensional Solid-State NMR and Polymers', Academic Press, London, 1994.
8. K. Schmidt-Rohr, W. Hu and N. Zumbulyadis, *Science*, 1998, **280**, 714.
9. P. G. Hatcher, D. L. VanderHart and W. Earl, *Org. Geochem.*, 1980, **2**, 87.
10. P. F. Barron and M. A. Wilson, *Nature*, 1981, **289**, 275.
11. C. M. Preston, M. Schnitzer and J. A. Ripmeester, *Soil Sci. Soc. Am. J.*, 1989, **53**, 1442.
12. J. A. Baldock, J. M. Oades, A. G. Waters, X. Peng, A. M. Vassallo and M. A. Wilson, *Biogeochemistry*, 1992, **16**, 1.
13. J. Mao, W. -G. Hu, K. Schmidt-Rohr, G. Davies, E. A. Ghabbour and B. Xing, submitted to *Environ. Sci. Technol.*
14. M. A. Wilson, *J. Soil Sci.*, 1981, **32**, 167.
15. A. E. Tonelli, 'NMR Spectroscopy and Polymer Microstructure: the Conformational Connection', VCH, New York, 1989.
16. D. L. Tzou, K. Schmidt-Rohr and W. H. Spiess, *Polymer*, 1994, **35**, 4728.
17. I. Kögel-Knabner, P. G. Hatcher, E. W. Tegelaar and J. W. de Leeuw, *Sci. Tot. Environ.*, 1992, **113**, 89.
18. R. H. Newman and L. M. Condron, *Solid State Nucl. Magn. Reson.*, 1995, **4**, 259.
19 P. G. Hatcher, E. C. Spiker, N. M. Szeverenyi and G. E. Maciel, *Nature*, 1983, **305**, 498.
20. E. W. Tegelaar, J. W. de Leeuw and C. Saiz-Jimenez, *Sci. Tot. Environ.*, 1989,

81/82, 1.
21.   E. W. Tegelaar, H. Kerp, H. Visscher, P. A. Schenck and J. W. de Leeuw, *Paleobiology*, 1991, **17**, 133.
22.   E. W. Tegelaar, J. W. De Leeuw, S. Derenne and C. Largeau, *Geochim. Cosmochim. Acta*, 1989, **53**, 3103.
23.   J. E. Potts, in 'Aspects of Degradation and Stabilization of Polymers', H. H. G. Jellinek, (ed.), Elsevier, New York, 1978, p. 617.
24.   D. J. Blundell, A. Keller, I. M. Ward and I. J. Grant, *J. Polym. Sci., Polym. Lett. Ed.*, 1966, **4**, 781.
25.   P. J. Lemstra, R. Kirschbaum, T. Ohta and H. Yasuda, in 'Developments in Oriented Polymers-2', I. M. Ward, (ed.), Elsevier, New York, 1987, p. 39.
26.   R. S. Porter and L. -H. Wang, *J. Macromol, Sci.-Rev. Macromol. Chem. Phys., C*, 1995, **35**, 63.
27.   E. A. Baker, in 'The Plant Cuticle', D. F. Cutler, K. L. Alvin and C. E. Price, (eds.), Academic Press, New York, 1982, p 139.
28.   R. A. J. Pacchiano, W. Sohn, V. L. Chlanda, J. R. Garbow and R. E. Stark, *J. Agric. Food Chem.*, 1993, **41**, 78.

CHANGES OF COLLOIDAL STATE IN AQUEOUS SYSTEMS OF HUMIC ACIDS

E.Tombácz[1] and J. A. Rice[2]

[1] Department of Colloid Chemistry, Attila József University, H-6720 Szeged, Hungary
[2] Department of Chemistry & Biochemistry, South Dakota State University, Brookings, SD 57007, USA

1 INTRODUCTION

Humic substances (HSs) are inherently composite materials in both chemical and structural points of view. They are considered to be multifunctional aromatic components linked chemically and physically by a variety of aliphatic constituents.[1] Their most probable molecular structure is a self-similar fractal, cross-linked aromatic network with different functional groups and side-chains. The formation of humic materials in a random polymerization process results in a large variety of polymers in chemical and dimensional points of view. A broad, polydisperse size distribution is characteristic of HSs. Even a power law-like distribution may well be probable, as suggested by Beckett et al.,[2] who found very broad molecular weight distributions from field-flow fractionation data.

According to the accepted principles of colloid science, systems are considered to exhibit colloidal properties when the dimensions of the dissolved or dispersed components (in hydrophilic or hydrophobic colloids, respectively) are in the range of 1 to 1000 nm (e.g. Everett[3]). Depending on the chemical composition of the dissolved or dispersed components and on the molecular interactions with the solvent molecules, a given type of colloidal system can generally be formed. For example, a macromolecular solution results if a polymer is dissolved in an appropriate solvent, or a micellar solution is formed when the surface active solute concentrations is above a critical value (critical micelle concentration, CMC); or dispersions result in which solid particles or immiscible liquid droplets are dispersed. Colloidal character (i.e. specific properties and changes of colloidal state) can be expected for aqueous humic systems because of the size ranges that they display in either the dissolved or dispersed (precipitated) state.

Various solution condition dependent intra- and intermolecular interactions can be supposed in aqueous systems of HSs. Because of their random formation, any uniform behavior of the polydisperse product can not be expected and only trends of properties can be predicted.

Hydrophobic moieties such as long alkyl side-chains from fatty acid residues provide amphiphilic character in humic molecules. Therefore, analogous properties to those of surface active agents can be expected.[4] For example, humic molecules can accumulate in interfacial layers, resulting in a decrease in the surface tension of their aqueous solutions[5] and in the interfacial tension at water-oil interfaces.[4] Also, the formation of self-assembled systems due to hydrophobic interactions can be expected, as for example in the formation of oriented bilayers in membrane-like structures when HS form coatings on mineral surfaces in soils[6] or as micellar solutions that can solubilize otherwise water-insoluble organic compounds.[7,8]

The dissociation of acidic HSs functional groups in aqueous media leads to the spontaneous formation of an electric double layer (edl). Therefore, electrostatic interactions play the determining role in the conformational changes of humic macroions as well as in the colloidal stability of humic solutions and in the aggregation of the individual humic macroions. The compensating charge in the edl consists of counterions that penetrate into the aromatic network and of co-ions that are expelled from it.

At high pH and low ionic strength the functional groups are fully ionized and the chemically linked charges endeavor to situate themselves as far apart as possible. However, expansion of the aromatic network is limited by the chemical cross-linkages. Therefore, the extent of expansion and compression of cross-linked humic nanospheres is much smaller than for linear polyelectrolytes.

The expanded or collapsed networks are totally or partially penetrable by water molecules, and the formation of water impenetrable units can be supposed after precipitation of humic materials. The macroscopic aggregation of humic acid (HA) particles (referred to as precipitation, coagulation, flocculation or phase separation) is mainly determined by the screened electric field developed from the dissociation of acidic groups bound chemically to the cross-linked polymeric network of humic molecules. Like any other aqueous colloidal dispersion,[9] the colloidal stability of an aqueous HA solution prior to the onset of macroscopic phase separation is controlled chemically and electrostatically. A theoretical approach has been introduced to describe the charge-induced stabilization and coagulation[10] and the conformational changes that occur during particle aggregation[11] of humic acids (HAs). A site-binding model is used to explain the concomitant effects of pH and neutral salts on the colloidal stability of HA solutions.

The inherent polydispersity in the size distribution of humic samples and the spontaneous changes in their conformational and aggregation states that take place under even slightly changing solution conditions are the reasons for the differences (often of orders of magnitude) in their average molecular weight or size distributions. Thus, some of the differences observed may be the result of slight differences in the experimental conditions of methods used to characterize the sizes of dissolved humic materials. The influences of polydispersity and fractal character on the evaluation of dynamic light and small angle X-ray scattering data have been analyzed,[11-14] and a characteristic example was given[10] of the orders of magnitude differences in the number average molecular weight values for a fractionated HA sample measured at low and high pH values and at different neutral electrolyte concentrations. We have analysed the role of the polydispersity in aggregation[20] and concluded that the sensitivity of humate solutions to electrolytes strongly depends on the size and acidic group density of humic nanospheres. Based on this feature, electrolyte induced spontaneous fractionation of multicomponent natural solutions can be predicted.[20]

To summarize, it can be stated that polyelectrolyte-, surfactant- and charged nanoparticle-like colloidal behaviour can be predicted for aqueous systems of HAs. In an aqueous medium, humic materials can be dissolved or precipitated, they can accumulate at interfaces, self-assemble, solubilize organic compounds and exist in different colloidal states depending on the solution conditions. The main external factors that influence their colloidal state are the pH, ionic strength, the presence of di- or multivalent metal ions, cationic organic compounds, organic liquids and solid particles. A schematic illustration in Figure 1 shows different aqueous systems of HSs that are supposed to exist under different solution conditions and the spontaneous processes in which their colloidal states change.

Aqueous systems of humic substances

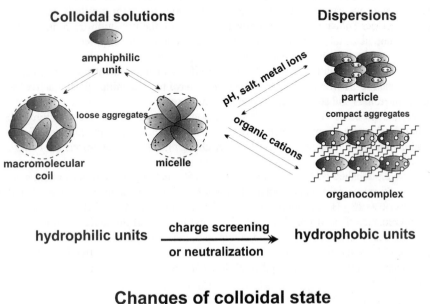

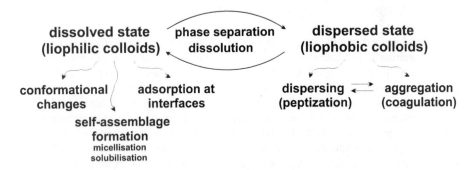

Figure 1 *Schematic illustration of different aqueous systems of HSs, and summary of characteristic colloidal processes and colloidal state changes*

It can be presumed that the building blocks of humic nanospheres are amphiphilic units containing hydrophilic groups linked chemically to an aromatic and/or aliphatic hydrophobic part. These amphiphilic units are associated to form loose aggregates similar to either macromolecular coils or micelles of liophilic colloids.[3] The loose nanospheres have polar, mainly negatively charged groups and are hydrophilic. So they are penetrated fully by water. In dilute solution at appropriate pH, FAs and HAs their alkali metal salts are considered to be real colloidal solutions, showing the properties of both cross-linked polyelectrolyte and surfactant solutions. Characteristic colloidal processes of dissolved HSs are conformational changes,[11-14] adsorption at different interfaces (mainly at the water surface[4,5,7] and at clay or oxide/water interfaces),[15-18] and self-assembly in solution[4,5,7,8] and at solid interfaces.[6] Changes in solution conditions such as a decrease in pH or addition of multivalent metal ions or organic cations induces the formation of compact aggregates. Since the charges are screened or neutralized, water molecules can not penetrate inside and so compact, hydrophobic nanospheres form similarly to those in liophobic colloids.[3] Changes of colloidal state of HSs are the transition from dissolved to dispersed state and the opposite process is dissolution. These processes should be reversible, which is obvious for example in pH and neutral electrolyte induced precipitation and coagulation.[10,11] However, the effects of multivalent metal and organic cations[4,19] seem to be irreversible, since these ions have high affinities for the active sites of humic macroions and so their displacement is hindered.

Colloidal processes involving HSs are spontaneous and the colloidal state of HSs continuously changes in environmental systems. Adsorption on mineral surfaces, dissolution and precipitation of humic materials under changing solution conditions are common in soils and natural waters.

We have published[4,5,8,10-14,16,17] adsorption, surface activity, micelle formation, solubilization, conformational changes, fractal features and a theoretical approach to explain the combined effects of pH and neutral salts on the colloidal stability of HA solutions. We have analyzed the effect of polydispersity[12-14,20] and the mutual influence of simultaneous colloidal processes such as conformational changes and aggregation[11,12] or adsorption and coagulation.[16,17] In the present work we summarize the characteristic trends in colloidal state changes of humic materials and how they correlate with the composition of humic materials.

2 DISCUSSION

The colloidal properties and the spontaneous colloidal state of HAs change under variable solution conditions. The effect of pH and neutral electrolytes and the influence of HAs' polydispersity have been systematically studied. The HA samples were isolated with traditional methods[4,5,8,10-14,16,17,19,20] from soil, compost, peat and brown coal.

We have discussed the surfactant-like character of aqueous solutions of HAs of different origins and of alkali metal humates.[5] Investigations of the accumulation of amphiphilic units at the surface of water (that is at the gas/liquid interface), the surface activity of HAs and alkali metal humate solutions, and micelle formation equilibria in solution (including the dependences on salt concentration and temperature) showed a complete analogy between aqueous solutions of surfactants and humic compounds. The extents of decreases in surface tension were different, depending on the origins of the HAs, and these decreases were greater when H^+, Li^+, Na^+ and K^+, in that order, were the compensating cations. The lowest surface tension values, in the range 40 and 42 mN m^{-1}, were obtained for the potassium salt of a compost HA, while the corresponding values for HAs from the brown coal were between 52 and 54 mN m^{-1}. The concentrations for critical micelle formation (CMC) were between 5 and 10 g L^{-1} humate. These values were relatively high compared to the values for surfactants. Both the surface tension lowering effects and the CMC values were significantly dependent on the origins of the HAs, which presumably can be attributed to the different chemical compositions of the samples. The salt concentration dependence of the CMC for K^+-humate samples, however, was the same as that for an ionic surfactant, and the enthalpy change of micellisation could be calculated from the linear plot of log CMC vs. log cK^+. The salting-out phenomenon was also observed above 0.2 M KCl. The heats of micellisation calculated from the temperature dependence of CMC were of the same order of magnitude as those for surfactants. The temperature dependence of the heat of micellisation focuses on the significance of hydrophobic interactions in the micellisation process arising from the presence of long alkyl chains.

The roles of the longer or shorter alkyl chains in conferring on HA solutions properties analogous to those of surfactants were studied using complexes of humates with a homologous series of synthetic n-alkylammonium cations.[4] It was found that the longer the hydrocarbon chain, the greater was the tendency for amphiphilic units to accumulate both at water surfaces and organic liquid/water interfaces, and hence to lower the surface and interfacial tension.

Further analogous properties of HAs and surfactants have been observed.[8] It was found that a hydrophobic organic compound like DDT that is water insoluble can be solubilized to levels many times greater than its solubility in water in a relatively concentrated humate solution (7.4 g L^{-1}), which is above the CMC where micelles with hydrophobic cores are present. This phenomenon is solubilization, which is particularly characteristic of micellar solutions of surfactants.

We have investigated the surface tension lowering effects of ultrafiltered fractions of a peat HA.[21] Spectroscopic investigations[20] (UV-visible, fluorescence, CPMAS ^{13}C-NMR) showed systematic differences in the chemical compositions of the different size fractions. Although aliphatic character predominated for the unfractionated peat HA, it became more pronounced for the largest size fraction and it was seen that aromaticity increased significantly with decreasing molecular size. A similar tendency was observed with regard to the abundance of functional groups; oxygen containing groups increased as the molecular sizes decreased. The measured surface tension data for dilute, slightly alkaline solutions showed a relatively high surface activity effect for all the samples, presumably as a result of the aliphatic character of the peat HA. In contrast to the extents of the surface tension decreasing effects of the samples, it could be seen that the unfractionated HA and the larger sized fractions, which had more pronounced aliphatic character, had larger effects than did the smaller fractions, where the relative amounts of aromatic constituents and oxygen containing functional groups were larger.

These experimental facts are in harmony with the expected trend. It should be noted that the influence of the most effective surfactant dominates in any mixture of surface active materials.[9] The whole unfractionated HA sample can be considered to be a mixture containing the most effective, largest sized components, and therefore the surface tension lowering in solutions of the whole sample is almost the same as that for the larger size fractions.

Dynamic light (DLS) and small angle X-ray (SAXS) scattering methods have been used to study changes in conformation and aggregation state of HA in the course of its transition from a colloidal solution to a phase-separated, heterogeneous system.[11,13] The DLS behavior of HA solutions showed a very unusual scattering angle dependence.[12] The initial relaxation rate (Γ_i) was shown to be a fractional power of the scattering vector q = (2.5±0.1) for all the samples measured under different solution conditions instead of exhibiting the expected quadratic dependence typical of translationally-driven particle dynamics. The explanation for this unusual behavior resides in both the polydispersity and the fractal character of the HA particles responsible for the observed scattering.[13,14] A detailed analysis of the DLS data with dynamic scaling theory for an ultrafiltered HA fraction at pH ~9 showed compaction of expanded particles (from a hydrodynamic radius of 95 nm down to a hydrodynamic radius of 48 nm) when the salt concentration was raised from 0.001 to 0.1 M, apparently due to charge screening effects. The rigid inner structure of the scatterers and the moderate power-law polydispersity demonstrated for this sample allows the mass fractal dimension $D_m = 2.1$ of the humate particles obtained from a SAXS measurement to be used without correction for polydispersity effects. A mass fractal dimension of 2.1 also suggests that a reaction-limited cluster aggregation model that typically produces aggregates with $D_m \sim 2.0$ might be appropriate to describe the aggregation behavior of HAs. This type of aggregation is a slow process in which clusters must overcome the repulsive forces between negative charges of humic colloids before aggregation and the formation of larger aggregates can occur.

Detailed analysis of the small angle X-ray scattering pattern allowed us to calculate the correlation length, the average size of inhomogeneities in electron density. Its values increased from 1.9 nm to 2.9 nm with increasing dilution of humate solutions at pH~9. By comparing these data with the radius value (95 nm) from the DLS measurement, we concluded that the humic nanoparticles in dilute alkaline solutions at low salt concentration have a loose, open inner structure containing inhomogeneities ~2 nm in electron density. These may be identified as aromatic rings linked together randomly by chemical bonds and highly penetrated by water.

The observed particle-size decrease just prior to the onset of macroscopic aggregation has been explained by the significant decrease in outer layer potential from surface complexation model (TLM) calculations.[11,13] This reduces the repulsion between charged sites in the humate nanoparticles. Further decreases in repulsive potential with either decreasing pH or increasing salt concentration can be predicted via the model calculation. The DLS measurement of several series of dilute HA solutions showed that a decrease in pH from 10 to 3 at very low salt concentration (0.001 M) caused a decrease in particle hydrodynamic radius from 95 nm to ~25 nm. At pH~3 even a slight increase in neutral electrolyte concentration initiates intraparticle aggregation of the now condensed nanoparticles, which results in the formation of very large particles with apparent hydrodynamic radii as large as several hundred nm. One of the more interesting observations is that the salt concentration at which the interparticle aggregation process starts is different by two orders of magnitude in acidic and alkaline solutions of the same HA fraction: it is 0.01 M at pH~3 while at pH~9 it is ~1 M. The repulsive potential values in the TLM calculation[11,13] predict this because the depressed ionization of functional groups at low pHs can provide only a weak electric field. By contrast, because of the enhanced dissociation at high pHs, a much lower salt concentration can effectively screen the particles than is needed to screen the stronger electric field existing at higher pHs.

In dilute systems of HAs, two subprocesses take place in the course of the transition from a stable colloidal solution to a phase-separated heterogeneous system. An intraparticle contraction or shrinkage is followed by an interaction between the collapsed nanoparticles. This produces an unstable colloidal system, which then coagulates and precipitates. Both subprocesses are chemically and electrostatically controlled. The pH and ionic strength dependent ionization of acidic functional groups creates an electrostatic field inside and around the particles that determines the aggregation state of the system.

Testing the sensitivity of different size fractions of HAs to pH and ionic strength resulted in noteworthy observations.[20] Besides the difference in size of the ultrafiltered fractions, their chemical composition also was different. The aromaticity of samples and the amounts and dissociation constants of acidic functional groups increased definitely with decreasing size of the fraction.[20,21] Since the stabilizing electrostatic field inside and around the humic nanoparticles is developed from the dissociation of acidic groups and the stability of an electrostatically stabilized, conditionally charged particle is influenced by both pH and ionic strength,[9] a characteristic difference between the HA fractions was predicted. The sensitivity of each fraction to pH and ionic strength was significantly different. At pH~3 complete phase-separation was induced by 0.005 M NaCl in a solution of the largest-size fraction, while the smallest-size fraction was only partially coagulated in the presence of 0.2 M NaCl. Coagulation values at constant pH for the size fractions increased with decreasing size and increasing acidity. The colloidal stability of all HA size fractions increased dramatically with increasing pH and the smallest-size fraction remained dissolved even in 5 M NaCl at pH~5.[9,20,21]

Thus, the sensitivity of HA solutions to salt strongly depends on their size and acidic group density. Any traditionally prepared sample is a mixture of HA molecules with different colloidal stability, so the phase separation of components will take place at different salt concentration at constant pH. Models describing the electrostatically influenced dissociation of HAs can predict the change in their colloidal stability under variable pH and electrolyte conditions. Because of HA's polydispersity in natural systems, any electrolyte-induced phase separation results in a spontaneous fractionation with respect to the macroion's size and chemical composition.

The dependence of the interaction of montmorillonite, a typical swelling clay mineral, with HSs on pH and neutral electrolyte concentration has been studied.[16,17] Two significantly different HAs and FAs were chosen in respect to their size and total acidity. As expected, their electrostatic stabilization also was different. It was found at very low electrolyte concentration (~0.001 M) that adsorption of dissolved material on the solid surface takes place. The dissolved humic polyions can be adsorbed in small amounts on the amphoteric edge sites (Al-OH) of montmorillonite lamellae. The relatively low adsorption density significantly increases the colloidal stability of aqueous clay suspensions because the adsorbed, negatively charged humic macroions recharge the edges.[22] However, in systems containing more electrolyte than ~ 0.01 M, coaggregation of the components takes place together with adsorption due to the joint effects of pH and electrolyte. Tarchitzky et al.[18] proposed the term mutual flocculation or heteroflocculation for the same process some years later. However, this term is questionable from a theoretical colloid stability point of view.

These systems[22] were new examples of adsorption systems in which coagulation processes due to the colloidal destabilization of components take place in parallel with adsorption, but experimental separation of these simultaneous processes is not possible.[23] At low pH, the sensitivity of adsorption systems to electrolyte was much larger: 0.01-0.02 M NaCl had an extremely large effect on the apparent adsorption of HA. At neutral pHs, however, a significantly greater NaCl concentration (~0.5 M) was needed to reach the same apparent extent of adsorption.

The effect of pH on the complex interaction can only be evaluated if it is considered along with the effect of neutral electrolytes. The combined effect of pH and electrolyte was interpreted by calculation based on the application of a surface complexation model and the electrostatic theory of colloid stability (DLVO theory) for montmorillonite and humic particles. Different intrinsic dissociation constants and planar geometry were presumed for fulvic acids and HAs. The probability of the proposed complex mechanism consisting of homocoagulation of clay and HAs and heterocoagulation of dissimilar particles was supported by the model calculation.[19,20,24] It revealed a new possibility of interaction between montmorillonite and HSs. In an aqueous medium (pH between 6 and 8, 0.2-0.5 M NaCl) heterocoagulation of humate or fulvate and montmorillonite particles is more pronounced than coagulation of identical particles. Under acidic conditions, however, coagulation of humic or fulvic acid has virtually the same probability as heterocoagulation with montmorillonite.

In summary, we can state that in any environmental system the probability of forming mixed aggregates containing both fine clay and humic particles is high. These mixed aggregates then separate from the aqueous phase and form sediment.

3 CONCLUSIONS

Colloidal character, that is specific properties and colloidal state changes, can be expected for aqueous humic systems because of the size ranges that they display in either the dissolved or dispersed (precipitated) state.

Amphiphilic building blocks of cross-linked humic nanospheres that contain hydrophilic groups linked chemically to an aromatic and/or aliphatic hydrophobic part can be presumed. These amphiphilic units are associated to form loose aggregates that are similar to either macro-molecular coils or micelles of liophilic colloids.

In dilute solution at appropriate pH, fulvic and humic acids and their alkali metal salts are considered to be real colloidal solutions that have the properties of cross-linked polyelectrolyte and surfactant solutions. Characteristic colloidal processes of dissolved HSs are conformational changes, adsorption mainly at the water surface and clay or oxide/water interface, and self-assembly in solution and at solid interfaces.

Changes in solution conditions such as a decrease in pH, addition of multivalent metal ions or organic cations induce the formation of compact aggregates. Since the charges are screened or neutralized, water molecules can not penetrate inside, and so compact, hydrophobic nanospheres form similarly to those in liophobic colloids. Changes of colloidal state of HSs are the transition from dissolved to dispersed state and the opposite process is dissolution. These processes should be reversible, which is obvious from the pH and neutral electrolyte induced precipitation or coagulation. Two subprocesses take place in the course of the transition from a stable colloidal solution to a phase separated heterogeneous system in dilute systems of HAs. An intraparticle contraction or shrinkage is followed by an interaction between the collapsed nanoparticles that produces an unstable colloidal system, which then coagulates and precipitates. Both subprocesses are chemically and electrostatically controlled. The pH and ionic strength dependent ionization of acidic functional groups creates an electrostatic field inside and around the particles that determines the aggregation state of the system.

In systems that contain clay mineral particles and dissolved HSs and have electrolyte concentrations greater than ~0.01 M, coaggregation of the components takes place due to the combined effect of pH and electrolyte. This process is accompanied by humic adsorption on the clay. The probability of homocoagulation of clays and HSs and heterocoagulation of dissimilar particles is supported by model calculations that reveal a new mode of interaction between montmorillonite and HSs.

Humic substances are the most abundant natural organic materials. In the environmental systems spontaneous colloidal processes involving HSs take place frequently, and the colloidal state of HSs continuously changes. Adsorption on mineral surfaces and dissolution and precipitation of humic materials under changing solution conditions are common in soils and waters. Because of HA's polydispersity in natural systems, any electrolyte-induced phase separation or coagulation results in a spontaneous fractionation with respect to the macroion's size and chemical composition. The probability of the formation of mixed aggregates containing both clay and humic particles is high. These heteroaggregates separate from the aqueous phase and form sediments.

ACKNOWLEDGEMENTS

This work was supported by grants OTKA T022436 and FKFP 0587/1999.

References

1. M. H. B. Hayes, P. MacCarthy, R. L. Malcolm and R. S. Swift, in 'Humic Substances II. In Search of Structure', M. H. B. Hayes, P. MacCarthy, R. L. Malcolm and R. S. Swift, (eds.), Wiley, Chichester, 1989, Ch. 24, p. 689.
2. R. Beckett, Z. Jue and J. C. Giddings, *Environ. Sci. Technol.*, 1987, **21**, 289.
3. D. H. Everett, 'Basic Principle of Colloid Science', Royal Society of Chemistry, London, 1988.
4. E. Tombácz and I. Regdon, in 'Humic Substances in the Global Environment and Implications on Human Health', N. Senesi and T. M. Miano, (eds.), Elsevier, Amsterdam, 1994, p. 139.
5. E. Tombácz, S. Sipos and F. Szántó, *Agrokémia Talajtan*, 1981, **30**, 365.
6. R. L. Wershaw, *Environ. Sci. Technol.*, 1993, **27**, 814.
7. N. Shinozuka and C. Lee, *Marine Chem.*, 1991, **33**, 229.
8. T. F. Guetzloff and J. A. Rice, *Sci. Total Environ.*, 1994, **152**, 31.
9. R. J. Hunter, 'Foundations of Colloid Science', Clarendon Press, Oxford, 1989.
10. E. Tombácz and E. Meleg, *Org. Geochem.*, 1990, **15**, 375.
11. E. Tombácz, J. A. Rice and S. Z. Ren, in 'The Role of Humic Substances in the Ecosystems and in Environmental Protection', J. Drozd, S. S. Gonet, N. Senesi and J. Weber, (eds.), PTSH, Wroclaw, Poland, 1997, p. 43.
12. S. Z. Ren, E. Tombácz and J. A. Rice, *Phys. Rev. E*, 1996, **53**, 2980.
13. E. Tombácz, J. A. Rice and S. Z. Ren, *ACH Models in Chem.*, 1997, **134**, 877.
14. J. A. Rice, E. Tombácz and K. Malekani, *Geoderma*, 1999, **88**, 251.
15. W. Stumm, 'Chemistry of the Solid-Water Interface', Wiley, New York, 1992.
16. E. Tombácz, M. Gilde, I. Ábrahám and F. Szántó, *Appl. Clay Sci.*, 1988, **3**, 31.
17. E. Tombácz, M. Gilde, I. Ábrahám and F. Szántó, *Appl. Clay Sci.*, 1990, **5**, 101.
18. J. Tarchitzky, Y. Chen and A. Banin, *Soil. Sci. Soc. Am. J.*, 1993, **57**, 367.
19. E. Tombácz, K. Varga and F. Szántó, *Colloid Polym. Sci.*, 1988, **266**, 734.
20. E. Tombácz, É. Mádi, M. Szekeres and J. A. Rice, in 'Humic Substances Downunder: Understanding and Managing Organic Matter in Soils, Sediments and Water', *Proceedings of IHSS-9*, Adelaide, Australia, 1999, in press.
21. E. Tombácz, *Soil Sci.*, 1999, in press.

22. E. Tombácz, G. Filipcsei, M. Szekeres and Z. Gingl, *Colloids Surfaces A*, 1999, **151**, 233.
23. T. W. Healy, 'Adsorption from Aqueous Solutions', D. E. Yates, (ed.), Royal Australian Chemical Institute, Parkville, 1978, p. 73.
24. E. Tombácz, I. Ábrahám and F. Szántó, *Appl. Clay Sci.*, 1990, **5**, 265.

HUMIC ACID PSEUDOMICELLES IN DILUTE AQUEOUS SOLUTION: FLUORESCENCE AND SURFACE TENSION MEASUREMENTS

R. von Wandruszka, R. R. Engebretson and L. M. Yates III

Department of Chemistry, University of Idaho, Moscow, ID 83844-2343, USA

1 INTRODUCTION

The interactions of dissolved humic acids (HAs) with hydrophobic organic matter (HOM) in aqueous media can be understood by consideration of the surfactant properties of HAs.[1,2] These properties dictate that the amphiphilic humic polymers arrange themselves in solution to form structures that have a relatively hydrophilic surface and a hydrophobic interior. This micelle-like organization is thought to exist at low concentrations (in the ppm range) and is not typified by a critical micelle concentration (CMC), as it is in synthetic surfactants.[3] The reason for this lies in the large molecular sizes (MW 10 to >1,000 kDa) found in humic materials, which allows them to "aggregate" intramolecularly by coiling and folding their polymer chains. This could, in principle, happen with a single HA molecule, but it should not be inferred that intermolecular associations are precluded. These are likely to become more important at higher concentrations, especially since the polydispersity of HAs can lead to entanglement of smaller fragments in larger ones. The term "pseudomicelle" has been used[4] to describe the micellar humic structures, setting them apart from customary surfactant micelles.

Humic pseudomicelles display distinct detergent properties in their interactions with HOM. As is the case with normal micelles, the relatively nonpolar interior of the pseudomicelles provides a favorable microenvironment for hydrophobic species. By partitioning into these domains, HOM can become sequestered and solubilized by dissolved HAs. Effective pseudomicellar HA/HOM interactions require that the humic polymers be sufficiently long, flexible and amphiphilic. Previous work has shown that molecular size is an important determinant in these interactions.[5] At the low end of the humic molecular size range, this is illustrated by the observation that small fulvic acid molecules show less evidence of aggregation at low concentrations. The role of molecular flexibility has been inferred from structural features, such as the degree of aromaticity and conjugation in humic polymers. Leonardite HA, for instance, which is known to be highly aromatic[6] and probably not very flexible, has been shown to aggregate poorly.[7] The amphiphilic nature of HAs strongly depends on solution conditions, including pH and ionic strength. Thus in a solution of high pH and low ionic strength, the humic polymers are polyanionic and have little surface activity.

Various pieces of evidence that support the HAs model described above have been reviewed recently.[3] The discussion below focuses on two types of fluorescence measurement and on the interpretation of surface tension data in terms of the pseudomicellar model.

2 MATERIALS AND METHODS

Water used in all solutions was deionized and treated with a 0.22μ Millipore filter system to a resistivity of 16 MΩcm. All measurements were taken in triplicate at a temperature of 22°C. HA was isolated from Latahco silt loam soil (Argiaquic Xeric Argialbolls) maintained as pasture for at least 20 years. The soil obtained from a depth of up to 30 cm was air-dried and crushed to pass a 2.0 mm sieve. It contained 41.5 g organic C, 159 g clay, 121 g silt, and 3.9 g total N/kg soil. The HA was extracted and deashed with the International Humic Substances Society (IHSS) procedure.[8] Stock HA solutions (1000 ppm) were prepared with at least 1 hr of sonication in a 0.01 M NaOH/1.0×10^{-7} M pyrene solution and stored in the dark at 4°C. Subsequent dilutions were stored in the dark at room temperature and allowed to equilibrate for at least 24 h before use. The pH of the 10 ppm HA solutions used for fluorescence measurements was in the range 7.8 - 8.2.

Pyrene (Sigma, 98%) was recrystallized from absolute ethanol and sublimed onto a cold finger. A 0.02 M pyrene stock solution was prepared in ethanol. Aqueous pyrene solutions (1.0×10^{-7} M) were prepared by placing the appropriate amount of stock solution in a dry volumetric flask, evaporating the ethanol, adding water, and then sonicating the solution for at least 5 h. All working pyrene solutions were stored in glass in the dark at room temperature. Analytical grade $MgBr_2$ and $MgCl_2$ (J. T. Baker) were used as received.

Fluorescence spectra were measured with a Hitachi F-4500 fluorescence spectrophotometer and a Perkin-Elmer MPF-66 fluorescence spectrophotometer, both equipped with thermostated cuvet holders. Fluorescence excitation was at 240 nm and the emission was measured at 373 nm. Surface tension measurements were made with a Fisher Surface Tensiomat Model No. 21 (Fisher Scientific, Pittsburgh, PA) fitted with a $^3/_4$-inch diameter platinum-iridium ring. Samples were prepared for measurement by adding the appropriate amounts of $MgBr_2$ or $MgCl_2$ to 10 mL of HA solution and shaking for a period of 24 h on a reciprocating shaker. Solutions were adjusted to pH 6.0 with HCl prior to salt addition. For measurement, they were placed in a shallow glass dish of >2-inch diameter and the platinum-iridium ring was inserted in the middle of the container to avoid edge effects. The ring was raised through manual operation of the torsion mechanism and the tension reading at the instant of surface detachment was noted.

3 RESULTS AND DISCUSSION

3.1 Fluorescence

The fluorescence of pyrene is strong in aqueous media despite the low solubility of the probe in water (~5×10^{-7} M). The intensity of the emission is quenched somewhat in the presence of HAs, although it remains clearly observable at HA concentrations up to ca. 100 ppm. At higher concentrations, measurements become difficult because of increasingly severe inner filter effects caused by HAs. These effects were relatively minor at the lower humic concentrations used in this work, since the 240-nm excitation wavelength used for pyrene was not strongly absorbed by HAs. The pyrene fluorescence at 372 nm was strong and suffered little interference in the presence of 10 ppm aqueous HA. A significant increase in pyrene emission intensity was observed when $MgBr_2$ was added to this solution (Figure 1). This was noteworthy because Br^- is a strong quencher of pyrene fluorescence. The fluorescence increase and the quenching effect are illustrated in Figure 1, the only difference being that the former solution contained HAs while the latter did not. The possibility that trivial effects such as increased pyrene solubility in the presence of HA or its desorption from cell surfaces were responsible were carefully considered and eliminated.

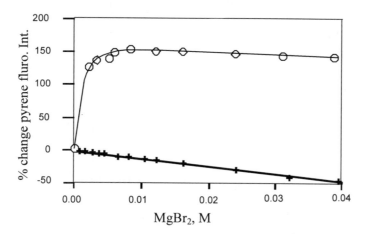

Figure 1 *Change of pyrene (10^{-7} M) fluorescence intensity with added $MgBr_2$;* x *in water;* o *with 10 ppm aqueous HA*

These results suggest that the presence of dissolved HA isolated the probe molecule from the solution borne bromide quencher. This can be rationalized on the basis of the pseudomicellar model: in the slightly alkaline HA solution (pH 7.5 - 8.0), the humic molecules were polyanionic and had little tendency to aggregate in either an inter- or intramolecular fashion. The addition of Mg^{2+} ion, however, led to both charge neutralization and bridging of carboxyl groups on different sections of the humic polymer. As a consequence, these chains were drawn together, forming micelle-like structures that served as sequestration sites for a hydrophobic species such as pyrene. The Br^- ion, on the other hand, was totally excluded from these domains because of polarity considerations.

This mechanism is consistent with the fact that bromide quenching of pyrene fluorescence was of no consequence in solutions containing HAs and a cation such as Mg^{2+}. However, the observations described above include an actual *increase* in pyrene emission. This must be ascribed to a decrease in probe fluorescence quenching by HAs when the salt was added. It required that at least part of the HA/pyrene quenching process was dynamic in nature, that is, consisting of encounters between the excited probe and quenching (probably heteroatomic) moieties on the humic polymer. Again invoking the sequestration mechanism, pyrene contained within the hydrophobic (hydrocarbon) regions of a HAs aggregate was held relatively immobile and protected from interactions with polar quenching groups on the polymer. Thus the fluorescence intensity increased.

The proposed partitioning of pyrene into the interior of humic pseudomicelles implies that the probe experienced a less polar environment under these circumstances. This was further explored through the use of the pyrene emission spectrum as a "spectroscopic ruler".[9] It is well established that the ratio of the first and third vibronic peaks of this spectrum (I_1/I_3) changes monotonically with the polarity of the microenvironment of the molecule. Thus in water the I_1/I_3 value is as high as 1.8, while it drops below 1.2 in hydrocarbon solvents. Figure 2 shows the I_1/I_3 variation in a 10 ppm HA solution with the addition of Mg^{2+}. Although the decrease in I_1/I_3 was fairly small it was consistent and reproducible, indicating that pyrene found itself in less polar surroundings as the concentration of the cation was increased. This is again in agreement with a process of increased pseudomicelle formation in HA as metal ions were added.

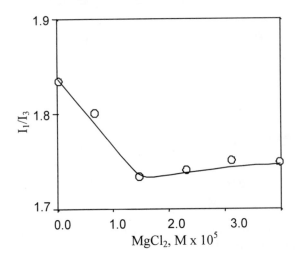

Figure 2 *Change of the pyrene (10^{-7} M) fluorescence I_1/I_3 ratio with added MgCl$_2$ in a 10 ppm HA solution.*

3.2 Surface Tension

Several other pieces of corroborating evidence for intra- and intermolecular micelle formation in dilute aqueous HA solutions have been reported recently.[10-12] Surface tension measurements of such solutions should in principle provide further insights into the processes that are operative under various solution conditions. However, meaningful results cannot be obtained with 10 ppm HA solutions and it was necessary for the present work to increase the HA concentration to 500 ppm. While this is still low compared to solutions used in some HA surface tension work,[13,14] it is possible that additional concentration-driven effects are present under these circumstances.

The influence of the Mg^{2+} concentration on the surface tension, γ, of an aqueous HA solution is illustrated in Figure 3. The value of γ in the absence of the salt, which cannot be displayed on the log scale of Mg^{2+} concentration, was 68 dynes cm^{-1}.

The surface tension decreased as more cation was added. This can be explained in terms of the amphiphilic nature of the humic polymer. At the pH of 6 used in the measurements, the HA molecules were largely anionic and fairly hydrophilic. As the salt was added, partial neutralization of the negative charges by the cation rendered the species more amphiphilic and caused them to partition to the surface. The surface tension therefore decreased. It must be understood that the effect measurable through surface tension was in all likelihood one among several, albeit the dominant one. It should not be inferred that humic pseudomicelle formation does not happen at these lower Mg^{2+} concentrations, but merely that it does not become immediately evident through surface tension effects. It is thus implied that this type of micellization is not a critical event and is not characterized by a CMC.

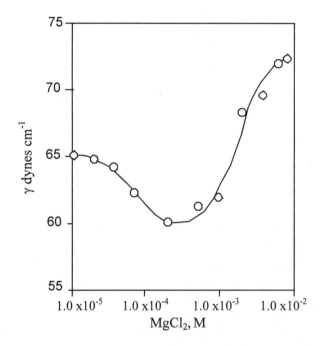

Figure 3 *Change of surface tension with added MgCl$_2$ in a 500 ppm HA solution*

After sufficient Mg^{2+} had been added the surface tension reached a minimum and then increased. In terms of the pseudomicelle model, this was an indication of the increasing prevalence of these aggregates among the dissolved HA molecules. The formation of pseudomicelles involves the association of humic polymers in self-contained 3-dimensional structures that have a relatively polar surface and a nonpolar core. This type of body is lyophilic and has little tendency to partition to the surface. Its increasing presence therefore caused the HAs to be removed from the surface, allowing γ to increase. The difference between the process described here and normal micelle formation is that the latter is entirely concentration driven. When the CMC of the surfactant in question is reached in such a case, micelles begin to form but the surfactant monomer concentration in solution remains approximately at the CMC. These monomers continue to aggregate at the solution-air interface and the surface tension reaches a constant low value. In contrast, when Mg^{2+} was added to a HAs solution in the present work, the amphiphilicity of the humic polymers was continuously changed. Thus the concentration of species organized in pseudomicellar aggregates increased while the humic equivalent of the surfactant "monomer" decreased in concentration. This circumstance allowed the surface tension to increase rather than remain a low value. It also manifested itself in the observation that the surface tension at higher Mg^{2+} concentrations was >70 dynes cm^{-1} that is well above that of the HAs solution without the salt. In view of the fact that the surface tension of pure water is 73 dynes cm^{-1} (marginally higher for the electrolyte solution used here) this suggests that a significant portion of HAs in the 500 ppm solution was included in pseudomicellar structures and that relatively little remained in the "monomer" form.

4 CONCLUSIONS

The fluorescence and surface tension data presented above are consistent with enhanced pseudomicelle formation upon the addition of Mg^{2+} to an HA solution. In this interpretation, the minimum in the curve shown in Figure 3 corresponds to the "turn-around" point at which the formation of pseudomicelles becomes dominant *vis-à-vis* HAs accumulation at the surface. This is likely to be a characteristic of the HA used in this work, especially with regard to its polar group content and distribution. As such it may be used as a feature in HAs description.

In earlier work from this laboratory[4] it was noted that HAs solutions undergo clouding under certain solution conditions. The clouding phenomenon is familiar from the chemistry of nonionic surfactants.[15] It involves the phase separation of such compounds from the aqueous solvent at a characteristic, elevated temperature (the cloud point). It is thought to be due to the decreased dielectric constant of water at higher temperatures and the lesser degree of hydrogen bonding that occurs with the hydrophilic portion of the surfactant. At the cloud point the micelles grow and/or coalesce to form larger aggregates that constitute a separate macroscopic phase. In the case of HAs, this occurs in solutions at 10 ppm, at a temperature of 40°C and with $[Mg^{2+}]$ 0.07 M.[4] The requirement for the metal ion is consistent with the discussion presented above: the HA polymer must be rendered sufficiently amphiphilic by the cation to assume a configuration that makes elimination of water from the aggregate energetically favorable at the cloud point. Since clouding only happens with nonionic surfactants and not with ionic ones, it is tempting to extend the analogy with HAs by proposing that the humic polymers in clouding HAs solutions have largely lost their ionic character. This is, however, probably not justified. The clouding mechanism and the molecular parameters required for it are not sufficiently well understood to warrant such a conclusion.

ACKNOWLEDGEMENT

The author gratefully acknowledges the financial support of the EPA (R82-2832-010) and the NSF Idaho EPSCoR program.

References

1.   W. Rochus and S. Sipos, *Agrochim.*, 1978, **22**, 446.
2.   R. L. Wershaw, *Environ. Sci. Technol.*, 1993, **27**, 814.
3.   R. von Wandruszka, *Soil Sci.*, 1998, **163**, 921.

4.   R. R. Engebretson and R. von Wandruszka, *Environ. Sci. Tech.*, 1994, **28**, 1934.

5.   R. R. Engebretson and R. von Wandruszka, *Org. Geochem.*, 1997, **26**, 759.

6.   K. A. Thorn, D. W. Folan and P. MacCarthy, in 'Water-Resources Investigations Report 89-4196', U. S. Geological Survey, Denver, CO, 1989.

7.   R. R. Engebretson, T. Amos and R. von Wandruszka, *Environ. Sci. Technol.*, 1996, **30**, 390.

8.   International Humic Substances Society Product Literature, Golden CO, published January 25, 1985, see <www.ihss.gatech.edu>.

9.   D. C. Dong and M. A. Winnik, *Photochem. Photobiol.*, 1982, **35**, 17.

10.   C. Ragle, R. R. Engebretson and R. von Wandruszka, *Soil Sci.*, 1997, **162**, 106.

11.   R. von Wandruszka, C. Ragle and R. Engebretson, *Talanta*, 1997, **44**, 805.

12.   L. M. Yates, R. R. Engebretson, T. M. Haakenson and R. von Wandruszka, *Anal. Chim. Acta,* 1997, **356**, 295.

13.   M. Tchapek and C. Wasowski, *Geochim. Cosmochim. Acta*, 1976, **40**, 1343.

14.   Y. Chen and M. Schnitzer, *Soil Sci.*, 1978, **125**, 7.

15.   W. L Hinze and E. Pramauro, *Crit. Rev. Anal. Chem.*, 1993, **24**, 133.

ATOMIC FORCE MICROSCOPY OF pH, IONIC STRENGTH AND CADMIUM EFFECTS ON SURFACE FEATURES OF HUMIC ACID

C. Liu and P. M. Huang

Department of Soil Science, University of Saskatchewan, Saskatoon, SK S7N 5A8, Canada

1 INTRODUCTION

Humic substances (HSs) are a series of high molecular weight, yellow to black organic materials formed by secondary synthesis reactions. HSs exist in all terrestrial and aquatic environments.[1,2] Humic acids (HAs) and other organic colloidal materials play important roles in a wide range of environmental and agricultural reactions and processes.[2-4] The size, shape and molecular weight (MW) characteristics of HAs have a profound effect on their interactions with other soil components such as clays, metal ions and biota.[2] The concentration of HA, the pH of the system and the ionic strength of the medium control the molecular characteristics of HAs.[5]

Humic acids have a substantial capacity to complex metal ions. The complexation process may affect the solubility of the metal ions bound.[6] Cadmium is very toxic and can cause severe human health problems such as bronchitis and itai-itai disease.[7,8] Cadmium is receiving increasing international attention because of its association with health problems associated with the kidney and particularly the renal cortex.

A study of the speciation of particulate-bound Cd in 16 surface soils of Saskatchewan indicates that the organic complex-bound Cd is highly correlated to the Cd bioavailability index.[9] Humic acids are a complicated mixture of colloidal macromolecular species whose state of aggregation and behavior may be affected by metal binding to specific coordination sites. However, the mechanisms of Cd binding by HSs and the impact on molecular conformations of HAs and fulvic acids (FAs) and Cd dynamics in nature remain obscure.

Transmission and scanning electron microscopy (TEM and SEM) have been used to determine the morphological conformation of FAs and HAs at various pH values and salt concentrations.[10-12] The structures of the complexes formed by FAs when reacted with Cu^{2+}, Al^{3+}, Fe^{2+} and Fe^{3+} was also studied by SEM.[10] However, the TEM and SEM operation is under vacuum condition that does not exist in natural soil environments. The high vacuum, a condition may affect the conformation of materials observed. Furthermore, the sample preparation is complicated when these two techniques are used to image FAs and HAs.

Atomic force microscopy (AFM) is a new surface-sensitive analytical technique that can image nearly any surface in a vacuum, in air or in solution with unprecedented resolution and can also measure three-dimensional profiles.[13-15] The sample preparation for

AFM is very simple.[16] Although the morphological conformation of FAs was recently observed by employing AFM,[15,16] the surface features of HAs and the influence of pH, salt and metal ions on the molecular conformation of FAs and HAs have not been studied by AFM.

In the present study, the influence of pH, an inert electrolyte ($NaNO_3$) and bound Cd ions on the surface features of HAs was investigated at the subnanometer-scale by AFM under ambient conditions. The Cd distribution in the Cd-humate complexes was also observed by electron microprobe analysis.

2 MATERIALS AND METHODS

2.1 Samples for the Study of pH and Salt Effects

Five mg of a soil HA standard (1S102H) from the International Humic Substances Society (IHSS) were dispersed in 8 mL deionized distilled water using ultrasonification (Model 350 Sonifier) at 150 watts for 2 min in an ice bath. The HA suspension was adjusted to pH 4.0, 6.0, 8.0 and 10.0 with 0.01 M NaOH and HNO_3 and diluted to 10 mL. Then 5 mL of deionized distilled water or $NaNO_3$ solutions of 0.003 M and 0.03 M at the respective pH was mixed with the HA suspension. The total volume of the final suspension was 15 mL. The HA concentration was 0.33 g L^{-1}. The final pH values of the HA suspensions were 4.0, 6.0, 8.0 and 10.0. The final $NaNO_3$ concentration of the suspension was 0, 0.001 and 0.01 M. A portion of the suspension was used for the AFM analysis. Another portion of the suspension was filtered with a 0.025 μ Millipore membrane and the electrical conductivity of the filtrate was measured.

2.2 Samples for the Study of Cadmium Effects

Fifty mg of the standard HA were dispersed in 40 mL deionized distilled water by the above-mentioned method. The HA suspension was adjusted to pH 5.0 with 0.01 M NaOH and diluted to 50 mL. Fifty mL of 0, 2×10^{-7}, 2×10^{-6}, 2×10^{-5}, 2×10^{-4} or 2×10^{-3} M $Cd(NO_3)_2$ containing 0.02 M $NaNO_3$ at pH 5.0 was then mixed with the HA suspension, bringing the final volume of the mixed suspension to 100 mL. The final Cd concentration was 0, 10^{-7}, 10^{-6}, 10^{-5}, 10^{-4} and 10^{-3} M and the $NaNO_3$ concentration was 0.01 M. The Cd-HA suspension was shaken for 0.5, 1, 4, 8 and 24 hrs at 298 K. The suspension at the end of each reaction period was filtered through a 0.025 μ Millipore membrane. The concentration of Cd remaining in the filtrate was determined by inductively coupled plasma (AXIAL ICAP 61E VACUUM, Thermo Instruments). The solid sample was washed with deionized distilled water until the conductivity of the washing water was less than 5×10^{-3} dS m^{-1}. After the sample was freeze-dried, 5 mg of the HA or Cd-HA complex was dispersed in 15 mL deionized distilled water using ultrasonification at 150 watts for 2 min in an ice bath. The suspension was used for the AFM analysis as described below. The freeze-dried sample was used for electron microprobe analysis.

2.3 Atomic Force Microscopy and Electron Microprobe Analysis

One drop of the HA or Cd-HA complex suspension was deposited on a watch glass and air-dried overnight at room temperature (296.5 ± 0.5 K). The watch glass was then fastened to a magnetized stainless steel disk (diameter 12 mm) with double-sided tape. The 3-dimensional AFM images were recorded in air and at room temperature with a

NanoScopeTMIII atomic force microscope (Digital Instruments). The imaging areas were 5 x 5 μ^2 and 15 x 15 μ^2. The scanner type was 1881E and the scanner size was 15 μ. A silicon nitride cantilever with a spring constant of 0.12 N/m was used in the contact mode. The scanning rate was 18.31 - 27.47 Hz. The AFM cantilever was changed frequently to prevent experimental artifacts. Furthermore, the scanning area and scanning angle were often changed by entering different area and angle parameters to detect artifacts caused by adhesion of HA particles to AFM tips. The diameter and thickness of particles were measured by section analysis of the AFM based on 50 HA particles.

X-ray mapping analysis of Cd distribution in HA samples was conducted with a JEOL JXA 8600 SUPERPROBE spectrometer at 15 kV.

3 RESULTS AND DISCUSSION

3.1 pH Effects on the Surface Features of Humic Acid

The subnanometer-scale surface features of the HA varied greatly with the solution pH as illustrated in Figure 1. The AFM images at the higher magnification (5x5 μ^2) (Figure 1E-H) show that the HA was basically small spheroids. The diameter of the spheroids ranged from 50 to 150 nm with an average diameter of 85 nm. The thickness was 9-30 nm with an average thickness of 16 nm.

These small spherical particles aggregated into different shapes at various pH values as shown by the AFM images at lower magnification (15x15 μ^2) (Figure 1A-D). At pH 4.0, the HA aggregates consisted of small spheroids with irregular shape (Figure 1A). At pH 6.0, the spheroids dispersed more evenly compared with those at pH 4.0 (Figure 1A and B). The dispersion of discrete HA particles was even more evident at pH 8.0 than at pH 6.0 (Figure 1B and C). The conformation of the HA changed considerably at pH 10.0. Bundles of fiber-shaped and elongated fiber-shaped HA aggregates were formed (Figure 1D). The width of the individual fiber-shaped particles was around 50-150 nm. It appears that the fiber-shaped HA aggregates consists of small spheroids (Figure 1H). The HA particles at pH 10.0 were more aggregated than those at pHs 6.0 and 8.0. The addition of Na^+ or NO_3^- to the suspension on adjusting the pH with NaOH or HNO_3 substantially increased the ionic strength (Table 1) and also enhanced HA aggregation.

Table 1 *Electrical conductivity and ionic strength of HA solutions*

pH	NaNO₃ concentration (M)					
	0		*0.001*		*0.01*	
	EC (dS/m)[a]	I[b]	EC (dS/m)	I	EC (dS/m)	I
4.0	0.057	0.0007	0.145	0.0019	0.950	0.0121
6.0	0.062	0.0008	0.181	0.0023	1.958	0.0249
8.0	0.075	0.0010	0.244	0.0031	3.258	0.0414
10.0	0.091	0.0012	0.307	0.0039	3.560	0.0452

[a] The standard error of the electrical conductivity measurements was <5%; [b] Ionic strength was calculated from the equation I = 0.0127 EC[17]

The atomic force micrographs in the present study show that the basic structure of HA is spheroidal with a diameter of 50 to 150 nm. Oades[18] proposed that the configuration of an extended HA polymer with a molecular weight of 150 kDa would have a diameter of

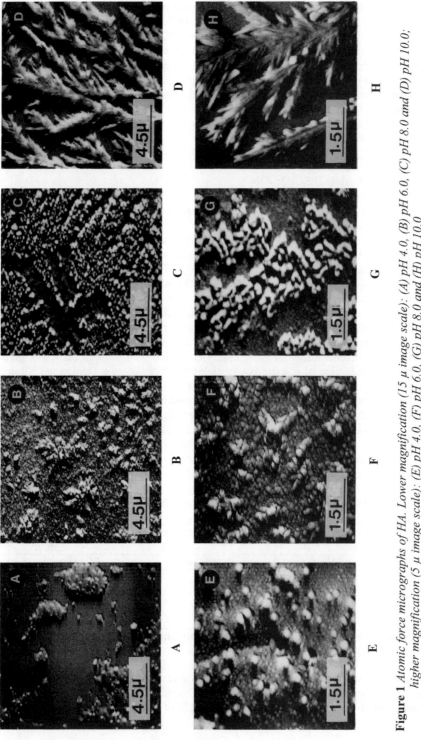

Figure 1 *Atomic force micrographs of HA. Lower magnification (15 μ image scale): (A) pH 4.0, (B) pH 6.0, (C) pH 8.0 and (D) pH 10.0; higher magnification (5 μ image scale): (E) pH 4.0, (F) pH 6.0, (G) pH 8.0 and (H) pH 10.0*

about 50 nm. The size of the HA particles observed in the present study is larger than the size of FA particles (10-50 nm) observed by AFM.[16] This is apparently due to the larger molecular weight of HAs than that of FAs. Although it is felt that tapping-mode AFM is better for imaging soft materials such as humic acids,[16] the AFM results of the present study indicate that good images of HA also can be obtained by contact-mode AFM at an HA concentration of 0.33 g HA L^{-1}.

The dark shadow in the AFM images (Figure 1) was caused by the thickness of the HA particles. Nevertheless, the dark shadow does not affect the morphological conformation of HA observed by AFM. The AFM analysis was attempted on a more diluted HA suspension (0.07 g L^{-1} HA). However, artifacts were observed in such a dilute HAs system. Some HA particles adhered to the AFM tip and the image at the same spot changed from scan to scan.

The atomic force micrographs in this study also show that the aggregation of HAs decreased with increasing solution pH (Figure 1). This trend is consistent with the previous findings for FAs and HAs.[12,19] At low pH, the aggregation of HAs is enhanced by strong hydrogen bonding.[10] As the solution pH increases, the hydrogen bonding becomes weaker and ionization of HA functional groups such as carboxylic acids and phenols increases. The charged functional groups of HAs make the particles repel each other electrostatically and the aggregation of HAs decreases.

Some individual spherical HA particles were observed at a low HA concentration (0.1 g HA L^{-1}) with TEM.[12] However, the TEM and SEM images of HAs and FAs in previous studies[10-12,19] did not show that HA aggregates consist of spheroids. The present AFM images clearly show that HA aggregates consist of spheroids (Figure 1). The present study demonstrates that AFM can be used to investigate the surface features of HA at the subnanometer-scale at much higher resolutions than TEM and SEM. The HA particles observed by TEM at 0.1 g HA L^{-1} ranged from 12 to 50 nm,[12] which is much smaller than the HA particles observed by AFM in our study. This is probably because different HA samples with different molecular weights were used in the two studies. Another reason is that the AFM was operated under ambient condition in this study while the TEM was operated in high vacuum conditions. High vacuum conditions could result in dehydration of materials and thus the HA particle size observed was smaller. In the study of Chen and Schnitzer,[10] the fiber structure of HAs was observed at pH 6.0 and the thin sheet structure was imaged at pH 10 by SEM. Compared to these results,[10] the different surface features observed in the present study at the same pH value are attributable to the different HA concentration used in the two studies. Ten g L^{-1} of HA was used before,[10] whereas a much lower HA concentration (0.33 g L^{-1}) was used in our work. The concentration of HA can significantly change the conformation of HA and FA observed.[12,16] Further, HA, which is a buffer, requires more NaOH to adjust pH to 6.0 and 10.0 at a higher HA concentration than at a lower HA concentration. The added ions may influence the observed conformation of HA through an ionic strength effect. Since the lowest pH at which the HA is soluble in H$_2$O was 6.0, the conformation of HA at pH <6.0 has not been studied by TEM and SEM. In this study, the subnanometer-scale surface features of HA at pH 4.0 were investigated by AFM after dispersing HA in deionized distilled water with ultrasonification.

3.2 Ionic Strength Effects on the Surface Features of Humic Acid

The subnanometer-scale conformations of the HA at pH 6.0 and various NaNO$_3$ concentrations are presented in Figure 2. The atomic force micrographs of HA at pH 6.0 and various NaNO$_3$ concentrations also show that the HA was spheroidal. As discussed

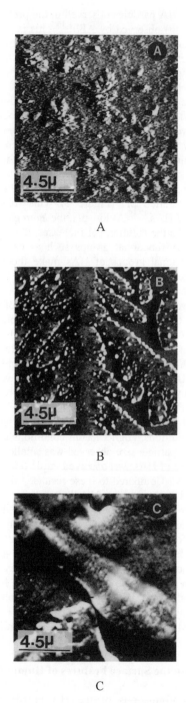

Figure 2 *Atomic force micrographs of HA at pH 6.0 with (A) 0 M NaNO₃, (B) 0.001 M NaNO₃ and (C) 0.01 M NaNO₃*

above, the HA occurred as small spherical particles at pH 6.0 and 0 M $NaNO_3$ and the spheroids were relatively well dispersed (Figure 2A). Some spheroids with larger diameters aggregated together. When the $NaNO_3$ concentration was increased from 0 to 0.001 M, the ionic strength increased by about 3 times (Table 1) and the HA aggregates formed pine leaf-shaped particles that consisted of spheroids (Figure 2B). When the $NaNO_3$ concentration was increased to 0.01 M the ionic strength increased by about 38 times (Table 1); the aggregation of the HA was very significantly enhanced and the spheroids aggregated into a thick and massive structure (Figure 2C). The effect of $NaNO_3$ on the surface features of the HA investigated is very similar to the influence of NaCl on the conformation of HAs observed by SEM.[11]

The pH of the HA dispersed in deionized distilled water at 0.33 g L^{-1} was about 3.98. When the system pH was adjusted to 6.0, the H-bonding was decreased and the dissociation of some functional groups of HA such as carboxylic acids or phenols was bound to take place, resulting in repulsion between HA particles. When the $NaNO_3$ salt was added to the system, the Na^+ ions of the salt were apparently adsorbed by the negatively charged functional groups of HA and neutralized the charge of HA. Further, according to the Poisson equation, the surface potential of the particles decreased with increase in the salt concentration, which resulted in the compression of the double layer. Thus, repulsion between HA particles decreased and HA aggregation increased. Effects of increasing the salt concentration on the configuration of HAs are apparently similar to those observed when the pH was decreased,[11] since protons and salts suppress the repulsion between HA particles.

According to the data of Griffin and Jurinak,[17] the ionic strength of soil solutions and river waters is usually between 0 and 0.45; the ionic strength of more than half of them is between 0 and 0.04. The ionic strength of the systems investigated here ranged from 0.0007 to 0.0452 (Table 1), which is very common in natural environments. The pH ranged studied was from 4.0 to 10.0, which is close to the limits of the natural environment in terms of pH.[20] The results show that ionic strengths and pHs common in soil solutions and natural waters greatly influence the surface features of HAs at the subnanometer-scale. Increase in ionic strength and decrease in pH significantly enhanced the aggregation of HA spheroids. The higher the degree of aggregation of HA spheroids, the lesser the functional groups of HA that would be accessible to react with nutrients and pollutants in soil and related environments. Extensive aggregation of HA particles may also suppress the release of nutrients and pollutants adsorbed by HA. Thus, the ionic strength and pH of a system should greatly affect the behavior of HAs in natural environments through modifying the molecular conformation of HAs.

3.3 Subnanometer-scale Surface Features of Cadmium-Humate Complexes

After mixing the Cd solution with the HA at pH 5.0, the concentration of Cd remaining in solution decreased rapidly with reaction time (Figure 3). At the end of a 4 hr reaction period, the adsorption of Cd by the HA reached equilibrium when the initial Cd concentration was 10^{-7} M (Figure 3A), whereas Cd adsorption approached equilibrium after an 8 hr reaction period at the 10^{-3} M initial Cd concentration (Figure 3B). The amount of Cd adsorbed by the HA at the end of a 24 hr reaction period increased with increased initial Cd concentration (Table 2).

The atomic force micrographs of Cd-humate complexes were taken after the samples had been washed to remove the free Cd salt in the solution. The different amounts of Cd complexed by the HA resulted in great variation of surface features of the HA at the subnanometer-scale (Figure 4). The HA without Cd treatment was present as spheroids

(Figure 4A). When the initial Cd concentration was increased to 10^{-7} and 10^{-6} M, the conformation of the HA did not change very significantly (Figure 4B and C). At an initial Cd concentration of 10^{-5} M, both spheroids and some thick, bean-like particles were observed (Figure 4D). As the initial Cd concentration was increased to 10^{-4} M, more bean-like particles were imaged (Figure 4E). When the Cd concentration was further increased to 10^{-3} M, the spheroids aggregated into massive, bean-like structures (Figure 4F). The length of the bean-like particles was around 1.7-2.5 μ and the width ranged from 620 to 800 nm.

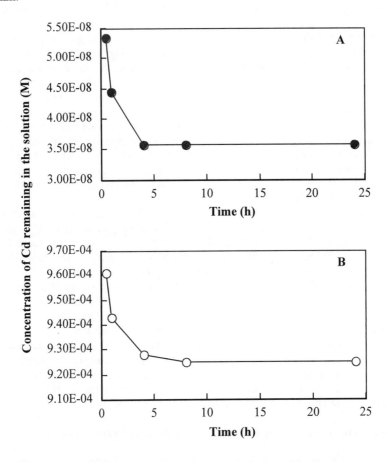

Figure 3 *The concentration of Cd remaining in the solution versus time in Cd-HA systems at initial Cd(NO₃)₂ concentrations of (A) 10^{-7} M and (B) 10^{-3} M*

Table 2 *Amount of Cd adsorbed by the HA at various initial Cd concentrations at pH 5.0 at the end of a 24 hr reaction period*

Initial Cd Conc. (M)	10^{-7}	10^{-6}	10^{-5}	10^{-4}	10^{-3}
g Cd kg $^{-1}$ HA	0.014	0.094	0.56	2.70	19.56

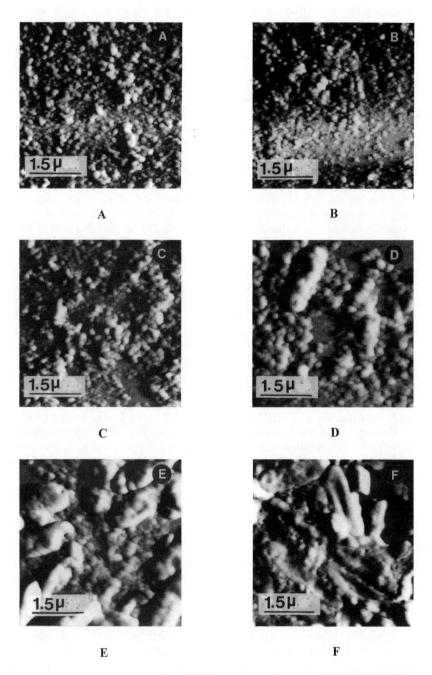

Figure 4 *Atomic force micrographs of Cd-humate complexes at pH 5.0 and NaNO$_3$ concentration of 0.01 M. (A) 0 M Cd, (B) 10^{-7} M Cd, (C) 10^{-6} M Cd, (D) 10^{-5} M Cd, (E) 10^{-4} M Cd and (F) 10^{-3} M Cd*

It was found[10] that different metal ions affect the aggregation of FA to different degrees, resulting in the formation of long and thin fibers of Cu^{2+}-FA complexes, thick and multi-layered aggregates of Al^{3+}-FA complexes, fibrous structures with many inter-connections of Fe^{3+}-FA complexes, and finger-like protrusions of Fe^{2+}-FA complexes. The Cu^{2+}-FA complex appears to have the highest orientation followed by the Fe^{3+}-FA complex.

The results of the present study (Figure 4) show that the conformation of Cd^{2+}-HA complexes is quite different from those of the above-mentioned metal-FA complexes[10] and that increasing in the amount of Cd complexed with the HA increased HA aggregation. Cadmium ions coordinated with the functional groups of the HA (Figure 5) and raveled the HA particles into a bean-like structure. The free Cd^{2+} and the inert electrolyte in the solution were removed by washing with deionized distilled water. Therefore, the atomic force micrographs obtained (Figure 4) show the conformation of Cd-humate complexes that were free of the effects of Cd ions in solution and of ionic strength. Electron microprobe analysis was employed to obtain the Cd X-ray map images and the corresponding electron backscattered images (Figure 6). The samples prepared by the same procedure as for AFM analysis were subjected to electron microprobe analysis. However, the backscattered electron image was not clear and the Cd in the Cd-humate complexes could not be detected since the electron microprobe resolution is much lower than that of AFM. Therefore, freeze-dried HA samples are directly used for electron microprobe analysis. The shapes of HA and Cd-humate complexes observed by electron microprobe were quite different from those observed by AFM. The Cd X-ray mapping image of the HA without Cd treatment was blank (Figure 6B), showing that no detectable Cd was present in the sample. The Cd^{2+} in the Cd-humate complexes formed at the initial Cd concentrations of 10^{-7} to 10^{-5} M could not be imaged due to the low Cd concentration in the complexes (Table 2). The Cd distribution in the Cd-humate complexes formed at initial $Cd(NO_3)_2$ concentrations of 10^{-4} and 10^{-3} M is shown as the white spots in the Cd X-ray mapping images (Figures 6D and F). The Cd X-ray image of the Cd-humate complexes formed at 10^{-3} M $Cd(NO_3)_2$ was substantially more intense compared with that of the Cd-humate complexes formed at 10^{-4} M $Cd(NO_3)_2$. The Cd-humate complexes with various degrees of aggregation had different surface features at the subnanometer-scale (Figure 4), which may affect the release of Cd from Cd-humate complexes through a molecular topological effect.

4 CONCLUSIONS

Atomic force micrographs of HA under ambient conditions show that HA was basically spherical particles with diameters of ca. 50-150 nm (average diameter 85 nm) and thickness ca. 9-30 nm (average thickness 16 nm). The spheroids aggregated into particles with various shapes at various solution pH values and ionic strengths. The aggregation of spheroids into massive particles has not been reported before. The aggregation of HA was enhanced by decreasing the pH and increasing the ionic strength due to hydrogen bonding and/or suppression of the repulsion between HA particles, which is in accord with previous reports. The data obtained in the present study indicate that higher degrees of aggregation of HA should result in fewer functional groups exposed at the HA surface, which then are less accessible to react with nutrients and pollutants. Therefore, ionic strength and pH should greatly affect the behavior of HAs in natural environments through modifying the molecular conformation of HA. Humic acid aggregation also was increased with increase in the amount of Cd complexed with the HA. The aggregation state of Cd-

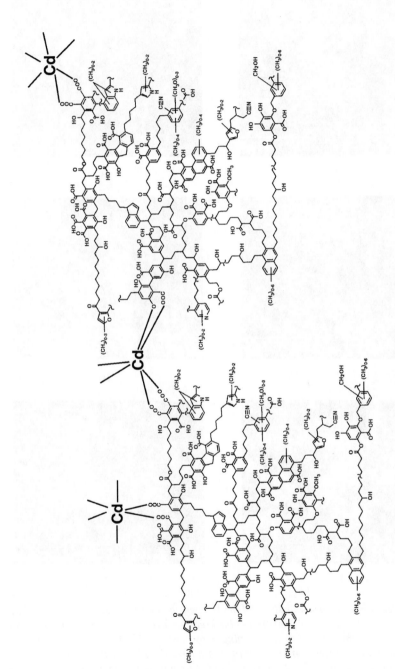

Figure 5 *Proposed chemical structure of Cd-humate complexes*

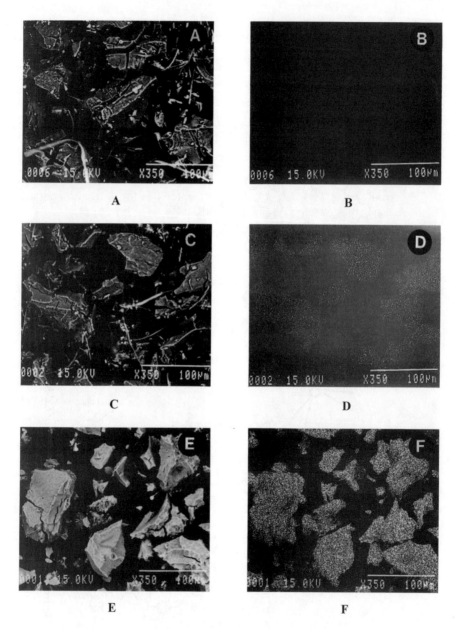

Figure 6 *Electron microprobe analysis of the HA before and after treating with Cd(NO₃)₂ at pH 5.0 and 0.01 M NaNO₃. (A) backscattered electron image of the HA before Cd treatment, (B) Cd X-ray mapping image of the HA before Cd treatment, (C) backscattered electron image of the HA after 10⁻⁴ M Cd treatment, (D) Cd X-ray mapping image of the HA after 10⁻⁴ M Cd treatment, (E) backscattered electron image of the HA after 10⁻³ M Cd treatment and (F) Cd X-ray mapping image of the HA after 10⁻³ M Cd treatment*

humate complexes may affect the release of Cd from Cd-humate complexes through a molecular topological effect and thus influence Cd dynamics and bioavailability in soil and related environments.

ACKNOWLEDGEMENTS

This research was supported by Research Grants GP2383- and EQP156628-Huang of the Natural Sciences and Engineering Research Council of Canada.

References

1. M. H. B. Hayes, P. MacCarthy, R. L. Malcolm and R. S. Swift, 'Humic Substances II: In Search of Structure', Wiley, New York, 1989.
2. F. J. Stevenson, 'Humus Chemistry. Genesis, Composition, Reactions', 2nd edn., Wiley, New York, 1994.
3   M. Schnitzer, *Soil Sci.*, 1991, **151**, 41.
4   J. S. Gaffney, N. A. Marley and S. B. Clark, 'Humic and Fulvic Acids. Isolation, Structure, and Environmental Role', ACS Symposium Series 651, 1996.
5   K. Ghosh and M. Schnitzer, *Soil Sci.*, 1980, **129**, 266.
6   Y. J. Zhang, N. D. Bryan, F. R. Livens and M. N. Jones, in 'Humic and Fulvic Acids. Isolation, Structure, and Environmental Role', J. S. Gaffney, N. A. Marley and S. B. Clark, (eds.), ACS Symposium Series 651, 1996, Ch. 12, p. 194.
7   B. J. Alloway, 'Heavy Metals in Soils', Blackie, London, 1995.
8   M. Webb, 'The Chemistry, Biochemistry and Biology of Cadmium', Elsevier, Amsterdam, 1979.
9   G. S. R. Krishnamurti, P. M. Huang, K. C. J. Van Rees, L. M. Kozak and H. P. W. Rostad, *Analyst*, 1995, **120**, 659.
10 Y. Chen and M. Schnitzer, *Soil Sci. Soc. Am. J.*, 1976, **40**, 682.
11 K. Ghosh and M. Schnitzer, *Geoderma*, 1982, **28**, 53.
12 I. L. Stevenson and M. Schnitzer, *Soil Sci.*, 1982, **133**, 179.
13 D. Rugar and P. Hansma, *Physics Today*, 1990, **43**, 23.
14 K. L. Nagy and A. E. Blum, 'Scanning Probe Microscopy of Clay Minerals', Clay Mineral Society Workshop Lectures, Clay Mineral Society, Boulder, CO, 1994, Vol. 7.
15 P. A. Maurice, in 'Structure and Surface Reactions of Soil Particles', P. M. Huang, N. Senesi and J. Buffle, (eds.), Wiley, Chichester, 1998, Ch. 4, p. 109.
16 K. Namjesnik-Dejanovic and P. A. Maurice, *Coll. Surf.*, 1997, **120**, 77.
17 R. A. Griffin and J. J. Jurinak, *Soil Sci.*, 1973, **116**, 26.
18 J. M. Oades, in 'Minerals in Soil Environments', 2nd edn., J. B. Dixon and S. B. Weed, (eds.), Soil Sci. Soc. Am., Madison, WI, 1989, Ch. 3, p. 89.
19 M. Schnitzer and H. Kodama, *Geoderma*, 1975, **7**, 93.
20 L. G. M. Baas Becking, I. R. Kaplan and O. Moore, *J. Geol.*, 1960, **68**, 243.

MOLECULAR SIZE DISTRIBUTION OF HUMIC SUBSTANCES: A COMPARISON BETWEEN SIZE EXCLUSION CHROMATOGRAPHY AND CAPILLARY ELECTROPHORESIS

Maria De Nobili,[1] Gilberto Bragato[2] and Antonella Mori[2]

[1] Dipartimento di Produzione Vegetale e Tecnologie Agrarie, Università di Udine, 33100 Udine, Italy

[2] Istituto Sperimentale per la Nutrizione delle Piante, Ministero per le Politiche Agricole, 34170 Gorizia, Italy

1 INTRODUCTION

In capillary zone electrophoresis (CZE), the addition of polyethylene glycols (PEG) capable of forming entangled solutions in the electrophoretic buffer causes an increase in the migration times of humic substances (HSs) when migrating through uncoated capillaries. This increase is linearly proportional to the apparent molecular weight (MW) of the fractions as deduced from the cut off limits of the ultrafiltration membranes used to prepare them.[1] A linear dependence of migration times on MW had already been observed in polyacrylamide gel electrophoresis (PAGE) and size exclusion chromatography (SEC) of HSs fractions,[2] suggesting that both techniques could in principle be employed to determine the molecular size distribution of humic substances under experimental conditions that either suppress or minimize charge density differences. The two techniques, in fact, have several common features that allow the description of the behavior of macromolecules in both systems through a unified model[3] that has been generalized to non spherical molecules.[4]

Sieving effects such as those observed for HSs in the presence of PEG have been observed in different electrophoretic systems and with different hydrophilic polymers for other biological macromolecules[5] as well as for substituted and unsubstituted benzoic acids.[6] The mechanism of separation differs according to the kind and molecular size of the analyte. For benzoic acids, interactions attributable to hydrogen bond formation represent the effective factor in determining the CZE behavior of these molecules in physical gels.[7] Molecular size based separations of DNA fragments are explained through the existence of a transient entanglement coupling mechanism.[8] SDS-protein complexes are much shorter than the long filaments of DNA and in PAGE are separated by a purely sieving mechanism. Because of the high electric field strength involved in CZE, the proposed mechanism of separation for the migration of SDS-protein complexes through entangled polymer solutions generated by addition of hydrophilic polymers to the electrophoretic buffer is either reptation with stretching[9] or hydrogen bond interactions.[10]

According to the reptation mechanism, very large polyions such as DNA, when travelling through gels with pores smaller than their molecular radius, orient themselves along the direction of the electric field. The long macromolecules therefore migrate, taking a tortuous path along the gel fibers in such a way that size differences are minimized. For DNA fragments, for instance, this means that the electrophoretic mobility becomes inversely proportional to the number of base pairs, that is to the length of the molecules and no longer depends on the molecular surface or volume.

Work from our laboratory on the behavior of HSs in the presence of short chain (400 g M^{-1}) PEG at concentrations varying from 2.5% to 12.5% shows that at a given PEG concentration all fractions migrate at about the same velocity. This seems to exclude the possibility of describing the CZE behavior of HSs in a PEG matrix in terms of hydrogen bonding interactions[11] and suggests that some kind of sieving is actually involved in the separation.

In the present work we examined the capillary electrophoresis (CE) and SEC behavior of HS fractions of different origin. Comparison of results obtained by either technique can in principle provide information on interrelationships between mobilities, partition coefficients and molecular radii of HSs. However, the mechanism of the separation must be correctly understood in order to apply the appropriate mathematical model.

2 MATERIALS AND METHODS

The HSs were extracted from commercial samples of sphagnum Baltic and Finnish peat, which had been air dried and milled to pass a 0.5 mm sieve, from the A1 horizons of an Inceptisol and a Spodosol, from Leonardite and from Aldrich humic acids. The dried sources were extracted for 1 hr at room temperature with 0.5 M NaOH under nitrogen flux. Extracts were filtered through a 0.2 µ cellulose nitrate filter (Whatman) and treated with Amberlite IR 120 H+ (Carlo Erba) to lower the pH to 7 and to remove excess sodium.

The extracts were ultrafiltered on Amicon Diaflo YM and XM membranes with 0.1 M sodium pyrophosphate adjusted to pH 7.1 with concentrated H_3PO_4. Five fractions of the following MW ranges, deduced from the 95% cut-off limits of the membranes, were obtained: 1-5, 5-10, 10-30, 50-100 kDa. The ultrafiltration of each fraction ended when the effluent coming out of the cell appeared completely colorless. All fractions were diafiltered on the same membrane with distilled water until the effluent had a conductance of about 30 µS cm^{-1} to eliminate pyrophosphate, concentrated to a final volume of about 10 mL and stored at -18°C.

CZE was performed with an Applied Biosystem 270A-HT unit equipped with a uv-visible detector and a software system for data acquisition on a PC. Uncoated capillaries of 55 cm (length to the detector 30 cm) × 75 µm I. D. (Composite Metal Services) were used. The electrophoretic buffer was a solution of 0.05 M tris-hydroxymethylaminomethane (TRIS) and 0.05 M sodium dihydrogen phosphate adjusted to pH 8.3 and filtered through a Whatman 0.2 µ filter. PEG (BDH) was added to the buffer at concentrations ranging from 1 to 20% (w/v). CZE was performed at a constant temperature of 30°C. Before each analysis, the capillary was rinsed with 0.1 M NaOH for 5 min (uncoated capillaries only) and with the buffer, eventually containing PEG, for 5 min.

In uncoated capillaries humic molecules are driven to the cathode by the electroosmotic flow (EOF). Samples were introduced at the anodic end of the capillary for 3 sec and electrophoretic separations were performed at a voltage of +14 kV. The EOF was determined using mesityl oxide (MSO) detected at 210 nm at the end of each series of measurements. All measurements were made in triplicate.

High performance size exclusion chromatography (HPLC SEC) was performed on a 30 cm long Biosil SEC 250 column (Biorad Laboratories) connected to a Waters 590 pump and a 480 uv-visible detector set at 400 nm. Samples were run in 0.05 M TRIS-phosphate buffer at pH 8.3.

3 RESULTS AND DISCUSSION

In size exclusion chromatography, HS fractions of reduced molecular size polydispersity extracted from the same soil show a linear relationship between the logarithm of the apparent MW of the fractions and the elution volume.[2] HS fractions of different origin show a much larger scatter of data as compared to regressions obtained for HS extracted from the same soil (Figure 1). Large errors can be expected in the determination of MW by SEC because of the effect of different charges and partial specific volumes among HSs extracted from different sources. Previous work has shown that the behaviour of HSs in CZE is much less sensitive to charge differences[1] than is SEC. However, the scatter of data for CZE in the same 0.05 M TRIS-phosphate buffer used for SEC in the presence of 5% w/v PEG 4000 is similar (Figure 2). In both cases the gel matrix has a sieving effect that causes a separation of HSs mostly based on molecular size differences. This suggests the possibility of applying the Ogston model for a random meshwork of fibers to describe the separation of HSs. This model allows prediction of the behavior of macromolecules in both gel chromatography and gel electrophoresis. It also provides equations for interrelationships between mobility, partition coefficients, gel concentration and molecular radius.

We have verified the possibility of applying this model to HSs in CZE. According to theory, for macromolecules behaving according to the Ogston model a plot of the logarithm of the effective electrophoretic mobility (μ_{eff}) against MW should be linear with slope of two. In the case of HSs (Figure 3) the plot is not linear, suggesting that reptation could occur. In the double logarithm plot of log μ_{eff} against log MW (Figure 4), the

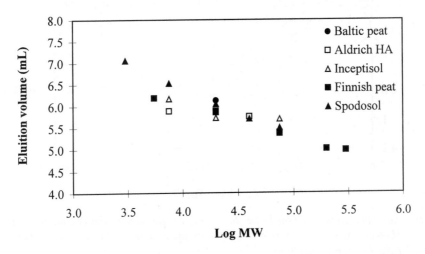

Figure 1 *Size exclusion chromatography of HS fractions extracted from different sources*

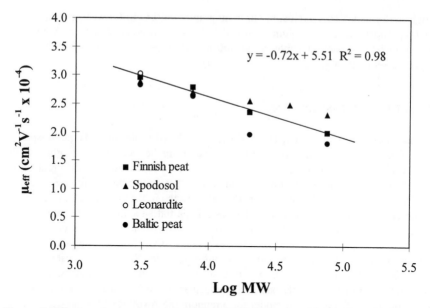

Figure 2 *Effective electrophoretic mobility of HS fractions of different origin in 5% w/v PEG versus the logarithm of their apparent molecular weight*

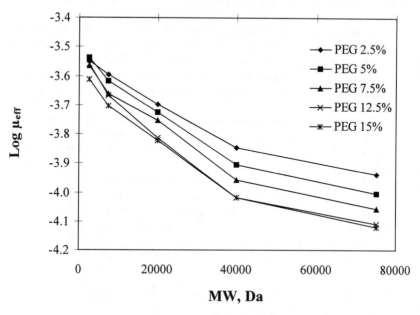

Figure 3 *Relationship between the logarithm of the effective electrophoretic mobility of HS fractions from the Spodosol and their apparent molecular weight at different PEG 4000 concentrations*

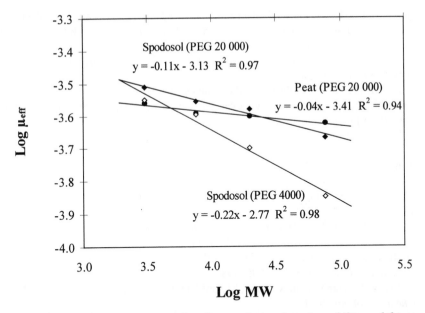

Figure 4 *Double logarithm plot of the effective electrophoretic mobility and the apparent molecular weight of HS fractions from the Baltic peat in the presence of 5% (w/v) PEG 4000 and PEG 20 000*

linearity is maintained but the slopes of the regression lines are much smaller than the theoretical value of -1 for pure reptation. This means that HSs, like SDS-protein complexes, are deformed by the high field strength employed in CZE and tend to orient themselves in the direction of the electric field.

A separation mechanism of this kind is called reptation with stretching and normally occurs for very large polyelectrolytes migrating under high electric fields. Further proof for a mechanism of this kind should be sought by plotting the mobility against the electric field. This was not possible for the system investigated here. At electric fields larger than that applied the Joule effect is no longer negligible. Excessively long migration times, associated with band broadening, limit the range that can be investigated at lower field strength. In view of the wide range of molecular weights examined and of the fact that for small size HSs the reptation with stretching mechanism seems unlikely, it is still theoretically possible to consider only the linear part of the single logarithm plot (Figure 3), and apply the Ogston model to HS fractions of molecular weight up to 20 kDa. According to this model, in a porous network of fibers much longer than the molecule, there should be a linear relationship between the square root of the retardation coefficient (K_r) and the sum of the radius of the gel fibers and the radius of the analyte.[4] This model applies also to non spherical molecules (cylinder, prolate ellipsoid, random coil, etc.) in which case the derived radius is that of an equivalent sphere with the same probability of "no contacts" with the gel.

The retardation coefficients were obtained as the slopes of the regression lines of the Ferguson plots (log μ_{eff} vs. PEG concentration) for low molecular size HSs (data not reported in Figure 4). As predicted by the model, we found a linear dependence ($R^2 = 0.99$)

of the square root of K_r vs. the equivalent radius of the HS fractions calculated from the apparent MW, assuming a partial specific volume of 0.63 mL/gm.[11] The intercept of the regression line gives an intercept value for the gel fibers radius (7.2 nm) that is in surprisingly good agreement with that calculated from the polymer molecular weight and entanglement threshold (6.3 nm).[12]

Because of the lack of absolute MW values for HSs, these results should be considered with caution. They are, however, indicative of an interesting trend that is certainly worth further investigation.

References

1.   M. De Nobili, G. Bragato and A. Mori, *Acta Hydrochim. Hydrobiol.*, 1998, **26**, 186.

2.   M. De Nobili and F. Fornasier, *Europ. J. Soil Sci.*, 1996, **47**, 223.

3.   A. G. Ogston, *Trans. Faraday Soc.*, 1958, **54**, 1754.

4.   D. Roadbard and A. Crambach, *Proc. Natl. Acad. Sci.*, 1970, **65**, 970.

5.   C. Heller, *J. Chromatogr. A*, 1995, **698**, 19.

6.   Y. Esaka, Y. Yamaguchi, K. Kano, M. Goto, H. Haraguchi and J. Takahashi, *Anal. Chem.*, 1994, **66**, 2441.

7.   Y. Esaka, M. Goto, H. Haraguchi, T. Ikeda and K. Kano, *J. Chromatogr. A*, 1995, **771**, 305.

8.   A. Barron, D. S. Soane and H. W. Blanch, *J. Chromatogr. A*, 1993, **652**, 3.

9.   E. Simò-Alfonso, M. Conti, P. G. Righetti and C. Heller, *J. Chromatogr. A*, 1994, **689**, 85.

10.   M. De Nobili, G. Bragato and A. Mori, *J. Chromatogr. A.*, 1999, in press.

11.   M. N. Jones, J. W. Birkett, A. E. Wilkinson, N. Hesketh, F. R. Livens, N. D. Bryan, J. R. Lead, J. Hamilton-Taylor and E. Tipping, *Anal. Chim. Acta*, 1995, **314**, 149.

12.   L. Mitnik, L. Salomá, J. L. Viovy and C. Heller, *J. Chromatogr. A*, 1995, **710**, 309.

CHARACTERIZATION OF HUMIC ACIDS BY CAPILLARY ZONE ELECTROPHORESIS AND MATRIX ASSISTED LASER DESORPTION/ IONIZATION TIME-OF-FLIGHT MASS SPECTROMETRY

L. Pokorná, D. Gajdošová and J. Havel

Department of Analytical Chemistry, Faculty of Science, Masaryk University, 61137 Brno, Czech Republic

1 INTRODUCTION

Humic acids (HAs) are the most widespread complexing ligands occurring in nature. They exhibit structural complexity and polyelectrolyte properties but still are of unknown structure.[1-3] Many general but strictly hypothetical models of HAs structure have been proposed.[4]

Characterization of HAs has been the focus of intense research for many years because soil organic matter contributes more to the quality of soil than the other constituents. HAs have properties similar to weak acid polyelectrolytes, with a wide range of molecular weights, solubilities and acidic strengths. They are sparingly soluble in acidic solutions but solubility increases with pH. HAs are thus fully soluble only in alkaline solutions.

As HAs play an important role in environmental chemistry, they have great importance with regard to the quality and productivity of a soil and the retention of metal ions and pollutants in the environment. Dissolved humic materials interact with metal ions or pollutants that may alter fates and transfer in the soil. The mechanisms involved in these interactions are not clear and may vary depending on the physico-chemical properties of the compounds, soil pH, redox status and the heterogeneous structures of HAs. The effect of pH, for example, is related to the degree of humic R-COOH and R-OH dissociation. Deprotonation increases the polarity of the humic material and alters its structure.[3] Complexation of lead and cadmium was recently studied by anodic stripping voltammetry,[5] and the acid-base behavior and complexation of Cu(II) and other metal ions by potentiometry.[6]

Electrophoretic methods have been applied to HAs studies since the sixties. The first work on HAs separation by a capillary technique was performed by Kopáček, et al. in 1991 using capillary isotachophoresis.[7]

Capillary zone electrophoresis (CZE) was perhaps applied for the first time by Rigol, et al.[8] However, few really suitable methods for the separation of HAs were developed in spite of valuable contributions of several authors.[9-11] Difficulty in HAs separation by CZE is due to the fact that HAs are sorbed on the surface of the quartz capillary and that they oligomerize, increasing their concentration in solution.[12,13] Conditions and additives to the background electrolyte (BGE) that prevent the sorption have been found.[12] An extensive search for the most suitable background electrolyte was performed and a detailed review is available.[14] The most efficient HSs separation achieved is perhaps that by Fetsch and Havel.[15] In that work separation was achieved by applying concentrated boric acid as the BGE. It was suggested that reaction of HA components with boric formed esters. Oligomerization of HA constituents to high molecular aggregates was suppressed and thus the separation into 25-30 fractions was observed.[15]

This work continues a systematic search for new highly efficient CZE separation procedures. In addition, MALDI-TOF mass spectrometry is applied to obtain information about molecular weights of humic acids components and to further characterize HAs. We also examined the possibility of using other reagents and/or derivatization procedures for humic acids in order to improve analysis of these compounds.

2 MATERIALS AND METHODS

2.1 Chemicals

All reagents used were of analytical grade. Urea, $NaHCO_3$, Na_2HPO_4, NaH_2PO_4, $Na_2B_4O_7$, Na_2CO_3, and some other chemicals were from Lachema (Brno). NaOH, H_3PO_4 and N-(3-dimethylaminopropyl)-N'-ethylcarbodiimide hydrochloride were from Merck, and caffeic, gentisic, 5-chlorosalicylic, sinapinic and α-cyano-4-hydroxy-cynnamic acids were from Aldrich. Mesityl oxide used as an EOF marker for CZE was from Fluka. Deionized water used to prepare all solutions was double distilled from a quartz Heraeus Quartzschmelze apparatus.

2.2 Humic Acids

The HA samples used in this work were Fluka HA preparation No. 53680 (analysis No. 38537/1 293) (Fluka I), Fluka HA preparation No. 53680 (analysis No. 38537/1 594) (Fluka II) and coal-derived Czech HA samples with origin and preparation as described elswhere:[13-15] MAR 193 (from oxyhumolite Bílina), MAR 329 and MAR 346 (from tschernozem obtained from a soil near Chotěšov Village, Czech Republic).

IHSS standards were purchased from the International Humic Substances Society (IHSS). The following samples were used: IHSS Peat HA Standard (1R103H), IHSS Leonardite HA Standard (1S104H), IHSS Soil Summit Hill HA Reference (1R106H) and IHSS Soil HA Standard (1S102).

2.3 Instrumentation

Mass spectra were measured with a Kompact MALDI III mass spectrometer (Shimadzu) either in linear or in reflectrone mode. The instrument was equipped with a nitrogen laser (wavelength 337 nm, pulse duration τ =10 nsec, energy pulse 200 μJ). The energy of the laser applied was in the range of 0-180 units. One hundred laser shots were used for each sample, the signals of which were averaged and smoothed. Insulin was used for mass calibration.

Spectrophotometric measurements were made with a UNICAM/UV2 uv-visible spectrophotometer. A Beckman Model P/ACE System 5500 equipped with a diode array detection (DAD) system, automatic injector, a fluid-cooled column cartridge and System Gold Data station was utilized for all CZE experiments. Fused silica capillary tubing of 47 cm (40.3 cm length to the detector) x 75 μ I. D. was used. The normal polarity mode of the CZE system (cathodic pole at the detection side) was applied.

The pH was measured using a glass G202C electrode, standard calomel electrode K401 (Radiometer) and a precision digital pH-meter OP-208/1 (Radelkis, Budapest). Standard buffer solutions from Radiometer and/or Radelkis were used for calibration.

3 RESULTS AND DISCUSSION

Several reagents or derivatization agents were examined with the aim of enhancing the low solubility of HAs. As found with boric acid, oxo-anions like germanate, molybdate and tungstate facilitate dissolution of HAs even in acid solution due to the interaction of hydroxo and carboxylic groups. However, the results of reaction mixture analysis by CZE or MALDI-TOF MS are complicated and are not given here. Similarly, derivatization of HAs with carbodiimide and tetramethyl- or tetrabutylammonium hydroxides gave complicated chromatographs. Therefore, only results with some dissolution agents are given below.

3.1 Dissolution of HA Samples

We need aqueous HAs solutions for capillary electrophoresis. It was necessary to find a suitable way to dissolve the samples. The use of several different agents such as urea, NaOH, KOH, $NaHCO_3$, H_2SO_4, dimethylsulphoxide (DMSO), dimethylformamide and so on for this purpose was examined. 96% H_2SO_4 was found to be the best solvent. However, after dilution the solution precipitated HA. DMSO and HAs mass spectra show similar peaks and so DMSO was not suitable for mass spectrometry.

Dissolution in $NaHCO_3$ solution was also not complete for most of the samples. Therefore, solutions in sodium hydrogencarbonate were always filtered before use.

3.1.1 Sample Preparation for MALDI-TOF MS. Aqueous solutions of HAs were prepared in the following way: 1 mg of the sample was dissolved in 50 μL of 1 M NaOH and the solution was completed with double distilled water to 1 mL, to give 1 mg/mL stock solutions of HA samples.

All HAs can be quite efficiently dissolved in 10 M urea. After dissolution the solutions were diluted to a concentration of 0.5 mg/mL in 1 M urea.

For mass spectrometry 0.5 μL of the sample solution was pipetted onto the spot of a sample slide, dried in a stream of air at room temperature and analyzed.

3.1.2 Sample Preparation for Capillary Zone Electrophoresis. The aqueous NaOH and urea stock solutions of HAs were prepared as described above. The aqueous $NaHCO_3$ stock solutions of HAs were prepared in a similar way: 1 mg of the HA was dissolved in 1 mL of 1 M $NaHCO_3$. However, total dissolution of the Leonardite sample in urea and of almost all HA samples in $NaHCO_3$ was not observed. Therefore, these stock solutions were always filtered before analysis using a sintered G4 porous glass filter.

3.2 MALDI-TOF Mass Spectrometry

The following matrices were examined for MALDI-TOF MS: gentisic, sinapinic, caffeic,

5-chlorosalicylic and α-cyano-4-hydroxy cinnamic acids. However, it was observed that the ionization of HAs was poor with these matrices. Peaks of matrices are situated in the range m/z 100 – 500 but the formation of clusters interfering with peaks of HAs and the fragmentation and clustering of matrix peaks was observed. Therefore, the mass spectra were measured in Laser Desorption/Ionization (LDI) mode, i.e., no matrix was used. In the other words, humic acids themselves were used as self-matrices.

The mass spectra of the IHSS Peat standard (samples dissolved in 1 M urea solution, 0.5 mg HA/mL) measured in linear positive mode are presented in Figure 1 as a function of laser energy.

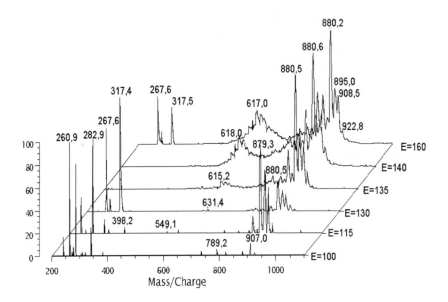

Figure 1 *Mass spectra of IHSS standard Peat HA dissolved in 1 M urea (HA concentration 0.5 mg/mL) in LDI linear positive mode as a function of laser energy*

With increasing laser energy we observed a difference in m/z values of about a multiple of 14 in the group of peaks around m/z 800. The mass spectra measured in reflectrone positive mode under the same conditions are given in Figure 2. In this mode the peaks with m/z values around 800 were not observed. This is an indication of the low stability of the ions observed in linear positive mode around m/z 800.

The other soil HA standards dissolved in different ways were investigated with the same laser energy of 130 units. The mass spectra of HA standards dissolved in diluted NaOH solution (HA concentration 1 mg/mL) in linear positive mode are given in Figure 3. The spectra are very similar. There are some differences between the spectra for different solvents. The mass spectra of the HA standard dissolved in 1 M aqueous urea (HA concentration 0.5 mg/mL) at laser energy = 130 units are given in Figure 4. The major peaks were observed around m/z 800 but significant differences were not observed there.

For the analysis of HA coded as MAR (193, 329 and 346) the same solvents as for standard HA samples were used. The mass spectra of the 346 MAR product as a function of laser energy in linear positive mode are given in Figure 5. These HAs were dissolved in

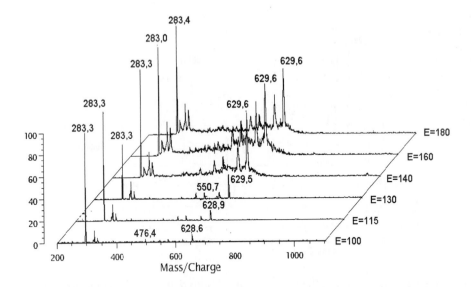

Figure 2 *Mass spectra of IHSS standard Peat HA dissolved in 1 M urea (HA concentration 0.5 mg/mL) measured in LDI reflectrone positive mode as a function of laser energy*

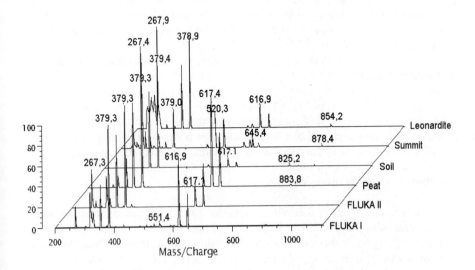

Figure 3 *The mass spectra of IHSS HA standard dissolved in diluted NaOH (HA concentration 1 mg/mL) measured in LDI linear positive mode*

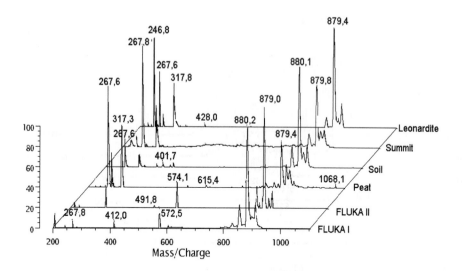

Figure 4 *The mass spectra of IHSS HA standard dissolved in 1 M urea (HA concentration 0.5 mg/mL) measured in LDI linear positive mode*

NaOH solution (HA concentration 1 mg/mL). The mass spectra are similar to those of HA standards with the same dissolution procedure, but the most suitable laser energy was between 130 and 140 units. Greater differences in mass spectra for different dissolution agents were observed for MAR samples (Figure 6). The mass spectra of the 346 MAR sample dissolved in 1 M urea (HA concentration 0.5 mg/mL) measured in linear positive mode are shown there. The same peaks at about m/z 267 and 317 were observed for most of the samples. The peaks around m/z 600 (observed for NaOH dissolved samples, Figure 5) were different in urea dissolved samples (Figure 6).

The mass spectra of Czech HA samples (MAR 193, 326 and 346) dissolved in diluted NaOH solution (concentration 1 mg/mL) were quite similar (Figure 7).

Comparing the mass spectra in Figure 1 and those in Figure 6, it is evident that the spectra of HA standards and of Czech HA samples are almost identical. There are two most probable explanations: (i) either the fractions are the same in all humic acids studied, or (ii) humic acids are fragmented by a laser to the same fragments (building blocks). It is also possible that only some of the components in a humic acid mixture are ionized.

3.3 Capillary Zone Electrophoresis

3.3.1 Separation of HA Dissolved in Aqueous Urea. The stock solutions were prepared by dissolving 5 mg of HA in 1 mL of 10 M urea solution. Because urea absorbs in the ultraviolet region, it was necessary to decrease its concentration. Thus, the stock solutions of HA were diluted to a concentration of 125 μg HA /mL in 25 mM urea.

The next step was optimization of the conditions for CZE analysis. Optimal conditions were found to be hydrodynamic injection for 22 sec, applied voltage 22 kV, detection at 210 nm and temperature 40°C. All HAs samples were analyzed under these conditions. 50 mM NaH_2PO_4 (pH 6.5) was used as a background electrolyte.

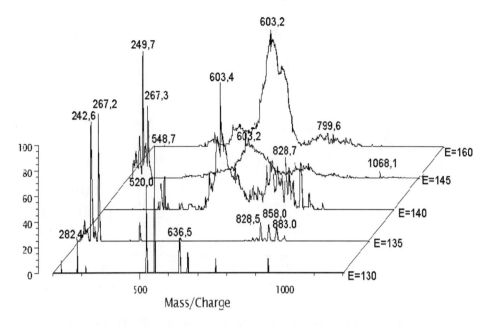

Figure 5 *Mass spectra of Czech HA sample 346 MAR dissolved in diluted NaOH (HA concentration 1 mg/mL) in LDI linear positive mode as a function of laser energy*

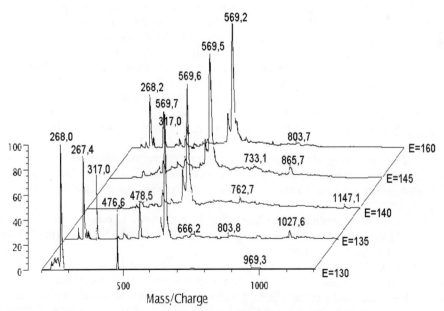

Figure 6 *Mass spectra of Czech HA sample 346 MAR dissolved in 1 M urea (HA concentration 0.5 mg/mL) in LDI linear positive mode as a function of laser energy*

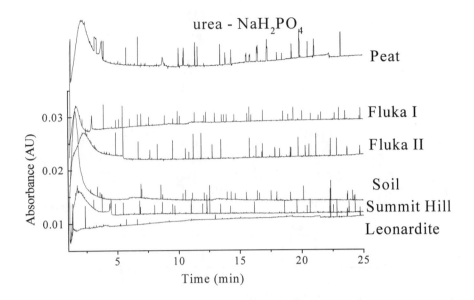

Figure 7a *Electropherograms of IHSS standards dissolved in 25 mM urea (HA concentration 125 µg/mL). BGE = 50 mM NaH$_2$PO$_4$ (pH 6.5), hydrodynamic injection 22 sec, applied voltage 22 kV, detection at 210 nm, temperature 40 °C*

For almost all HA standard samples we obtained similar electropherograms with a typical HA hump with a migration time near to 2 min and many smaller peaks. The part of the electropherogram for IHSS Peat HA standard at migration times from 8 to 22 min is shown in Figure 7b, where it is possible to see about 20 separate peaks. Only for IHSS Leonardite HA standard, which was not completely dissolved even in 10 M urea, was the electropherogram completely different. Two peaks with a migration time near to 2 min and some other peaks at high migration times were observed (Figure 7a).

For MAR 346 HA no significant peaks were observed and the electropherogram is similar to that of Leonardite HA, while MAR 193 and 329 show a characteristic humic hump. Different other peaks in the electropherogram (Figure 7c) were also obtained with MAR 193.

3.3.2 Separation of HA Samples Dissolved in Aqueous NaOH Solution. The stock solutions were prepared by dissolving of 1 mg of HA in 1 mL of 50 mM NaOH solution. CZE analysis was performed after further diluting to a concentration of 100 µg HA /mL in 5 mM NaOH.

In this case the optimal conditions were: hydrodynamic injection 22 sec, applied voltage 12 kV, detection at 210 nm and temperature 40°C. 50 mM Na$_2$HPO$_4$ (pH 11) was chosen as a background electrolyte.

Although all HAs samples were dissolved using sodium hydroxide, the results of electrophoresis were no better than in the previous case. Only for Fluka I was the electropherogram similar to that obtained with urea, but a different migration time (near to 12 min) for the hump was observed. For Fluka II peaks with migration times 7-10 min were found, but the rest of the electropherogram was practically empty. For the other IHSS

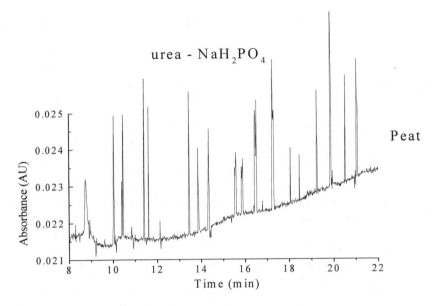

Figure 7b *Electropherogram of IHSS Peat HA standard dissolved in 25 mM urea (HA concentration 125 µg/mL). Conditions are the same as in Figure 7a*

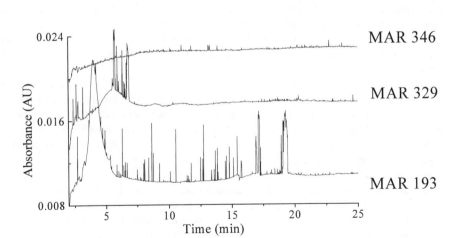

Figure 7c *Electropherograms of Czech HA samples dissolved in 25 mM urea (HA concentration 125 µg/mL). Conditions are the same as in Figure 7a*

standards a small hump with a migration time of about 7 min was observed and nothing more.

For Czech HA samples we obtained electropherograms similar to that of Fluka I for MAR 346 and 329 HAs. But for MAR 193 HA we obtained an electropherogram practically without peaks, which corresponded to data for other IHSS samples.

As another background electrolyte for HA samples dissolved in NaOH, 50 mM $Na_2B_4O_7$ (pH 9.5) was applied. Optimal conditions in this case were: hydrodynamic injection 22 sec, applied voltage 15 kV, detection at 210 nm, temperature 40°C.

For IHSS Leonardite, Summit Hill and Peat HA standards we obtained electropherograms typical of humic acids, which means a "hump" and a number of peaks. For Fluka I and Fluka II the electropherograms were similar, but the "hump" was not so distinct. For IHSS Soil standard HA a completely different electropherogram was obtained, where a peak with a migration time near to 5 min was observed and there were a few small peaks in the rest of electropherogram.

For Czech HA samples, we obtained almost the same electropherogram fingerprints for MAR 346 and 193, with only a hump, and a very similar one for MAR 329, where in addition to the hump we can see another peak at a migration time of about 20 min.

3.3.3 Separation of HA Samples Dissolved in Aqueous NaHCO₃. Stock solutions were prepared by dissolving of 1 mg of HA in 1 mL of 1 M NaHCO₃. For CZE analysis, further dilution to concentration 100 μg HA /mL in 0.1 M NaHCO₃ was applied. For 50 mM $Na_2B_4O_7$ (pH 9.5) BGE the optimal separation conditions were: hydrodynamic injection 22 sec, applied voltage 15 kV, detection at 210 nm, temperature 40°C.

The results for IHSS standards were quite different. For Fluka I and Peat HA we obtained similar electropherograms with several peaks at a migration time near to 5 min, two peaks around 10 min, a hump and some other peaks with longer migration times. The Fluka II electropherogram was without a hump, but peaks with migration times of about 5-6 min were evident, and several peaks with migration times from 18 to 27 min were observed. For other IHSS standards, electropherograms show only three or four peaks. For Czech HAs we obtained almost the same electropherograms for all samples, quite similar to the one of MAR 329 dissolved in NaOH and using tetraborate as BGE.

A mixture of 90 mM NaHCO₃ + 8 mM Na_2CO_3 (pH 9) was tested as a BGE with the following optimal conditions: hydrodynamic injection 22 sec, applied voltage 20 kV, detection at 210 nm, temperature 40°C.

Fluka I, Fluka II, Soil, Summit Hill and Peat HAs show very similar electropherograms with a hump and number of peaks. For Soil and Summit Hill HAs one outstanding peak with a migration time near to 26 min was observed. This peak was not present in the electropherograms for the other three humic acid samples. However, for IHSS standard Leonardite an electropherogram practically without peaks was obtained (Figure 8a). A part of the electropherogram for Fluka I HA at migration times 18 - 30 min is shown in Figure 8b.

Similar electropherograms were obtained for all three Czech HA samples, in which for migration times from 0 to 15 min it is possible to find almost the same peaks. But for MAR 193 HA there are other peaks with migration times from 15 to 25 min that are not present on the electropherograms for the other two samples (Figure 8c).

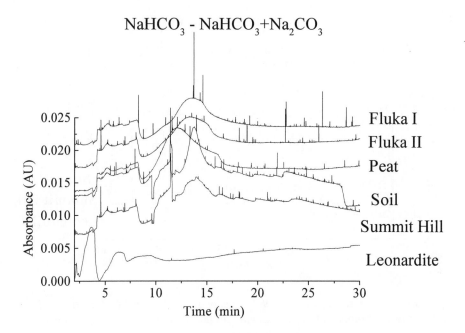

Figure 8a *Electropherograms of IHSS standards dissolved in 0.1 M NaHCO₃ (HA concentration 100 µg/mL). BGE = (90 mM NaHCO₃ + 8 mM Na₂CO₃) of pH 9, hydrodynamic injection 22 sec, voltage 20 kV, detection at 210 nm, temperature 40°C*

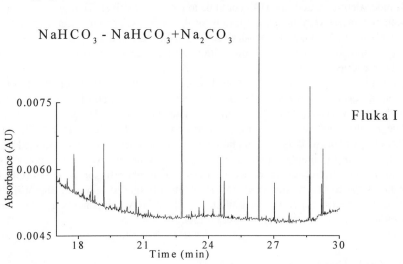

Figure 8b *Electropherogram of IHSS Peat standard dissolved in 0.1 M NaHCO₃ (HA concentration 100 µg/mL). Conditions are the same as in Figure 8a*

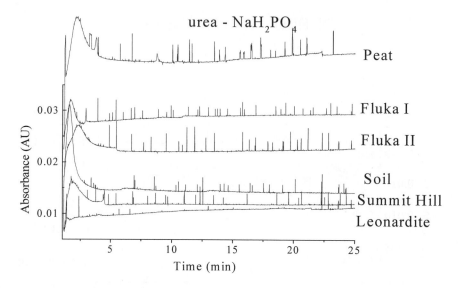

Figure 8c *Electropherograms of Czech HA samples dissolved in 0.1 M NaHCO₃ (HA concentration 100 µg/mL) Conditions are the same as given in Figure 8a*

4 CONCLUSIONS

Urea is a suitable dissolution agent for HAs. MALDI-TOF mass spectrometry performed in LDI mode demonstrated that HAs could be ionized to positively charged ions. Ions with the same m/z values were obtained for most of the samples. The most frequent m/z values for Czech soil (tschernozem), coal-derived and IHSS standard HAs were at 261, 267.5, 283, 317.5, and so on. A group of ions with m/z values around 600-880 was observed at higher laser energy.

By capillary electrophoresis the best separation patterns were obtained for HAs dissolved in aqueous urea and a phosphate buffer (pH = 6.5), and for HAs dissolved in sodium hydrogen carbonate and hydrogen carbonate-carbonate buffer (pH = 9).

The results confirm the previous conclusions from this laboratory[5,6,12-16] that (i) capillary electrophoresis is powerful and perhaps the best separation method for HAs, (ii) although HAs are a complex mixture of many low molecular weight compounds some similarities and differences in either capillary electrophoresis or MALDI-TOF mass spectrometry patterns are observed for HAs of different origin and that (iii) mass spectrometry strongly suggests that building blocks of HA components are the same or very similar even for HAs of very different origin.

ACKNOWLEDGEMENTS

This work is part of a research program within BARRANDE Czech-French bilateral scientific cooperation project no. 96006. The Czech Ministry of Education (Prague) and the French Ministry of Foreign Affairs are acknowledged for support of this work.

References

1. F. J. Stevenson, 'Humus Chemistry: Genesis, Composition, Reactions', Wiley, New York, 1982.
2. M. Remmler, A. Georgi and F. -D. Kopinke, *Eur. Mass Spectrom.*, 1995, **1**, 403.
3. G. Davies and E. A. Ghabbour, (eds.), 'Humic Substances: Structures, Properties and Uses', The Royal Society of Chemistry, Cambridge, 1998.
4. S. M. Shevchenko and G. W. Bailey, *Crit. Rev. Environ. Sci. Technol.*, 1996, **26**, 95.
5. J. Šenkýř, A. Ročáková, D. Fetsch and J. Havel, *Toxicol. Environ. Chem.*, 1999, **68**, 377.
6. P. Lubal, D. Široký, D. Fetsch and J. Havel, *Talanta*, 1998, **47**, 401.
7. P. Kopáček, D. Kaniansky and J. Hejzlar, *J. Chromatogr. A*, 1991, **545**, 461.
8. A. Rigol, M. Vidal and G. Rauret, *J. Chromatogr. A*, 1998, **807**, 275.
9. M. Nordén and E. Dabek-Zlotorzynska, *Electrophoresis*, 1997, **18**, 292.
10. R. Dunkelog, H. -H. Rüttinger and K. Peisker, *J. Chromatogr. A*, 1997, **777**, 355.
11. P. Schmitt-Koplin, A. W. Garisson, E. M. Perdue, D. Freitag and A. Kettrup, *J. Chromatogr. A*, 1998, **807**, 101.
12. D. Fetsch, M. Fetsch, E. M. Peña-Méndez and J. Havel, *Electrophoresis*, 1998, **19**, 2465.
13. D. Fetsch, M. Hradilová, E. M. Peña-Méndez and J. Havel, *J. Chromatogr. A*, 1998, **817**, 313.
14. D. Fetsch, A. M. Albrecht-Gary, E. M. Peña-Méndez and J. Havel, *Scripta Fac. Sci. Nat. Univ. Masaryk Brun. Chemistry*, 1997, **27**, 3.
15. D. Fetsch and J. Havel, *J. Chromatogr. A*, 1998, **802**, 189.
16. J. Havel, D. Fetsch, E. M. Peña-Méndez, P. Lubal and J. Havliš, in Abstracts, International Humic Substances Society 9th International Meeting, Adelaide, 1998, p. 103.

MALDI-TOF-MS ANALYSIS OF HUMIC SUBSTANCES – A NEW APPROACH TO OBTAIN ADDITIONAL STRUCTURAL INFORMATION?

G. Haberhauer,[1] W. Bednar,[1] M. H. Gerzabek[1] and E. Rosenberg[2]

[1] Department of Environmental Research, Austrian Research Centers, A-2444 Seibersdorf, Austria
[2] Institute of Analytical Chemistry, Vienna University of Technology, A-1060 Vienna, Austria

1 INTRODUCTION

The exact chemical structure of humic substances (HSs) is still unknown.[1-3] This biomaterial is much more diverse and complicated than DNA or proteins, which consist of a limited number of monomers. Elemental analysis of HSs has been carried out and we do have knowledge about the functional groups of HSs. What is still lacking is information about configuration and conformation (there are some hypotheses)[4-5] and how the functional groups are linked together to form three-dimensional structures.

Present knowledge of HSs structure is mainly derived from NMR,[6-8] vibrational spectroscopy[9-11] and pyrolysis mass spectrometry (py-MS) methods.[12] All these methods are powerful for determination of average amounts of certain functional groups or moieties and for studying HSs dynamic processes. Py-MS has been used extensively in the study of HSs, primarily as a means of obtaining information about fragments after thermal degradation.

Virtually all early mass spectrometric studies of HSs were characterized by the extensive fragmentation produced by conventional electron impact ionization, leading to fragment ions at almost every nominal mass below mass-to-charge ratio m/z 200 and a few ions above m/z 200. Soft ionization methods such as electrospray ionization (ESI), laser ablation and fast atom bombardment (FAB) MS[13] are successfully used for proteins and may be valuable for HSs analysis.

Matrix Assisted Laser Desorption/Ionization Time of Flight MS (MALDI-TOF-MS) is another soft ionization technique that was developed in the late 80's by Karas and Hillenkamp.[14] The combination of matrix assisted laser desorption with time of flight MS enables the analysis of high molecular weight substances. MALDI-TOF-MS is now routinely used for the analysis of proteins, RNA, DNA[15] and oligosaccharides.[16] The principle of MALDI-TOF MS is shown in Figure 1. The co-crystallized sample/matrix mixture absorbs the energy of the laser pulse. This leads to desorption and ionization of both sample and matrix molecules. The ions are accelerated and separated according to their mass-to-charge ratio by a time-of-flight mass analyzer. The mechanism of ion formation is still not clear but is thought to proceed via charge transfer processes after desorption. Another formation mechanism for ion production is desorption of already preformed ions from the matrix.[17]

Molecular ions and cluster ions are detected by MALDI-TOF-MS. Cluster ions can

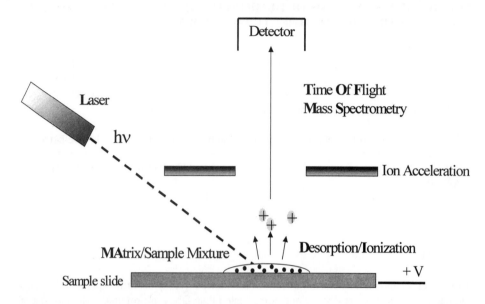

Figure 1 *Principle of MALDI-TOF–MS*

consist of molecules adhering to each other through either specific interactions such as hydrogen bonding or through non-specific interactions.[18,19]

Only a few applications of the above mentioned soft ionization MS techniques to HSs analysis have been published.[20,21] An investigation of fulvic acids (FA) using electrospray ionization MS demonstrates that changes of the analysis parameters change the number of charges per molecule and consequently the spectra.[21]

Thus, the advantages of MALDI-TOF-MS, which are soft ionization, formation of singly charged ions and high mass range detection, are opposed by several limitations in the analysis of HSs. These are (i) impurities such as high inorganic salt contents that are known to reduce the ion yield,[16] (ii) low solubility of mainly uncharged HSs and (iii) the nature of HSs themselves. HSs consist of a large number of macromolecules with a wide mass range distribution, which excludes the analysis of single molecules. Even if MALDI-TOF MS enables femtomol analysis of pure substances,[22] analysis of a mixture containing many different molecules with each at low concentration will certainly create a detection problem. The wide mass range distribution of HSs requires improvement of conventional sample preparation. Application of MALDI-TOF-MS to HSs could reveal information about the mass distribution and give fingerprint identification but it can not identify individual molecules. The aim of this work was to develop a suitable technique to investigate the potential of MALDI-TOF-MS for HSs analysis.

2 MATERIAL AND METHODS

MALDI-TOF-MS was applied to both HSs and solid samples from organic soil horizons in suspension.

Three different types of HSs were used. Two were extracted from the Ap-horizon of an Austrian agricultural soil (Chernozem) using either Chelex® or sodium pyrophosphate procedures as described in ref 23. The third HS was purchased from Aldrich. The milled HSs samples were mixed and suspended with the matrix solution and applied to the sample slides. After evaporation, the sample was placed in the MALDI-TOF-MS instrument.

For the organic soils, 5 individual soil profiles (humic podsol) of a site at Roundwood Forest near Dublin (Ireland) were collected down to a depth of approximately 10 cm from an undisturbed forest area. Organic horizons were divided into L (fresh litter), F (fermentation horizon) and H (humification horizon) layers, respectively. For thicker horizons the plots were divided vertically into several subsamples. The soil samples were air-dried and sieved to remove material >2 mm. Then the samples of sieved soil were powdered in an agate mill. The organic forest soils were milled, suspended with matrix solution, applied to the sample slide and after evaporation were placed to the MALDI-TOF-MS instrument.

All mass spectra were acquired on a Kratos Kompact MALDI III. The instrument was operated at an acceleration voltage of 20 kV, using linear mode. The samples were desorbed/ionized from the probe tip using a pulsed nitrogen laser (3 nsec pulse duration) with an output wavelength of 337 nm and a maximum output of 6 mW. Calibration was performed with peptides/proteins with known molecular mass (bovine serum albumin, ovalbumin, lysozyme and cytochrome C, Sigma) with an amount of 50 pmol. The matrices (sinapinic acid and gentisinic acid, Sigma) and standards were dissolved in 75% aqueous acetonitrile (Merck) with 0.1% trifluoroacetic acid (Fluka). The milled HSs were suspended in distilled water.

A mathematical reduction of resolution was applied for better comparison of the spectra. This was accomplished by simply adding up all abundances of certain mass ranges ($\Sigma 1000$ nominal masses = one parameter). The calculated parameters were then plotted against mass.

3 RESULTS AND DISCUSSION

The use of conventional sample preparation techniques for MALDI-TOF MS failed in our experiments and resulted in spectra with no significant peaks that could be assigned to HSs. However, suspensions of HSs in matrix solution allowed us to examine HSs samples of low solubility. A milled HS sample was directly suspended with the matrix solution and applied to the sample slide. After evaporation, the sample was analyzed. This simple sample preparation procedure enables the investigation of HSs by MALDI-TOF-MS and mass spectra of HSs can be obtained.

The modified technique can be regarded as a combination of ion formation either due to laser desorption from solid particles, which are still present in the suspension, or due to matrix assisted laser desorption. A typical set of spectra of these HSs is shown in Figure 2. HSs of different origin resulted in different MALDI spectra. It is not possible to assign one peak to a specific molecule because the recorded peaks are either due to molecular or cluster ions. However, the abundance pattern is repeatable, although the abundance of a specific mass is not always the same when repeat spectra of the same sample were compared. This might be a problem of both heterogeneity of the HS and inhomogeneity due to sample preparation, which leads to laser desorption of both matrix dissolved molecules and molecules from particles. Nevertheless, the abundance pattern of certain mass ranges was repeatable. A mathematical reduction of resolution was applied for better comparison of the mass distribution. This was accomplished by simply adding up all abundances of certain mass ranges. The calculated values were then plotted against mass.

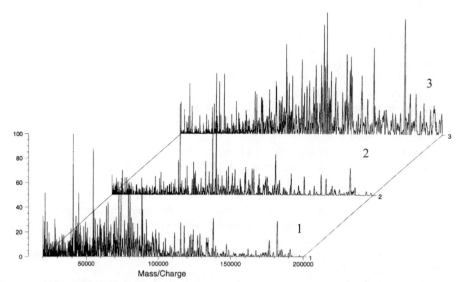

Figure 2 *MALDI-TOF-MS of HSs of different origin: 1, Na-pyrophosphate extracted, 2, Aldrich-HA and 3, Chelex® extracted*

The results obtained are displayed in Figure 3. The mathematical reduction of resolution emphasizes the differences between the spectra. High abundances were found for the pyrophosphate extracts in the mass range below 100 kDa while the spectra of the Aldrich humic acid showed high abundances up to 200 kDa. Predominantly molecular and cluster ions below 100 kDa were obtained for pyrophosphate extracted HSs.

A shift to higher masses up to 150 kDa was observed for the Chelex® extracts. Thus, from our MALDI observations it can be concluded that the Chelex® procedure extracts HSs-fragments and clusters of higher molecular weight than pyrophosphate or leads to less destruction of larger molecules during the extraction procedure itself. The Aldrich HS yielded mass to charge ratios in a wide range up to more than 200 kDa.

In addition to applications of MALDI to extracted HSs, the simple new sample preparation technique was used to apply MALDI to investigate solid samples from organic soil. Five forest soil profiles sampled at Roundwood Forest near Dublin were taken.[24] Applying FTIR and other methods litter, fermentation and humus are clearly distinguishable and the layers clearly differ in their decomposition status. The question was whether the differences observed with other analytical methods could also be detected by MALDI. The sample preparation was similar to that of the extracted HSs. The soils were milled, suspended with matrix solution, applied to the sample slide and after evaporation placed in the instrument.

A typical set of spectra is shown in Figure 4. Clear differences between the spectra were found. For the L-horizon only a few mass peaks are observed, while with increasing decomposition status (down to the H-horizon) the abundance of peaks in the mass range below 150 kDa increased. All five profiles were analyzed several times and the same abundance pattern always was obtained.

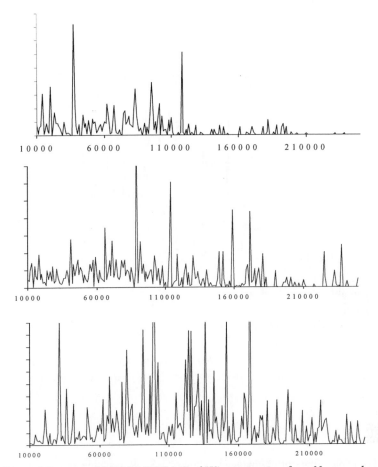

Figure 3 *Processed MALDI-TOF-MS of HSs originating from Na-pyrophosphate (top), Chelex® extractions (middle) and Aldrich HA (bottom)*

So even if the abundance at one mass varies, the abundance pattern remains constant over a certain mass range. It can be assumed that desorption of cluster and molecular ions from mainly intact plant material of the L layer is less pronounced in comparison to desorption from already decomposed and humified material of the H-layer.

This finding suggests that MALDI-TOF-MS allows monitoring of organic matter decomposition processes. Again, it is not possible to assign a specific peak to one molecule because both cluster and single molecules are detected.

Reduction of resolution makes this trend even more visible. Highly significant differences in the spectra are obtained for the L, F and H horizons. Examination of the abundances of every 1000 mass range segments (Figure 5) shows that the abundance of the mass range up to 150 kDa increases significantly with increased decomposition status from L to H.

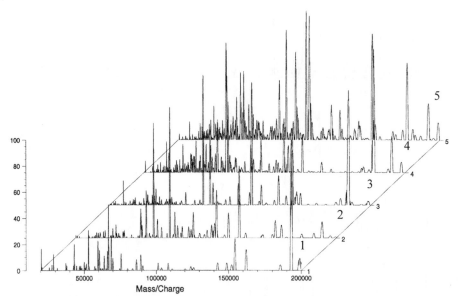

Figure 4 *MALDI-TOF MS of organic forest soil horizons of Roundwood humic podsol: 1,*
L- litter, 2, F - fermentation, 3-5 humic horizons H1-H3 with increasing soil
depth

4 CONCLUSIONS

MALDI-TOF-MS enables fingerprint characterization of HSs. Repeatable and distinctive mass distributions of cluster and/or molecular ions can be obtained. However, discrimination of clusters from single molecules is not possible due to the complexity of the analyzed material. Nevertheless, MALDI-TOF-MS examinations of forest organic soils suggest that decomposition can be followed and horizon identification can be accomplished. However, there are still drawbacks concerning the application of MALDI-TOF-MS to inhomogeneous materials like HSs. The high heterogeneity due to sample preparation and of the samples themselves might explain the varying abundances, which allows only abundance pattern comparison and no single peak comparison, and for the relatively low ion yield of HSs.

However, our results suggest that the application of statistical methods for data analysis together with further improvement in sample preparation and comparison of the MALDI-TOF-MS data with data obtained by other methods should make MALDI-TOF-MS a powerful additional information source for HSs analysis.

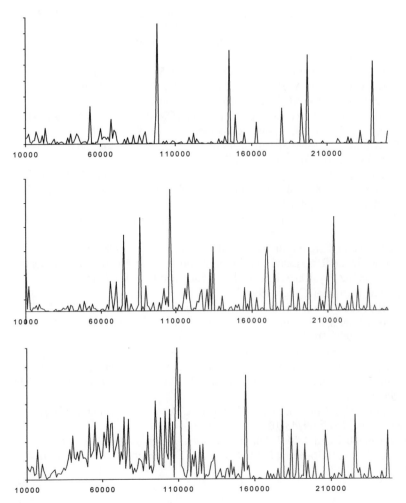

Figure 5 *Processed MALDI-TOF MS of litter (top), fermentation (middle) and humic (bottom) horizons from Roundwood Forest (humic podsol)*

ACKNOWLEDGEMENTS

This research was partially supported by the Austrian Nationalbank (Nationalbank fonds project No.6761), which is greatfully acknowledged.

References

1.   G. Davies and E. A. Ghabbour, 'Humic Substances: Structures Properties and Uses', Royal Society of Chemistry, Cambridge, 1998.
2.   R. L. Wershaw, *Environ. Sci. Technol.*, 1993, **27**, 814.
3.   M. A. Schlautman and J. J. Morgan, *Environ. Sci. Technol.*, 1993, **27**, 961.

4.   L. T. Sein, J. M. Varnum and S. A. Jansen, *Environ. Sci. Technol.*, 1999, **33**, 546.
5.   H. R. Schulten and M. Schnitzer, *Soil Science*, 1997, **162**, 115.
6.   C. M. Preston, R. Hempfling, H. R. Schulten, M. Schnitzer, J. A. Trofymow and D. E. Axelson, *Plant and Soil*, 1994, **158**, 69.
7.   C. A. Fox, C. M. Preston. and C. A. Fyfe, *Can. J. Soil Sci.*, 1993, **74**, 1.
8.   Y. Inbar, Y. Chen and Y. Hadar, *Soil Sci. Soc. Am. J.*, 1989, **53**, 1695.
9.   F. J. Stevenson and K. M. Goh, *Geochim. Cosmochim. Acta*, 1971, **35**, 471.
10.   G. Haberhauer and M. H. Gerzabek, *Vibrational Spectroscopy*, 1999, **19/2**, 415.
11.   J. Niemeyer, Y. Chen and J. -M. Bollag, *Soil Sci. Soc. Am. J.*, 1992, **56**, 135.
12.   H. R. Schulten, in 'Mass Spectrometry of Soils', T. W. Boutton and S. Yamasaki, (ed.), Dekker, New York., 1995, p 373.
13.   T. L. Brown, F. J. Novotny and J. A. Rice, in 'Humic Substances: Structures Properties and Uses', G. Davies and E. A. Ghabbour, (eds.), Royal Society of Chemistry, Cambridge, 1998, p. 91.
14.   F. Hillenkamp, M. Karas, R. C. Beavis and B. T. Chait, *Anal. Chem.*, 1991, **63**, 1193 A.
15.   S. Berkenkamp, F. Kirpekar and F. Hillenkamp, *Science*, 1998, **281**, 260.
16.   J. A. Carroll and R. C. Beavis, in 'Laser Desorption and Ablation: Experimental Methods in the Physical Sciences', J. C. Miller and R. F. Haglund, (eds.), Academic Press, San Diego, 1997, Vol. 30.
17.   V. Karbach and R. Knochenmuss, *Rapid Comm. Mass Spectrom.*, 1998, **12**, 968 .
18.   R. Knochenmuss, F. Dubois, M. J. Dale and R. Zenobi, *Rapid Comm. Mass Spectrom.*, 1996, **10**, 871.
19.   E. Lehmann and R. Zenobi, *Angew. Chem.*, 1998, **110**, 3600.
20.   M. Remmler , A. Georgi and F. -D. Kopinke, *Eur. Mass Spectrom.*, 1995, **1**, 403.
21.   A. Fievre, T. Solouki, A. G. Marschall and W. T. Cooper, *Energy & Fuels*, 1997, **11**, 554.
22.   D. P. Little, T. J. Cornish, M. J. O´Donnell, A. Braun, R. J. Cotter and H. Köster, *Anal. Chem.*, 1997, **69**, 4540.
23.   M. H. Gerzabek and S. M. Ullah, *International Agrophysics*, 1989, **5**, 197.
24.   G. Haberhauer, B. Rafferty, F. Strebl and M. H. Gerzabek, *Geoderma*, 1998, **83**, 331.

LASER SPECTROSCOPY OF HUMIC SUBSTANCES

C. Illenseer, H. -G. Löhmannsröben, Th. Skrivanek and U. Zimmermann

Institute of Physical and Theoretical Chemistry, University of Erlangen-Nürnberg, D-91058 Erlangen, Germany

1 INTRODUCTION

Humic substances (HSs) play an important role in chemical, physical and geological processes occurring in natural waters and soils.[1-3] For example, the binding of organic contaminants such as polycyclic aromatic compounds (PACs) to HSs can have significant consequences on the fate and properties of these environmental contaminants. The investigation of HSs properties and of their complexation behavior is therefore of considerable interest.

Optical spectroscopy has the distinct advantage that it allows an *in-situ*, non-intrusive analysis of HSs and HSs complexes in waters and soils. Spectroscopic research on HSs has always been situated between two extreme positions: on the one hand, the ubiquity and importance of HSs has prompted essentially every new and promising spectroscopic or analytical technique to be applied to HSs studies. This holds also for a large variety of laser spectroscopic techniques. On the other hand, due to their extremely complex, diverse and heterogeneous nature, HSs are elusive to a comprehensive characterization of their structural or physico-chemical properties. It seems appropriate, therefore, to elucidate to what extent laser spectroscopy can help to improve our understanding of structures and properties of HSs and their interactions with xenobiotics.

Laser spectroscopic techniques applied to the investigation of HSs include laser Raman spectroscopy,[4,5] laser light scattering,[6,7] time-resolved emission and absorption measurements after laser excitation[8-11] and laser desorption studies.[12-14] A review of these studies is beyond the scope of this paper. Rather, we shall attempt to illustrate some principles and advantages of laser spectroscopy, assisted by conventional spectroscopy. We recently have employed laser-induced fluorescence (LIF), laser desorption followed by laser-induced multiphoton ionization of HSs and time-resolved triplet-triplet absorption (TTA) measurements of laser-excited PACs in the presence of HSs. The results presented here are parts of on-going projects on the spectroscopic investigation of HSs and HSs/PAC complexes[15-20] and for laser-based ion mobility spectrometry (LIMS).[14]

The unique properties of laser light have revolutionized many areas of spectroscopy and instrumental analysis. Some of these properties are (i) broad wavelength tunability, which can be achieved with all-solid state devices based on non-linear optical effects.

With so-called optical parametric oscillators or amplifiers (OPO or OPA), the whole uv-vis-NIR spectral range can be covered to the benefit of absorption and fluorescence excitation spectroscopy; (ii) laser pulses of extremely short durations with temporal pulse widths in the sub-picosecond regime can easily be achieved, for example with mode-coupling techniques and with fiber lasers. Such short pulses lend themselves to kinetic measurements. Moreover, the employment of laser beams with low divergence (less than 1 mrad) allows almost diffraction-limited focussing on sub-μ focal diameters. This provides excitation power flux densities in excess of terawatts cm^{-2} and thus facilitates multiphoton and ablation processes. Other intriguing features of lasers such as high coherence may prove of less importance for HSs research.

2 MATERIALS AND METHODS

2.1 Humic Substances and Aromatic Compounds

Most of the fulvic acid (FA) and humic acid (HA) fractions employed in this study were provided within the priority program ROSIG funded by Deutsche Forschungsgemein-schaft. A detailed description of the sampling, preparation and characterization of these HSs, which were isolated from a brown water lake (HO), soil seepage (BS), ground water (FG), a strongly anthropogenically-influenced secondary effluent (ABV) as well as from a brown coal waste water (SV) can be found elsewhere.[21] In addition, the standard HSs samples 1S101H to 1S104H from the International Humic Substances Society (IHSS) and a commercially available HS from Aldrich (AHS; Steinheim) were used. Other HSs investigated have been described elsewhere.[15] Aqueous HSs solutions were either prepared by dilution (liquid HSs samples) or dissolution (lyophilized solid HSs samples) of appropriate amounts of HSs in distilled deionized water (Millipore). Adjustments of pH were carried out by addition of 0.1 M NaOH or HCl.

The organic dyes eosine yellowish (EO^{2-}, Merck, dye content ~ 90 %), erythrosine B (EB^{2-}, Aldrich, dye content ~ 90 %), rose bengal (RB^{2-}, Sigma, dye content ~ 90 %), the aromatic ketone benzophenone (BP, Aldrich, sublimed 99+ %), and the PACs pyrene (PY, Aldrich, zone-refined 99.9+ %), tetracene (TET, Aldrich, 98%), chrysene (CHR, Aldrich, 98%), anthracene (ANT, Aldrich, 99+ %) and 9,10-dichloranthracene (DCA, Aldrich, 98%) were used as received or recrystallized if necessary. Fullerene (C$_{60}$) and the water soluble fullerene adduct C$_{60}$A were obtained from Prof. Hirsch (University of Erlangen-Nürnberg, Germany).[22] Solutions of the aromatic compounds were prepared either in water or methanol (Aldrich, spectroscopic grade).

For the detection of uv/visible absorption spectra (recorded on a Perkin-Elmer Lambda 2 spectrophotometer), stationary fluorescence spectra (recorded on a Perkin-Elmer LS 50 fluorescence spectrometer) and triplet-triplet absorption measurements, sample solutions were transferred into cuvettes (1 cm × 1 cm quartz or optical glass) if not otherwise stated, de-aerated by bubbling nitrogen through the solution (20-45 min) and sealed.

The absorbances of the HSs samples for time-resolved fluorescence measurements did not exceed A = 0.4 at the excitation wavelength λ_{ex} = 337 nm.

For triplet quenching experiments, sample series of constant aromatic compound concentration and increasing HSs concentration (typical mass concentrations β_{HS} = 0-30 mg L^{-1}) were prepared by mixing aliquots of corresponding stock solutions. Because of the low water solubility of PACs and the low methanol solubility of some HSs, triplet quenching experiments with these substances were performed either in methanol or in a methanol/water mixture (0.1-0.2 v/v of water). The absorbances of all aromatic compound stock solutions used were in the range A = 0.2-0.5 at the corresponding laser excitation wavelength λ_{ex}. All aqueous solutions were brought to pH 10 with 0.1 M NaOH.

For LIMS measurements of HSs, 60 μL of aqueous HSs solutions ($\beta_{HS} \approx$ 1.5-2.0 g L^{-1}) were dropped on the surface of the sample holder. The water was then allowed to evaporate for exactly 60 min. LIMS samples of toluene and fullerene were prepared analogously using pure or cyclohexane solutions with an evaporation time of 30 min. All steps of the LIMS sample preparation procedure were performed according to this strict protocol.

2.2 Lasers

The following lasers were used in the experiments: (i) A Nd:YAG laser (B.M. Industries, 5021 DNS/DPS 10, wavelength λ = 1064 nm, harmonic generation of 532, 355 and 266 nm respectively, laser pulse width $\tau_L \approx$ 6 nsec or 30 psec, maximal laser pulse energies E_L= 900/260/100/60 mJ in the nsec-mode or E_L = 90/36/17/10 mJ at 1064/532/355/266 nm in the psec-mode, beam diameter d = 7 mm and maximal repetition rate f = 10 Hz); (ii) A dye laser pumped by an XeCl excimer laser (Lambda Physik, EMG 500, λ = 308 nm, $\tau_L \approx$ 15 nsec, E_L = 100 mJ and f = 30 Hz); and (iii) a nitrogen laser (LaserTechnik Berlin, MSG 800-TD, λ = 337 nm, $\tau_L \approx$ 500 psec, E_L = 350 μJ and f = 30 Hz).

2.3 Time-Resolved Fluorescence Measurements

Time-resolved emission (TRE) spectra of HSs were recorded with an experimental setup using a nitrogen laser (see above) as the pulsed excitation source and a slow-scan CCD-camera system (PicoStar F, La Vision) equipped with a pulsed image intensifier as the detection unit. The CCD has 576 $\times$ 384 pixels, which results in combination with the emission spectrograph (01-001, LTI) in a spectral resolution of 0.3 or 0.6 nm pixel^{-1} depending on the grating used (300 or 600 lines mm^{-1}). The intensifier of the CCD camera can be gated with temporal widths in the range t_G = 150 psec to 5 nsec and the gate can be delayed relative to the excitation pulse with minimal steps of Δt = 50 psec. The photon signal obtained from the CCD camera is digitized in 14 bit (dynamic resolution 1:16384) and processed by a personal computer.

2.4 LIMS Measurements

The setup for LIMS measurements consists of a laser as desorption and ionization source, a home-built ion mobility spectrometer with a drift tube in which the separation of the ions occurs and a detection device for the time-resolved registration and amplification of the ion current. The drift tube used for our investigations consists of alternating rings made of teflon and stainless steel. This arrangement allows the application of a constant electrical field (E = 296 V cm^{-1}) along the drift distance (drift length d = 10 cm). The aluminum sample holder is located in the middle of the cathode plate. Desorption and ionization of the sample is performed in a single step using the light of the frequency-quadrupled Nd:YAG laser (λ_{ex} = 266 nm, E_L= 20 mJ), which enters the ion mobility spectrometer through a 1 mm diameter pinhole and a quartz window. The laser energy at the location of the sample ranges up to 1.2 mJ. Detection of the ions is made by a faraday plate made of stainless steel after they pass an aperture grid (mesh width 1 mm). After amplification and transformation of the ion current (10 GV A^{-1} amplifier, ISAS-Dortmund), the signal is digitized by an A/D-D/A card (DAS-1602, Keithley) and read out by a personal computer. Nitrogen (Linde) is used as drift gas and flows opposite to the ion drift direction with a flow rate of 10 L hr^{-1}.

2.5 Triplet-Triplet Absorption Measurements

The apparatus used is a standard setup for transient absorption spectroscopy in which a Nd:YAG or a XeCl-excimer laser pumped dye laser is used for photoexcitation of the sample. The monitoring light was provided by a xenon lamp (150 W continuous wave output) placed at right angles to the excitation light beam. Time-resolved detection of the change in absorbance ΔA of the sample before and after laser excitation is done with a photomultiplier tube in combination with a monochromator (time resolution ca. 250 nsec). The signal of the photomultiplier is read out by a digital storage oscilloscope (DSO, Gould 4072) and transferred to a personal computer for processing, storage and output.

3 RESULTS AND DISCUSSION

3.1 Stationary Absorption and Fluorescence Spectroscopy of HSs

Absorption spectra of HSs are characterized by a continuous increase from low absorbances in the NIR to strong absorbances in the uv. Traditionally, the broad and almost structureless HSs spectra are empirically described by the so-called E_4/E_6 ratios (ratios of absorbances at 465 nm and 665 nm), which typically are in the range 3-20.[2,15] Fluorescence spectra are mostly characterized by one broad emission band. Both fluorescence spectra and efficiencies are excitation wavelength dependent. Maximal emission intensities for excitation in the near-uv are in the wavelength range of 450-500 nm. Fluorescence efficiencies are small, varying between $\eta_F = 0.1$-1.3 % for different HSs investigated by us.[15]

As an example, the absorption spectra of AHS and HO13 FA together with corresponding fluorescence spectra are displayed in Figure 1 (upper part). The HSs absorption is expressed by effective extinction coefficients ε_{HS} given on a mass concentration basis. Extended long-wavelength absorption tails are often described in terms of the so-called Urbach phenomenology. This was initially an empirical concept based on the assumption that a thermally or structurally disordered system with a distribution of absorber sites can produce an absorption tail which depends exponentially on the photon energy E (the Urbach rule):[23]

$$\varepsilon(E) \propto \exp(E/E_0) \propto DOS(E) \tag{1}$$

Here, E_0 is an energy term consisting of thermal and structural components and DOS is the density of electronic states provided by the disordered system. The absorption spectra of a large variety of very diverse materials, such as solvated electrons,[24] liquid water,[25]

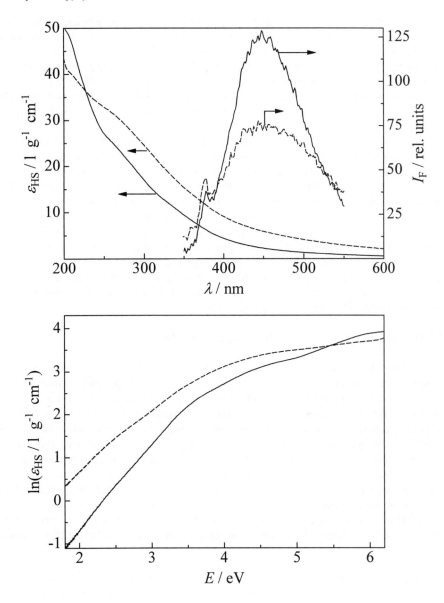

Figure 1 *Top: Absorption and fluorescence spectra ($\lambda_{ex} = 333$ nm, excitation and emission spectral bandwidths of 2.5nm, no quantum correction) of AHS (dashed line) and HO13FA (solid line) in aqueous solution at pH 8.5 (air-saturated samples). Bottom: Semilogarithmic plot of the effective extinction coefficient ε_{HS} versus energy E according to eq 1 for AHS (dashed line) and HO13 FA (solid line)*

amorphous semiconductors,[26] organic dyes,[27] crude oils[28] and asphaltenes[29] were treated with some success with this approach. Several theoretical approaches that will not be outlined here have been advanced for the explanation of the Urbach phenomenology.[26,27] The long-wavelength absorption tails of the HSs investigated here are consistent with the Urbach rule, as shown in exemplar Figure 1 (bottom) for AHS and HO13 FA by an exponential absorbance increase in the energy range of 1.8-3.0 eV (700-400 nm). The other HSs exhibited similar behavior but with somewhat varying slopes in the $\ln \varepsilon(E)$ vs. E representation.

The values of E_0 derived for 15 different HSs are in the range of 0.5-1 eV. This is significantly higher than the thermal energy at room temperature. The absorption spectra thus indicate that HS can be regarded as systems with high structural disorder encompassing a quasi-continuous distribution of electronic states, the density of which strongly increases with increasing energy. This situation is schematically depicted in Figure 2. The interpretation of high structural disorder in HSs derived here within the framework of the Urbach phenomenology is in agreement with the concept of HSs being highly complex substances and also is supported by our other spectroscopic studies (see below). The structural disorder of HSs treated by fractal approaches has been discussed repeatedly.[30-33] Future investigations and further analysis will show whether the energy parameter E_0 obtained from $\ln \varepsilon(E)$ vs. E plots is better suited to characterize HS absorbance properties than the E_4/E_6 ratios mentioned above.

3.2 Time-Resolved Fluorescence Spectroscopy of HSs

Information on the rates of deactivation of electronically excited HSs molecules can be obtained with time-resolved LIF measurements. For a single compound in homogeneous solution at room temperature, the fluorescence decay measured after a short pulsed laser excitation (ideally a δ-pulse) usually follows first-order kinetics. Since HSs are macromolecular compounds of non-uniform chemical constitution and conformation simple fluorescence decay behavior is not expected. In fact, time-resolved fluorescence measurements of different HSs reported[34,35] and performed by us with the method of time-correlated single-photon counting[15] have shown that the fluorescence decay cannot be described with a monoexponential decay law. Two different interpretations of this decay behavior are conceivable.

(i) HSs are composed of different types of fluorophores. The measured decay of the HSs fluorescence intensity $I_F(t)$ is then given by the sum of i exponential functions in eq 2

$$I_F(t) = \sum_{i=1}^{n} A_i \exp(-t/\tau_i) \tag{2}$$

where A_i and τ_i denote the corresponding amplitudes and fluorescence decay times, respectively. Ideally, the fluorescence decay times τ_i can be obtained from this model from a regression analysis. This evaluation procedure is known as the discrete component approach.

(ii) Even in the case where only one type of fluorophore contributes to the measured fluorescence, spatial heterogeneities of the fluorophore's environment within a macromolecule and different conformative microstates may be present (e. g., tryptophane in proteins[36-38]). Each of these microstates is characterized by a typical fluorescence decay

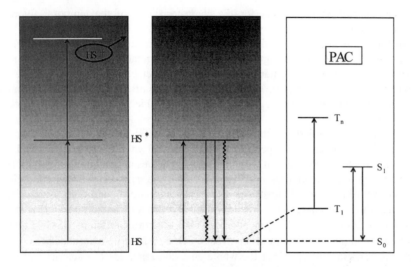

Figure 2 *Schematic representation of a simplified electronic energy level diagram of HSs (left and middle; HS and HS*: electronic ground and excited state, HS+ positive HS ion) and of aromatic compounds (right; S0: electronic ground state; S1: first excited singlet state, T1,Tn: first and higher excited triplet states). The increasing shading in the left and middle diagrams is to indicate the increasing density of electronic states with higher energy in HSs. The arrows indicate radiative (absorption and fluorescence) and non-radiative (wavy) transitions. The left diagram depicts the ionization of HS via two-photon ionization and the middle panel shows the transitions responsible for uv-vis-absorption as well as for fluorescence of HSs. The middle and right diagram symbolize the interaction of HSs with aromatic compounds in electronic ground states and the first excited triplet state*

time, which leads to the observed non-monoexponential fluorescence decay. Moreover, reactions during the lifetime of the fluorophore's excited state such as conformational relaxation processes can have the same effect. In the context of time-resolved fluorescence studies concerning proteins and membranes, the so-called continuous distribution method is often applied for data evaluation. This model assumes that the environmental complexity leads to a continuous Lorentzian or Gaussian distribution of the fluorescence decay times.

Both approaches were recently applied to time-resolved fluorescence measurements of different HSs and compared.[34] We decided to use the discrete component approach for the analysis of HS LIF decay curves and found that they are sufficiently described by the sum of four exponential functions. The fluorescence decay times τ_i were in the range $\tau_1 \approx$ 0.5, $\tau_2 \approx 2$, $\tau_3 \approx 5$ and $\tau_4 \approx 20$ nsec, with highest amplitudes for τ_1 and τ_2. In accordance with our values, fluorescence lifetime studies of HSs that applied mono- or biexponential regression models for data analysis also found fluorescence decay times in the range of ≤ 2 nsec connected with high amplitudes.[10,35,39] Since information on the compositions and structures of HSs are insufficient for a comprehensive photophysical interpretation, the extracted fluorescence decay times are not considered as 'real' fluorescence lifetimes but they can be used as parameters for the characterization of the temporal fluorescence behavior of HSs.

In contrast to conventional fluorescence lifetime measurements (detection of $I_F(t)$ at a single, fixed emission wavelength λ_{em}) for the measurement of so-called time-resolved emission (TRE) spectra, the fluorescence intensity is detected as function of time as well as of emission wavelength ($I_F(t, \lambda_{em})$). Such spectra are often graphically represented in a 3-dimensional way and contain all information available by time-resolved fluorescence spectroscopy.

Exemplary Figure 3 shows the TRE spectrum of an HS isolated from a bog water. The broad and structureless LIF spectra (sections parallel to the yz-plane) extend over a wavelength range of 350-630 nm with maximal intensities at around 480-500 nm. In order to analyze the temporal fluorescence behavior over the emission wavelength regime, we determined for individual fluorescence decay curves (sections parallel to the xz-plane) the time after which the fluorescence intensity decreases from its maximum to a value of 1/3e ($\tau_{1/3e}$, Figure 3). Applying this procedure to TRE spectra of different HSs we found that $\tau_{1/3e}$ depends on the emission wavelength for all HSs under investigation. The wavelength-dependent variations of $\tau_{1/3e}$ over the whole emission range are characteristic of different HSs and are in the range 2-3 nsec. The absolute values of $\tau_{1/3e}$ averaged over 480-490 nm vary between 6.0 and 6.8 nsec for different HSs.[15]

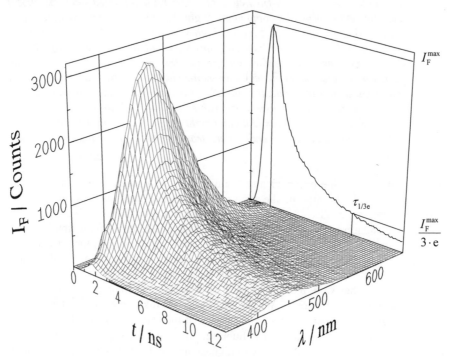

Figure 3 *Time-resolved emission spectrum of HO13 FA in aqueous solution at pH 10 (de-aerated sample, $\lambda_{ex} = 337$ nm, gate width $t_G = 250$ psec, time between two spectra $\Delta t = 100$ psec). No quantum correction of the emission spectra was performed. Some data were omitted for better graphical representation*

Only a few studies have dealt with the wavelength-dependence of HSs fluorescence decay times.[10,11,40] For example, Panne[11] and Kumke et al.[40] used an analogous evaluation procedure of TRE spectra and found behavior similar to that in this work. It has to be pointed out that the observation of non-monoexponential fluorescence decays kinetics and of decay time variation over the emission wavelength regime does not allow a photophysical interpretation. It is therefore not possible to draw conclusions from our results about the existence or properties of different fluorophores, different microstates or relaxation processes in HSs. Because of the complexity of HSs it is likely that combinations of various contributions lead to the observed complicated spectral and temporal behaviors.

3.3 Laser-Based Ion Mobility Spectrometry of HSs

Apart from optical methods (detection of photons), laser spectroscopic techniques in combination with the detection of molecular ions are promising tools for the investigation of HSs. In our experiments we use the combination of laser desorption and laser-induced multiphoton ionization with ion mobility spectrometry for the generation and detection of HS ions. The generation of positive HS ions via two-photon absorption is schematically depicted in Figure 2. In contrast to conventional mass spectrometric methods such as time-of-flight mass spectrometry, LIMS works under atmospheric pressure conditions. Because of collisions with neutral drift gas molecules (mass M, usually nitrogen or air), the separation of the positive ions by the electric field (electric field strength E) along the drift tube (length d) depends on the ion masses (m) and also on the collision cross sections (Ω_D). Analyte ions can be characterized by their gas-phase ion mobility K, which in the model of Mason et al.[41] is given in the simplest form by eq 3

$$K = \frac{3q}{16N\Omega_D} \sqrt{\frac{2\pi}{\mu k_B T}} \tag{3}$$

where q is the ion charge, N is the density of the neutral molecules, $\mu = mM/(m+M)$, the reduced mass of the ion-neutral collision pair, k_B is the Boltzmann constant and T is the temperature. It is notable that for M $\ll$ m the mobility becomes essentially independent of the ion masses m.

At low electric field strengths the mobilities are related to the gas-phase diffusion coefficients D via the Einstein relation, eq 4

$$K = \frac{qD}{k_B T} \tag{4}$$

Experimentally, K is determined from measured drift velocities v_D or drift times t_D

$$K = \frac{v_D}{E} = \frac{d}{t_D E} \tag{5}$$

and often empirically expressed as reduced mobility K_0

$$K_0 = \frac{d}{t_D E} \frac{T_0}{T} \frac{p}{p_0} \text{ cm}^2 \text{ V}^{-1}\text{sec}^{-1} \tag{6}$$

with $p_0 = 1.013 \times 10^5$ Pa, $T_0 = 273$ K, d in cm, E in Vcm^{-1}, T in K and p in Pa to make the mobilities comparable under different atmospheric pressure (p) and temperature (T) conditions.[42] A comprehensive overview of the principles of ion mobility spectrometry can be found in the monograph by Eiceman.[43]

Figure 4 shows the LIMS signals measured from different HSs and background signals. Shapes as well as total intensities of the ion signals varied for each HS under investigation. The signals obtained from direct laser illumination of the empty aluminum sample holder are referred to as background signals. Since the use of HSs samples changes the optical and surface properties of the sample holder, we do not subtract these background signals from the HSs LIMS signals. While the LIMS signals obtained for some HSs are only slightly larger than these background signals, other HSs produce significantly larger LIMS signals. We feel confident that the LIMS signals are qualitative characteristics of the HS investigated. The HS LIMS signals exhibit a broad distribution of drift times in the range $t_D = 15\text{-}40$ msec. This indicates that a large number of ion components or fragments are detected, as expected for polydisperse macromolecules.

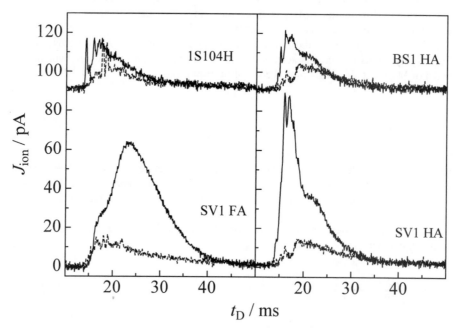

Figure 4 *Ion mobility spectra (plots of ion current J_{ion} vs. drift time t_D) of different HSs together with the corresponding background signals (signals of 1S104H and BS1 HA are displaced upward for better visualization). Samples were desorbed from an aluminum plug and ionized with the fourth harmonic of a Nd:YAG laser ($\lambda_{ex}=$ 266 nm and $E_L= 1$ mJ). HSs concentrations were in the range $\beta_{HS} \approx 1.5\text{-}2.0$ g L^{-1}*

Measurements of the HSs LIMS signal intensities with different power flux densities of the ionization laser showed no significant qualitative changes of the LIMS signals, suggesting that the HS ion fragments were ionized via a two-photon excitation process (results not shown). That the LIMS signal features remained unchanged under more intense irradiation might indicate that the HS undergo little fragmentation. In addition to the broad ion distribution, several sharp peaks in the range of t_D = 15-18 msec are observed. At present it is not clear if these sharp peaks, which may contain some contributions from the background signal, can be attributed to smaller ion components or fragments common to several HSs.

A comparison of the LIMS signal of SV1 FA with the LIMS signals obtained with toluene (C_7H_8, m = 92 Da) and fullerene (C_{60}, m = 720 Da) shows that most HS ions have drift times between those of the toluene and fullerene positive ions (Figure 5). This means that HS ions with a broad mass distribution were detected. Whereas the lower limit of this distribution can be positioned for small-sized ions at ca. 100 Da, due to the insensitivity of K for large ion masses the upper limit can only be estimated to be above ca. 750 Da. It is perhaps more instructive to consider the maximum of the SV1 FA LIMS signal located in the drift time region t_D = 26-28.5 msec, which corresponds to a reduced mobility of K_0 = 1.5-1.4 $cm^2V^{-1}sec^{-1}$. According to eqs 3 and 4, this mobility can be related to medium-sized ions for which m = 200-240 Da (assuming Ω_D to be constant in this small mass range) and diffusion coefficients of D ≈ 0.035 cm^2sec^{-1}.

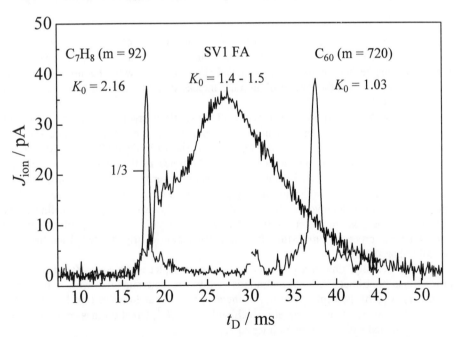

Figure 5 *Comparison of the ion mobility spectrum of HS (SV1 FA) with those of organic compounds (toluene and fullerene). The signal intensity of toluene was rescaled by a factor of 1/3. Laser parameters were λ_{ex} = 266 nm and E_L = 1 mJ. The concentration of SV1 FA was β_{HS} ≈ 1.5 g L^{-1}. The reduced mobilities K_0 are given in $cm^2V^{-1}sec^{-1}$*

To the best of our knowledge, this is the first application of LIMS to study HSs. Our work demonstrates the potential of LIMS to obtain HSs mass distributions. However, at present we have not performed a systematic mass calibration and cannot detect masses larger than ca. 750 Da with certainty. MALDI investigations, which are experimentally related in the desorption process but employ a different mass spectrometric detection, seem to yield unclear results concerning HS mass distributions. In one study using a fulvic acid, ions were detected in a mass range (100-1000 Da) similar to those in this work.[12] In another study using different ultrafiltered HSs, fractions up to extremely large masses (larger than 400 kDa) were reported.[13] It is unclear whether these are real different mass distributions of the different HSs investigated or whether varying experimental ionization conditions are responsible for the findings.

Further investigations are necessary to find out if LIMS investigations, with softer ionization and without expansion into a vacuum, can provide more reliable information about HSs mass distributions. Notable is the potential of LIMS to provide gas phase diffusion coefficients for HS ions. Diffusion coefficients for large, non-volatile substances in air can be found in the literature. For example, a value of $D = 0.07$ cm^2sec^{-1} is reported for benzo[a]pyrene (m = 252 Da).[44] This value is close to the diffusion coefficients obtained from intermediate drift times measured in the present work (Figures 4 and 5) and thus is consistent with our interpretation of medium-sized HS ions.

3.4 Time-Resolved Absorption Spectroscopy of Laser-Excited PAC in the Presence of HSs

Flash photolysis, a method to measure absorption spectra of transient species, was introduced by Norrish and Porter in 1949.[45] Since that time, tremendous progress in improving this technique has been achieved. For instance, with the use of lasers as excitation light sources the time resolution has increased dramatically from milliseconds to femtoseconds.[46]

Time-resolved absorption spectroscopy after laser flash irradiation can provide insight into the interactions between HSs and organic molecules in excited electronic states. A schematic representation of PAC/HS interactions is shown in Figure 2. For most PACs in non-viscous solvents at room temperature, the lifetimes in the first excited electronic singlet states (S_1) are typically 1-10 nsec, whereas the experimentally determined lifetimes in the lowest triplet states (T_1) are usually in the range of 0.1-10 msec. Therefore, TTA spectroscopy allows the investigation of PAC/HS interactions not only for non-fluorescent systems but also on much longer time scales. It is obvious that for photophysical or photochemical processes involving HSs the participation of triplet excited species is of central importance.

Experimentally, time-resolved measurements of triplet absorbances (assumed to be proportional to triplet-PAC concentrations) allow the determination with a first-order kinetic analysis of the triplet lifetimes in the absence (τ^0_T) and presence of HSs (τ_T). The bimolecular rate constants of triplet quenching by HSs (k_q^T, based on mass concentration) are then obtained with the Stern-Volmer eq 7.

$$\frac{\tau^0_T}{\tau_T} = 1 + K^T_{SV} \cdot \beta_{HS} = 1 + k^T_q \cdot \tau^0_T \cdot \beta_{HS} \qquad (7)$$

Here, K_{SV}^T is the so-called Stern-Volmer constant and β_{HS} denotes the HSs mass concentration.

As an example, Figure 6 shows the temporal behavior of EO^{2-} TTA signals in the absence and presence of HS (SV1 HA, = 20 mg L^{-1}) together with the corresponding fits to mono-exponential decay. As can be seen, the triplet lifetime of EO^{2-} decreases significantly with increasing HS concentration. From the Stern-Volmer plot shown in the inset of Figure 6, a bimolecular rate constant for the triplet quenching $k_q^T = (98\pm12) \times 10^6$ $Lkg^{-1}sec^{-1}$ was determined. For all HSs used in this study, decreases of EO^{2-} (in water at pH 10), CHR and TET (in methanol solution) triplet lifetimes with increasing concentration of HSs were also found. The quenching rate constants k_q^T determined for EO^{2-} varied from $(14\pm2) \times 10^6$ $Lkg^{-1}sec^{-1}$ for HO10 FA and the value for SV1 HA given above, k_q^T for TET varied between $(19\pm5) \times 10^6$ $L kg^{-1}sec^{-1}$ for AHS and $(30\pm5) \times 10^6$ L $kg^{-1} sec^{-1}$ for 1S101H, whereas for CHR k_q^T values of $(1.2\pm0.2) \times 10^9$ $Lkg^{-1}sec^{-1}$ for AHS and $(3.6\pm0.7) \times 10^9$ $Lkg^{-1}sec^{-1}$ for SV1 HA were found (framed boxes in Figure 7).[16,18,47] Thus, for EO^{2-} and CHR the triplet quenching efficiencies of the different HSs under investigation vary by less than one order of magnitude.

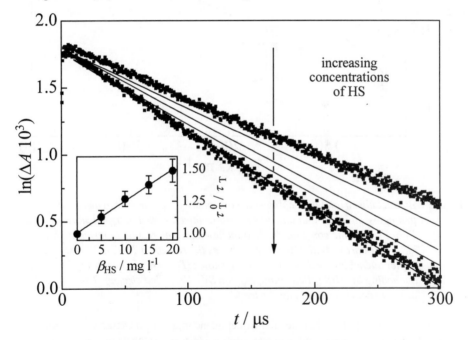

Figure 6 *Time-resolved triplet-triplet absorption signals ($\lambda_{ex} = 532$ nm, $\lambda_{det} = 565$ nm) of EO^{2-} in aqueous solution (pH 10) in the absence (top) and in the presence of HSs (bottom). Shown are the experimental data for $\beta_{HS} = 0$ and 20 mgL^{-1} and the fits to monoexponential decay. The inset shows the Stern-Volmer plot of the measured triplet lifetimes τ_T^0 and τ_T.*

To study the influence of the physicochemical and photophysical properties of aromatic compounds on the triplet quenching by SV1 FA in water at pH 10, EO^{2-} and the aromatic compounds RB^{2-}, EB^{2-}, BP and $C_{60}A$ were selected (for abbreviations see section 2.1). These substances have useful water solubilities and suitable photophysical properties (e.g., triplet quantum yields).[49-51] The T_1 state energies (E_T) of all investigated molecules range from 123 kJ mol[-1] for TET to 289 kJ mol[-1] for BP. Because there is no E_T of $C_{60}A$ in the literature, we approximated E_T of $C_{60}A$ as equal to E_T of buckminsterfullerene C_{60} in nonpolar solvents (151 kJ mol[-1]).[49]

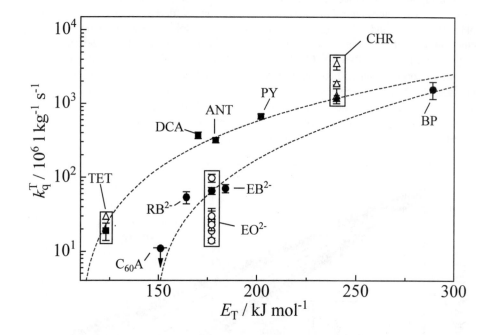

Figure 7 *Bimolecular triplet quenching rate constants k_q^T determined in methanol (triangles, squares) and aqueous solutions (circles) vs. the triplet state energies E_T of the organic compounds. Also included are the results reported by Kumke et al.[47,48] (squares). Within the framed boxes the results of TET, EO^{2-} and CHR triplet quenching by different HSs are shown. The other k_q^T values refer to triplet quenching by AHS (filled triangles) and SV1 FA (filled circles). The dashed lines are only drawn to insinuate the trend*

Figure 7 is a plot of the bimolecular triplet quenching rate constants k_q^T determined in methanol (triangles, squares) and aqueous solutions (circles) vs. the triplet state energy E_T. Because of the viscosity dependence of diffusional processes, the results obtained in methanol and aqueous solutions move a little closer if the different viscosities of these solvents are taken into account. For $C_{60}A$ no significant triplet quenching was observed so that only an upper limit of $k_q^T < 11 \times 10^6$ Lkg[-1]sec[-1] can be given. The highest triplet quenching efficiency in aqueous solution was found for BP ($k_q^T = (1.6\pm0.4) \times 10^9$ L kg[-1] sec[-1]), which is still significantly below diffusion control.[15] The trend of increasing k_q^T values with increasing E_T, which has earlier been suggested by Kumke et al.[47,48] for PAC in methanol solution and which is insinuated by the dashed lines in Figure 7, is obvious. In the only previous work on triplet quenching by HSs in aqueous solution, a value of k_q^T = (39±4) x 10^6 Lkg^{-1}sec^{-1} for the triplet quenching of RB^{2-} by a fulvic acid was reported.[52] This is very similar to our value of k_q^T = (54±10) x 10^6 Lkg^{-1}sec^{-1} for the triplet quenching of RB^{2-} by SV1 FA.

The results presented in Figure 7 show that over the triplet energy range from E_T = 123 kJ mol^{-1} for TET to E_T = 289 kJ mol^{-1} for BP the triplet quenching efficiencies of the different aromatic compounds under investigation by either SV1 FA or AHS vary by ca. two orders of magnitude. Evidently, the triplet state energy of the organic compounds is of great importance for the HS triplet quenching efficiencies. HSs properties seem to play a less significant role in the quenching process.

4 CONCLUSIONS

Important issues of HSs research extending from the characterization of HSs to the study of interactions with other organic compounds can be addressed with laser spectroscopy. Uv-visible spectroscopy, stationary and time-resolved LIF spectroscopy and LIMS are valuable tools for evaluation of the photophysical properties of HSs. Describing the typical uv/visible absorption behavior of HS within the framework of the Urbach phenonenology, these biological macromolecules can be regarded as highly disordered systems. This is reflected by an increase of the density of electronic states when going to higher energies. The excitation into high energy states, leading to efficient ionization of HSs, is possible via multi-photon absorption as shown by our LIMS studies. The simultaneous absorption of two photons reaching a large number of states of high energies accounts for this behavior.

Urbach behavior also becomes evident in the broad and unstructured fluorescence spectra and the complex fluorescence decay kinetics of HSs. Moreover, the experimental finding that the efficiency of PAC triplet quenching by HSs increases strongly with triplet state energy is also consistent with an Urbach-type description: the higher the PAC T$_1$-state energy, the larger is the number of HS electronic states available in the PAC/HS interaction as accepting states and the more efficient PAC triplet quenching becomes. It thus appears that the Urbach phenomenology is suited to describe the properties of HSs themselves and also allows a qualitative representation of the interactions between HSs and organic compounds (Figure 2).

It is hoped that the examples presented here demonstrate some of the potentials of laser spectroscopic studies of HSs. Our current work is directed towards investigation of the interactions between HSs and reactive organic compounds.

ACKNOWLEDGEMENTS

This work was financially supported by Deutsche Forschungsgemeinschaft within the priority program ROSIG (Refraktäre organische Säuren in Gewässern). The authors wish to thank Prof. Dr. F. H. Frimmel and Dr. G. Abbt-Braun (Engler-Bunte Institute, University of Karlsruhe), Dr. F.-D. Kopinke (UFZ-Umweltforschungszentrum Leipzig /Halle GmbH, Leipzig) and Prof. Dr. A. Piccolo (Dipartimento di Science Chimico-Agrarie, University of Naples "Frederico II") for the isolation and preparation of humic substances.

References

1. M. N. Jones and N. D. Bryan, *Adv. Coll. Interface Sci.*, 1998, **78**, 1.
2. F. J. Stevenson, 'Humus Chemistry - Genesis, Composition, Reactions', 2nd edn., Wiley, New York, 1994.
3. J. S. Gaffney, N. A. Marley and S. B. Clark, 'Humic and Fulvic Acids: Isolation, Structure, and Environmental Role', ACS Symposium Series 651, American Chemical Society, Washington, DC, 1996.
4. O. Francioso, S. Sanchez-Cortes, V. Tugnoli, C. Ciavatta and C. Gessa, *Appl. Spectrosc.*, 1998, **52**, 270.
5. Y. Yang and H. A. Chase, *Spectrosc. Lett.*, 1998, **31**, 821.
6. D. B. Wagoner, R. F. Christman, G. Cauchon and R. Paulson, *Environ. Sci. Technol.*, 1997, **31**, 937.
7. R. Kretzschmar, H. Holthoff and H. Sticher, *J. Coll. Interface Sci.*, 1998, **202**, 95.
8. F. H. Frimmel, H. Bauer, J. Putzien, E. Murasecco and A. M. Braun, *Environ. Sci. Technol.*, 1987, **21**, 541.
9. J. I. Kim, D. S. Rhee, H. Wimmer, G. Buckau and R. Klenze, *Radiochimica Acta*, 1993, **62**, 35.
10. S. L. Hemmingsen and L. B. McGown, *Appl. Spectrosc.*, 1997, **51**, 921.
11. U. Panne, 'Zeitaufgelöste, faseroptisch geführte, multi-dimensionale Fluoreszenz als Sensorprinzip zum Nachweis wassergelöster Fluorophore', Dissertation, Technical University München, 1994.
12. F. J. Novotny and J. A. Rice, *Environ. Sci. Technol.*, 1995, **29**, 2464.
13. M. Remmler, A. Georgi and F. -D. Kopinke, *Eur. Mass Spectrom.*, 1995, **1**, 403.
14. Th. Roch and J. I. Baumbach, *Int. J. Ion Mobility Spectrom.*, 1998, **1**, 43.
15. U. Zimmermann, H. -G. Löhmannsröben and Th. Skrivanek, in 'Remote Sensing of Vegetation and Water, and Standardization of Remote Sensing Methods', G. Cecchi, T. Lamp, R. Reuter and K. Weber, (eds.), *Proc. SPIE*, 1997, **3107**, 239.
16. U. Zimmermann, 'Fluoreszenzspektroskopische Untersuchung der Struktur-Wechselwirkungseigenschaften von Huminstoffen und polycyclischen aromatischen Kohlenwasserstoffen', Dissertation, FAU Erlangen-Nürnberg, 1999.
17. H. -G. Löhmannsröben, Th. Skrivanek and U. Zimmermann, in preparation, 1999.
18. H. -G. Löhmannsröben, Th. Skrivanek and U. Zimmermann, in preparation, 1999.
19. M. U. Kumke, H. -G. Löhmannsröben and Th. Roch, *Analyst*, 1994, **119**, 997.
20. M. U. Kumke, H. -G. Löhmannsröben and Th. Roch, *J. Fluoresc.*, 1995, **5**, 139.
21. F. H. Frimmel and G. Abbt-Braun, *Environ. Int.*, 1999, **25**, 191.
22. M. Brettreich and A. Hirsch, *Tetr. Lett.*, 1998, **39**, 2731.
23. F. Urbach, *Phys. Rev.*, 1953, **92**, 1324.
24. C. Houée-Levin, C. Tannous and J. -P. Jay-Gerin, *J. Phys. Chem.*, 1989, **93**, 7074.
25. F. Williams, S. P. Varma and S. Hillenius, *J. Chem. Phys.*, 1976, **64**, 1549.
26. M. V. Kurik, *Phys. Stat. Sol. (A)*, 1971, **8**, 9.
27. S. Kinoshita, N. Nishi, A. Saitoh and T. Kushida, *J. Phys. Soc. Japan*, 1987, **56**, 4162.
28. O. C. Mullins, S. Mitra-Kirtley and Y. Zhu, *Appl. Spectrosc.*, 1992, **46**, 1405.
29. O. C. Mullins and Y. Zhu, *Appl. Spectrosc.*, 1992, **46**, 354.
30. L. Beyer, *Z. Pflanzenernähr. Bodenk.*, 1996, **159**, 527.

31. N. Senesi, G. F. Lorusso, T. M. Miano, G. Maggipinto, F. R. Rizzi and V. Capozzi, in 'Humic Substances in the Global Environment and Implications on Human Health', N. Senesi and T. M. Miano, (eds.), Elsevier, Amsterdam, 1994, p. 121.

32. N. Senesi, in 'Humic Substances in the Global Environment and Implications on Human Health', N. Senesi and T. M. Miano, (eds.), Elsevier, Amsterdam, 1994, p. 3.

33. M. H. B. Hayes, in 'Humic substances: Structures, Properties and Uses', G. Davies and E. A. Ghabbour, (eds.), Royal Society of Chemistry, Cambrige, 1998, p. 1.

34. M. U. Kumke, G. Abbt-Braun and F. H. Frimmel, *Acta Hydrochim. Hydrobiol.*, 1998, **26**, 73.

35. C. H. Lochmüller and S. S. Saavedra, *Anal. Chem.*, 1986, **58**, 1978.

36. A. R. Holzwarth, *Methods Enzymol.*, 1995, **246**, 334.

37. D. M. Jameson and T. L. Hazlett, in 'Biophysical and Biochemical Aspects of Fluorescence Spectroscopy', T. G. Dewey, (ed.), Plenum, New York, 1991, p. 105.

38. J. R. Alcala, E. Gratton and F. G. Prendergast, *Biophys. J.*, 1987, **51**, 925.

39. P. J. Milne and R. G. Zika, *Mar. Chem.*, 1989, **27**, 147.

40. F. H. Frimmel and M. U. Kumke, in 'Humic Substances: Structures, Properties and Uses', G. Davies and E. A. Ghabbour, (eds.), Royal Society of Chemistry, Cambrige, 1998, p. 113.

41. H. E. Revercomb and E. A. Mason, *Anal. Chem.*, 1975, **47**, 970.

42. R. H. St. Louis and H. H. Hill, Jr., *Crit. Rev. Anal. Chem.*, 1990, **21**, 321.

43. G. A. Eiceman and Z. Karpas, 'Ion Mobility Spectrometry', CRC Press, Boca Raton, Florida, 1994.

44. R. P. Schwarzenbach, P. M. Gschwend and D. M. Imboden, 'Environmental Organic Chemistry', Wiley, New York, 1993.

45. R. G. Norrish and G. Porter, *Nature*, 1949, **164**, 658.

46. R. Bonneau, J. Wirz and A. D. Zuberbühler, *Pure. Appl. Chem.*, 1997, **69**, 979.

47. M. U. Kumke and F. H. Frimmel, in 'The Role of Humic Substances in the Ecosystems and in Environmental Protection', J. Drozd, S. S. Gonet, N. Senesi and J. Weber, (eds.), Polish Society of Humic Sbstances, Wroclaw, Poland, 1997, p. 525.

48. M. U. Kumke, 'Spektroskopische Untersuchungen der Wechselwirkung zwischen Huminstoffen und polyzyklischen aromatischen Kohlenwasserstoffen', Dissertation, Technical University Braunschweig, 1994.

49. S. L. Murov, I. Carmichael and G. L. Hug, 'Handbook of Photochemistry', Dekker, New York, 1993.

50. J. W. Arbogast, A. P. Darmanyan, C. S. Foote, Y. Rubin, F. N. Diederich, M. M. Alvarez, S. J. Anz and R. L. Whetten, *J. Phys. Chem.*, 1991, **95**, 11.

51. S. Reindel and A. Penzkofer, *Chem. Phys.*, 1996, **213**, 429.

52. D. P. Hessler, F. H. Frimmel, E. Oliveros and A. M. Braun, *J. Photochem. Photobiol. B: Biol.*, 1996, **36**, 55.

COMPREHENSIVE STUDY OF UV ABSORPTION AND FLUORESCENCE SPECTRA OF SUWANNEE RIVER NOM FRACTIONS

Gregory V. Korshin,[1] Jean-Philippe Croué,[2] Chi-Wang Li[1] and Mark M. Benjamin[1]

[1] Department of Civil and Environmental Engineering, University of Washington, Seattle, WA 98195-2700
[2] Laboratoire de Chimie de l'Eau et de l'Environnement ESIP, Université de Poitiers, Poitiers, France 86022

1 INTRODUCTION

The absorption of light by humic substances (HSs) in the UV region has been associated mainly with aromatic chromophores.[1,2] Deconvolution of the UV absorbance spectra of these substances into chromophore-specific components is, for all practical purposes, impossible. Nevertheless, the structure and reactivity of HSs can be probed by absorption spectroscopy.[3-5] Acquisition of the relevant information requires identification of the functionalities contributing to the absorbance and other relevant spectral phenomena such as fluorescence. In this paper we attempt to advance the understanding of absorbance and fluorescence spectra of HSs based on a comprehensive study of the structure and composition of a set of fractions of natural organic matter (NOM) extracted from Suwannee River water.

2 MATERIALS AND METHODS

Water was collected at the outlet of the Okeefenokee Swamp on October 18, 1995. The entire sample (453 L) was filtered on-site through two Balston glass-fiber cartridge filters in series (25 μ and 0.3 μ porosity) and then shipped to the USGS laboratory in Denver. The NOM was fractionated there into hydrophobic, transphilic, hydrophilic and ultrahydrophilic fractions using the comprehensive separation/isolation protocol developed by Leenheer.[6-10] These fractions were then characterized by ^{13}C CPMAS NMR, pyrolysis GC-MS, FTIR and elemental analyses. The corresponding data have been described by Croué et al.[11] The present work is primarily concerned with the UV and fluorescence spectra of these fractions, which were recorded at the University of Washington using a Perkin-Elmer Lambda-18 spectrophotometer and a Perkin-Elmer LS-50B fluorescence spectrophotometer, respectively. The excitation wavelength for the fluorescence analyses was 320 nm.

The fraction naming convention employed here differs from that used in previous publications on aquatic NOM fractionation.[8-10] Each fraction is denoted by a prefix and a one-letter suffix. A short explanation of the prefixes is given in Table 1. The suffixes A, N,

B correspond to the acid, neutral and base fractions, respectively. The only exception is the ultrahydrophilic humic acids fraction, which is denoted simply as uHA.

Table 1 *Fraction Name Conventions and Their Correlation with the Previously Used Terminology*

Fraction Name and Prefix	Fraction Isolation Method	Other Descriptors
Hydrophobic (HPO)	Adsorption on XAD-8	Humic material; fulvic and humic acid
Transphilic (TPH)	Adsorption on XAD-4 of material that does not adsorb to XAD-8	Acidic part of this fraction has sometimes been referred to as hydrophilic acid
Hydrophilic (HPI)	Adsorption on XAD-4 of material that does not sorb in first pass through media	Previously referred to as part of the non-adsorbable part of NOM
Ultrahydrophilic (uHPI)	Zeotrophic distillation with acetic acid	Previously referred to as part of the non-adsorbable part of NOM

3 RESULTS AND DISCUSSION

The mass contributions of various fractions of NOM to the total organic carbon are shown in Figure 1. The HPOA fraction dominates, but the HPIA, TPHA and TPHN fractions also contribute significantly. The contribution of all other fractions was < 4%. Approximately 20% of the organic carbon was lost during the isolation of the fractions. The reported losses of NOM were caused by spills or broken glassware.

The UV and fluorescence spectra of selected NOM fractions are shown in Figures 2 and 3, respectively. The spectra of some fractions are not presented because they were virtually identical to those shown. The spectra do not exhibit any obvious features other than broad shoulders between 200 and 250 nm and at >250 nm. Nevertheless, they can be

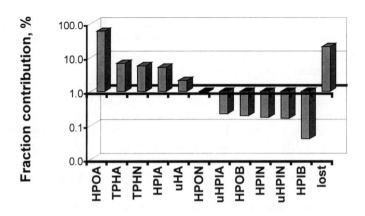

Figure 1 *Contribution of Suwannee River NOM fractions to the total dissolved carbon*

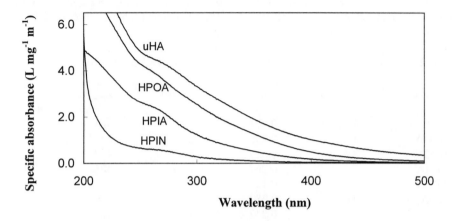

Figure 2 *UV-visible spectra of selected Suwannee River NOM fractions. All spectra are normalized to 1 mg/L DOC and a 1 cm cell length*

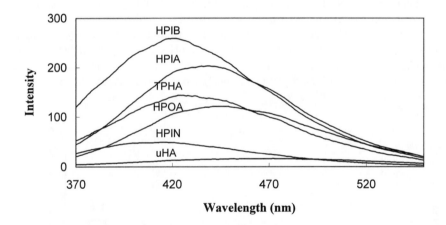

Figure 3 *Fluorescence emission spectra of selected Suwannee River NOM fractions. All spectra are normalized to 1 mg/L DOC*

characterized quantitatively and distinguished from one another based on their specific absorbance at 254 nm (SUVA$_{254}$) and by the band half-width of their electronic transition (ET) absorbance bands (Δ_{ET}). This latter parameter can be estimated based on the ratio of absorbances at 280 and 350 nm according to Equation 1, which is derived from the Gauss formula for the absorbance band shape for wavelengths > 250 nm.[12,13]

$$\Delta_{ET} = 2.18 \cdot \left(\ln\left(\frac{A_{280}}{A_{350}} \right) \right)^{-\frac{1}{2}} \tag{1}$$

The SUVA$_{254}$ values of the different fractions vary substantially, being largest for the uHA fraction and smallest for the HPIN and uHPIN fractions (Table 2). The trend for the Δ_{ET} values in Table 2 is very similar.

Table 2 *UV and fluorescence parameters for Suwannee River NOM fractions*

Sample	SUVA$_{254}$, L/(mg·m)	A_{280}/A_{350}	Δ_{ET}, eV	λ_{max} for fluorescence, nm
uHA	4.7	0.471	2.51	463
HPOA	4.2	0.392	2.25	446
HPON	2.4	0.317	2.03	437
TPHA	3.5	0.267	1.89	425
TPHN	0.6	0.128	1.52	
HPIA	2.6	0.282	1.93	439
HPIN	1.4	0.247	1.84	427
uHPIA	1.3	0.223	1.78	430
uHPIN	0.6	0.128	1.52	
HPIB	2.2	0.256	1.86	421

The intensities of the fluorescence spectra varied widely for the fractions, but in all cases the emission spectra consisted of a single broad band without visually perceptible sub-bands (Figure 3). The specific fluorescence yield (fluorescence intensity per mg/L DOC) varied widely (Figure 4), being lowest for the uHA fraction and highest for the HPIB fraction. The wavelength of maximum emission (λ_{max}) also varied considerably, from 467 nm for uHA to 417 nm for HPIN (Table 2). No values of λ_{max} are shown for the

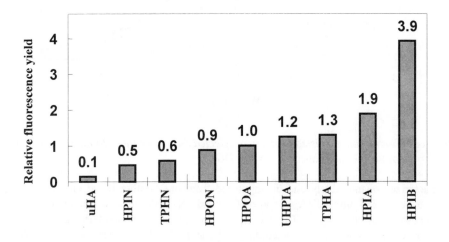

Figure 4 *Comparison of the relative fluorescence emission yields of Suwannee River NOM fractions. Excitation at 320 nm. The fluorescence intensity at the emission band maximum is normalized to 1 mg/L DOC. The emission yield for the HPOA fraction is normalized to 1*

uHPIN and TPHN fractions because of the low emission intensity of these fractions and the overlap of the bands with the Raman band of water. All the neutral fractions had low emission yields, although the emission intensity of the HPON fraction was comparable to that of the HPOA fraction.

Although the literature indicates the importance of aromatic groups to the spectroscopic properties of HSs,[1-5] the strength of the correlations between various spectral parameters (e.g., $SUVA_{254}$, Δ_{ET} and λ_{max}) and the aromatic carbon content of the samples has not been established. In addition, it is not currently possible to discern the contributions of individual chromophores and/or groups of chromophores to the overall absorbance/emission spectra of HSs.

The values of $SUVA_{254}$ and Δ_{ET} were strongly correlated with the aromaticity of the samples, as determined by integration of the ^{13}C NMR spectra (Figure 5); $SUVA_{254}$ and Δ_{ET} were also strongly inter-correlated ($R^2=0.89$). The interpretation of the result for the Δ_{ET} values is more complex, since this parameter probably reflects the combined effects of aromaticity and inter-chromophore interactions. For example, it has been shown[13] that Δ_{ET} is much more strongly correlated than $SUVA_{254}$ is with the gyration radii of HS molecules (and therefore their molecular weight).

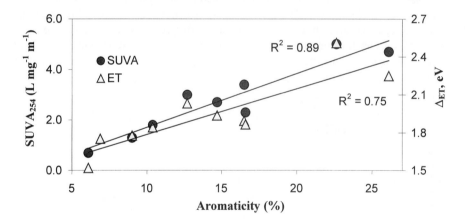

Figure 5 *Correlations of the aromaticity of the Suwannee River NOM fractions (from $SUVA_{254}$ and ^{13}C NMR data) and the half-width of the ET band*

The correlations between Δ_{ET} and HSs aromaticity and/or $SUVA_{254}$ can be hypothesized to result from two processes. First, the accumulation of aromatic carbon is likely to be accompanied by an increase in the number of individual ET bands contributing to the composite ET band, causing it to expand. Second, inter-chromophore interactions may be expected to become increasingly prominent as the number of chromophores increases and the distances between them decrease. These interactions are likely to generate absorbance bands that are red-shifted compared with the ET bands of the individual chromophores.

The wavelength of maximum emission in the fluorescence spectra was correlated with the corresponding aromaticities ($R^2 = 0.60$), although this relationship was weaker than those between either $SUVA_{254}$ or Δ_{ET} and the aromaticity. Generally, as the hydrophilicity of NOM fractions increases so does the fluorescence yield, although the neutral fractions tend to have lower emission yields. The HPIB fraction had a much higher emission yield than any other fraction. Since this fraction contained a substantial amount of organic nitrogen, it is suggested that the high emission yield might be a manifestation of emission from proteins and amino acids. However, no correlation between the total concentration of dissolved amino acids (TDAA) and the fluorescence yield was found ($R^2 = 0.01$). On the contrary, the lowest fluorescence yield was observed for the uHA fraction, for which the TDAA concentration was highest (53 mg of N per gram of organic carbon).The HPIB fraction, which had the highest fluorescence yield, had a much lower TDAA concentration (19.3 mg N/g C). The emission intensity of the uHA fraction is extremely low compared with that of all other fractions. The most likely explanation for this specific feature of the uHA fraction is that the larger size of organic molecules in that fraction (as indicated by its very high Δ_{ET} value) caused radiationless transitions to become much more prominent than in the fractions with smaller molecules.[14-17]

Because the TDAA includes both fluorescing and not-fluorescing amino acids, an attempt was made to correlate the emission yield with the concentration of the fluorescing amino acid species only. Although the amino acids tyrosine, phenylalanine and tryptophan all fluoresce, only the first two of these could be quantified because tryptophan decomposes when the sample is subjected to acidic hydrolysis in preparation for the TDAA analysis. Excluding the uHA fraction, a weak correlation was found between the fluorescence yield and the sum of the concentrations of tyrosine and phenylalanine ($R^2 = 0.48$) in the various fractions (Figure 6). Even this weak correlation must be viewed skeptically, because it is dominated by a single data point (the HPIB fraction). All other data points exhibit almost random scatter.

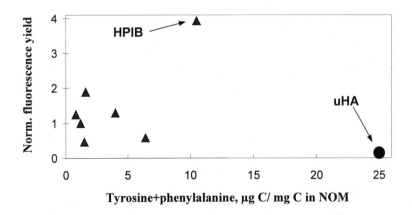

Figure 6 *Dependence of the fluorescence yield of the Suwannee River NOM fractions vs. the concentration of tyrosine and phenylalanine*

Next, an attempt was made to identify distinct sub-bands that contribute to the absorbance or emission spectra of HSs by calculating the derivatives of the spectra (d(intensity)/d(λ)). The basis for the interpretation of these derivative functions is that the contribution of any uniform group of chromophores i to the overall intensity is maximized at a wavelength $\lambda_{max,i}$, so $d(\text{intensity})/d(\lambda)$ is zero at $\lambda_{max,i}$. Even if the overall intensity of the spectrum contains contributions from several groups of chromophores ($i, j, k...$), the derivative of the intensity is likely to have a distinct feature at values of λ corresponding to $\lambda_{max,i}, \lambda_{max,j}, \lambda_{max,k}$, etc.

A semi-theoretical justification for the preceding statement can be developed by considering the absorption spectra of HSs. The energy of light (E) at any wavelength λ equals $1240/\lambda$, where E is expressed in electron-volts (eV) and λ in nanometers. Therefore, defining E_{i0} as $1240/\lambda_{max,i}$, spectral intensity is maximized at E_{i0}. It is common to represent the absorption spectrum of a uniform group of chromophores as a Gaussian function of E.[18] The first derivative of such a function is as follows:

$$\frac{dA_i}{dE} = -\frac{8(\ln 2)(E - E_{i0})}{\Delta_i^2} A_{i0} \exp\left[-\left(\frac{2(\ln 2)^{1/2}(E - E_{i0})}{\Delta_i}\right)^2\right] \tag{2}$$

where i identifies the band of interest, A_i is the absorbance contributed by band i at energy E, A_{i0} is the absorbance at the maximum of the band, Δ_i is the width of the absorption band at one half of its maximum intensity and E_{i0} is defined above. If only one band is present this derivative is zero at $E = E_{i0}$. If two or more bands are present, the derivative of the total spectrum is described by the summation of the contributions of individual bands:

$$\frac{dA}{dE} = \sum_i \left(-\frac{8(\ln 2)(E - E_{i0})}{\Delta_i^2} A_{i0} \exp\left[-\left(\frac{2(\ln 2)^{1/2}(E - E_{i0})}{\Delta_i}\right)^2\right]\right) \tag{3}$$

No analytical solution exists by which the values of E_{i0}, E_{j0}, E_{k0}, etc., can be found from the overall dA/dE function. Numerical simulation shows, however, that when two broad bands (i and j) are superimposed, the values of E_{i0} and E_{j0} are more apparent by visual inspection of the dA/dE vs. E curve than by inspection of the A vs. E curve (Figure 7). By extension, if an experimental spectrum comprises contributions from several sub-bands, calculations of the first and possibly higher order derivatives might help identify the E_0 values of the contributing sub-bands.

The first derivatives of the absorbance spectra of the Suwannee River NOM fractions are shown in Figure 8. The functions have distinct features at energies near 4.8, 4.2 and 3.7 eV (260, 300, 330 nm, respectively, as shown by arrows in the Figure). Similarly, the first derivatives of the emission spectra have noticeable features near 3.0, 2.7, 2.5 and 2.3 eV (410, 460, 490 and 530 nm, respectively).

Mathematical processing of the spectra shows the complex nature of the light absorption bands in the spectra of HSs. The features in the derivatized spectra have not been assigned to any specific functionality. Although calculation of higher order derivatives might allow detection of more subtle contributions, this would require measurement of a very high quality numerical signal. More work is needed to explore their chemical and/or spectroscopic identity. Nevertheless, the identification of any consistent features in the derivatized spectra is a promising development that might yield valuable information with further analysis.

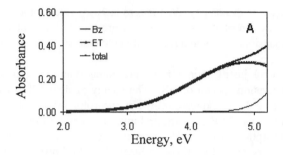

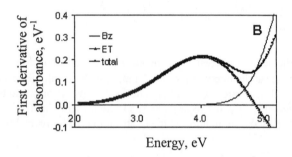

Figure 7 *Numerical calculation of absorbance spectra comprised by two Gauss-shape bands with maxima located at 6.14 and 4.88 eV, respectively. A. Zero-order spectrum with contribution components. B. First derivatives*

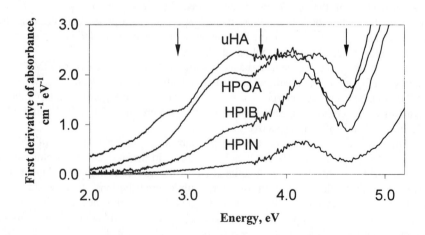

Figure 8 *First order derivatives of UV spectra of selected Suwannee River NOM fractions*

4 CONCLUSIONS

The specific absorbance at 254 nm ($SUVA_{254}$), width of the ET absorbance band (Δ_{ET}), and the position of the maximum fluorescence emission intensity (λ_{max}) of Suwannee River NOM fractions are well correlated with the aromaticity of the samples estimated by ^{13}C-NMR. No other functionality appears to affect the absorbance spectra significantly. The aromaticity and molecular weight of NOM both seem to affect their fluorescence spectra. The contribution of the nitrogen-containing aromatic species to the yield of NOM fluorescence needs further study.

Mathematical processing of the absorption spectra by calculation of their derivatives reveals the presence of several features that might be contributed by sub-bands with specific structural and/or spectroscopic identities that remain to be elucidated. The presence of the features indicates, however, that further numerical processing of the absorption and emission spectra of HSs coupled with their characterization by structure-sensitive methods is likely to provide more insight into the nature of HSs.

ACKNOWLEDGEMENTS

The authors acknowledge the major role of Dr. Jerry A. Leenheer (U. S. Geological Survey, Denver) in the isolation and characterization of the samples, as well as his active participation in the discussion and interpretation of the data. This work was funded by AWWA Research Foundation (grant # 159-94). Support from HDR Engineering, Inc. (Omaha, NE), SAUR (Maurepas, France) and DYNAMCO (West Sussex, UK) is greatly appreciated. We thank Steve Reiber (HDR Engineering) for keen interest and encouragement of this study.

References

1. S. J. Traina, J. Novak and N. E. Smeck, *J. Environ. Qual,* 1990, **19**, 151.
2. J. M. Novak, G. L. Mills and P. M. Bertsch, *J. Environ. Qual.,* 1992, **21**, 144.
3. P. R. Bloom and J. A. Leenheer, in 'Humic Substances II: In Search of Structure', M. H. B. Hayes, P. MacCarthy, R. L. Malcolm and R. S. Swift, (eds.), Wiley, Chichester, 1989, p. 409.
4. P. MacCarthy and J. Rice, in 'Humic Substances in Soil, Sediment and Water. Geochemistry, Isolation and Characterization', G. R. Aiken, D. M. McKnight, R. L. Wershaw and P. MacCarthy, (eds.), Wiley, New York, 1985, p.527.
5. G. V. Korshin, C. -W. Li and M. M. Benjamin, *Water Res.,* 1997, **31**, 1787.
6. J. A. Leenheer and E. W. D. Huffman, *J. Res. U. S. Geol. Surv.,* 1976, **4**, 737.
7. J. A. Leenheer, Proceedings of the Natural Organic Matter Workshop, Poitiers, France, 1997, **4**, 1.
8. J. A. Leenheer, *Environ. Sci. Technol.,* 1981, **15**, 578.
9. J. A. Leenheer, P. A. Brown and T. I. Noyes, in 'Aquatic Humic Substances: Influence on Fate and Treatment of Pollutants', I. H. Suffet and P. MacCarthy, (eds.), Advances in Chemistry Series 219, American Chemical Society, Washington, D.C., 1989, p. 25.
10. G. R. Aiken and J. A. Leenheer, *Chem. and Ecol.,* 1993, **8**, 135.

11.   J. P. Croué, G. V. Korshin, J. A. Leenheer and M. M. Benjamin, 'Isolation, Fractionation and Characterization of Natural Organic Matter in Drinking Water', to be published by the American Water Works Association.

12.   G. V. Korshin, C. –W. Li and M. M. Benjamin, in 'Water Disinfection and Natural Organic Matter', R. A. Minear and G. Amy, (eds.), ACS Symposium Series No. 649, American Chemical Society, 1996, Ch. 12, p. 182.

13.   G. V. Korshin, M. U. Kumke, C. -W.Li and F. H. Frimmel, *Environ. Sci. Technol.*, 1999, **33**, 1207.

14.   G. Jones and G. L. Indig, *New J. Chem.*, 1996, **20**, 221.

15.   T. M. Miano, G. R. Sposito and J. P. Martin, *Geoderma*, 1990, **47**, 349.

16.   P. L. Smart, B. L. Finlayson, W. D. Rylands and C. M. Ball, *Water Res.*, 1976, **10**, 805.

17.   A. J. Stewart and R. G. Wetzel, *Limnol. Oceanogr.,* 1980, **25**, 559.

18.   P. Pelikan, M. Ceppan and M. Liska, 'Applications of Numerical Methods in Molecular Spectroscopy', CRC Press, Boca Raton, 1994.

FLUORESCENCE BEHAVIOUR OF MOLECULAR SIZE FRACTIONS OF SUWANNEE RIVER WATER. THE EFFECT OF PHOTO-OXIDATION

T. M. Miano[1] and J. J. Alberts[2]

[1]Istituto di Chimica Agraria, Università di Bari, 70126 Bari, Italy
[2]Marine Institute, University of Georgia, Sapelo Island, GA 31327, USA

1 INTRODUCTION

The Okefenokee Swamp is a large (approx. 1,700 km^2) wetland in southeastern Georgia. The major source of freshwater to the swamp is via rainfall. Surface water leaves the swamp through two rivers: the Suwannee River, which drains to the southwest through Florida and into the Gulf of Mexico, and the St. Mary's River, which drains south and then northeast forming part of the Georgia-Florida border before entering the Atlantic Ocean. The Suwannee drainage represents approximately 75% of the total outflow, with the St. Mary's accounting for the remaining 25%. A more detailed description of the swamp, its vegetation, geology, and hydrology is given in ref 1.

Because of decomposition of swamp vegetation, both rivers have high organic carbon loads that are dominated by dissolved organic carbon (DOC).[2] The high DOC loading is primarily composed of humic substances (HSs), which lend a stained color to the water. They are referred to as "black waters" in the southeastern US. Black water rivers may have 75% or greater of the DOC in the form of humic and fulvic acids (HAs and FAs).[3]

As part of our ongoing investigations into the nature of DOC in southeastern waters, we are studying the Suwannee River, which contains high DOC content that has averaged 48 ± 8 ppmC/L from 1986-1996.[4] In this paper we discuss the fluorescence spectral characteristics of the DOC in this river and of various size fractions of that DOC as determined by ultrafiltration. Since the river is very dark in color and the southeastern US receives large inputs of solar energy, we were also interested in the potential alterations to the fluorescence spectral characteristics that might be produced by photo-oxidation. These findings are also discussed below.

2 MATERIALS AND METHODS

Approximately 10 L of water was collected from the Suwannee River (SUW) at Fargo, GA, where US highway 441 crosses the river. This site is located very close to the sill that forms the boundary of the Okefenokee Swamp and is fed by outfall from the swamp, which passes over a dam approximately ½ mile from the sampling site. Chemical characteristics of these waters have been described.[5] The sample was immediately taken to the Marine Institute laboratory on Sapelo Island and kept refrigerated in a cool room (10°C) prior to analysis (approx. 24 hrs).

The Suwannee River water sample was filtered through 0.45 μ membrane filters (Millipore Corp.). One liter of the filtered water was concentrated to a volume of approximately 200 mL over an Amicon XM100 ultrafilter in an Amicon Model 400 Stirred Cell Ultrafiltration Unit using approximately 8 psi N_2 gas pressure. Nominal molecular weights (MW) are expressed in Daltons (Da). The concentrate, a fraction of MW < 0.45 μ but > 100 kDa was kept refrigerated for further analysis. The filtrate was then concentrated to a volume of 200 mL over an Amicon XM10 ultrafilter under 18 psi N_2 pressure. The concentrate, MW < 100 kDa but > 10 kDa, and the filtrate, MW < 10 kDa were kept refrigerated for further analysis.

Thus, four sub-samples for the Suwannee River water samples were obtained: the whole water sample (whole); the fraction > 100 kDa but < 0.45 μ (fraction I); the fraction < 100 kDa but > 10 kDa (fraction II); and the fraction < 10 kDa (fraction III). All fractions were kept refrigerated for further analysis.

2.1 Fluorescence Spectroscopy

The whole water sample and the three water fractions were analyzed by fluorescence spectroscopy on a Hitachi Model F-3010 fluorescence spectrophotometer. Data were collected with Lab Calc and successively elaborated by Grams/32 and Grams/3D. All softwares are produced by Galatic Industries. All fluorescence spectra were measured with a scan time of 240 nm/min and a response of 2 sec; emission bandpass was set at 5 nm and excitation bandpass was set at 3 nm. Conventional emission and excitation spectra as well as synchronous-scan excitation and 3-D fluorescence spectra were measured on all samples. Emission spectra were taken from 380 to 550 nm with the excitation wavelength set at 360 nm. Excitation spectra were measured from 300 to 500 nm with the emission wavelength set at 520 nm. Synchronous spectra were taken from 290 and 550 nm by scanning simultaneously the excitation and emission wavelength and keeping a constant Δλ of 18 nm between them.[6-8] Finally, total luminescence (emission-excitation matrices) spectra and contour maps were obtained from 400 to 600 nm, while the excitation wavelength varied from 220 to 400 nm, increasing sequentially by 6 nm. All fluorescence spectra were directly measured at natural concentration without any isolation, extraction steps or adjustment of the solutions.

2.2 Photo-Oxidation

Aliquots (100 mL) of the whole water sample and the three water fractions were placed in open quartz tubes and irradiated with a medium pressure mercury lamp for 9-10 hours. Samples were left open to the atmosphere and were cooled by circulating cold water throught the immersion well (quartz) in which the lamp had been placed. After the photo-oxidation process (photox), all samples were closed and kept in a cool room prior to analysis. The pH of the whole water and three size fractions were taken before and after photo-reaction (Fisher Accumet Model 750 equipped with combination pH electrode, T = 22°C and calibration before and after measurements had been taken). The mean pH of the non photo-reacted samples was 3.77 ± 0.17 and the samples that were photo-reacted had pH = 4.09 ± 0.30.

Fluorescence spectra were also collected for the photo-oxidized samples and compared with their respective precursor data.

3 RESULTS AND DISCUSSION

3.1 Emission Spectra

Emission spectra of the Suwannee River whole sample and related fractions show typical bell-shaped emission curves (Figures not shown), with a maximum fluorescence intensity occurring between 450 and 465 nm (Table 1). The smaller molecular size fractions (II and III) contribute the greatest portion of the overall fluorescence intensity of the whole sample (11.8). The wavelengths of maximum emission in the spectra of the three fractions markedly decrease with decreasing of average molecular weight of the samples, confirming the close dependence of the emission wavelength on the size and possibly the molecular configuration of water dissolved organic matter. Further, dissolved organic molecules exhibiting the highest molecular weights (fraction I, > 100 kDa), apparently show the lowest relative fluorescence intensity (RFI = 4.1) together with the longest emission wavelength maximum (465 nm) (Table 1).

Table 1 *Fluorescence emission data of Suwannee River whole sample and its molecular size fractions*

Sample	Fraction	Emission wavelength (nm)			
		no photox		photox	
		λ_{max}	RFI	λ_{max}	RFI
Suwannee	whole	456	11.8	456	9.5
	I	465	4.1	463	4.9
	II	459	9.9	457	9.6
	III	450	9.5	451	4.5

Similar results were observed in the spectra of photo-oxidized samples (Table 1), where slight shifts of the wavelength of maximum emission towards shorter wavelengths and reductions of fluorescence intensity were generally observed. Only fraction III showed a marked decrease in fluorescence intensity, indicating possible structural changes of small molecular weight organic molecules as a result of photo-oxidation.

3.2 Excitation Spectra

Excitation spectra (Figure 1a) of the Suwannee River fractions show trends similar to those of the emission spectra. In general, Fraction II largely contributes to the overall fluorescence intensity of the whole water sample whereas Fraction I is significantly characterized by longer emission wavelengths but limited fluorescence intensity with respect to the other fractions. Maximum excitation wavelengths are observed around 350 nm for fraction I, 366 nm for the whole sample and fraction II, and 420 nm for fraction I.

Photo-oxidation generally reduces relative fluorescence intensity in all fractions (Figure 1b), especially fraction II and III, and apparently induces significant shifts towards shorter excitation wavelengths, as observed for the whole sample, fraction II and particularly fraction I (420 nm to 401 nm). Also, a slight narrowing of the overall excitation lineshapes is evident.

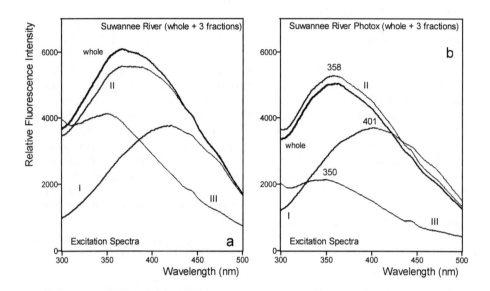

Figure 1 *Fluorescence excitation spectra of Suwannee River whole water sample and its molecular size fractions (a) before and (b) after photo-oxidation*

3.3 Synchronous Spectra

Synchronous spectra appear more informative and structured than the previous ones, exhibiting a greater number of peaks located around 350-360, 430 and 470-480 nm, and smaller signals and shoulders around 380, 414 and 490-500 nm (Figure 2a). In addition, a sharper difference among each of the fractions is observed. As in Figure 1, fraction I shows strong fluorescence signals only in the long wavelength range, indicating fluorophores with higher molecular weight and possibly molecular structures with extended chemical conjugation and/or an extended π system (highly conjugated phenolics and quinoids).[9-11] Fraction II (< 100kDa but > 10kDa) exhibits a more complex spectrum with fluorescence signals occurring mostly in the mid range (~ 430 nm) but also at longer wavelengths (475 nm). Fraction III probably contributes mostly to the short wavelength region of the whole Suwannee River spectrum, resulting in main peaks at 357 and 380 nm and minor shoulders at 414, 473 and 494 nm.

As previously noted, photo-oxidation markedly reduces the overall fluorescence intensity of the spectra, especially those of low and intermediate molecular weight fractions, and it induces a general slight shift of fluorescence signals towards shorter wavelengths (Figure 2b).

The effect of photo-oxidation is reported for each fraction in Figure 3 a-d. For the whole sample (Figure 3a), photo-oxidation dramatically reduces the overall fluorescence intensity, apparently depressing mid and long wavelength ranges of the spectrum. In Figure

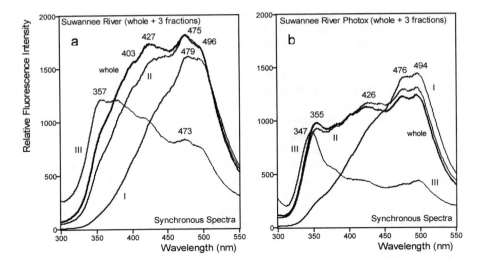

Figure 2 *Fluorescence synchronous spectra of Suwannee River whole sample and its molecular size fractions* **(a)** *before and* **(b)** *after photo-oxidation*

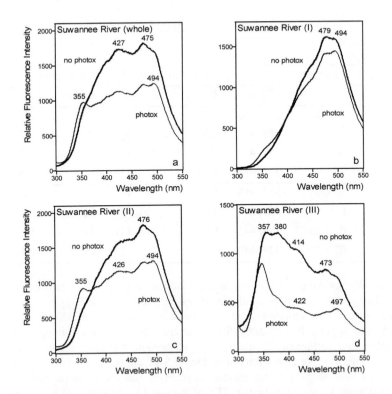

Figure 3 *Comparative fluorescence synchronous spectra of Suwannee River before and after photo-oxidation:* **(a)** *whole sample and* **(b)** *fractions I,* **(c)** *II and* **(d)** *III*

3b, fraction I clearly shows only a slight intensity decrease together with a quenching and a minor shift of the peak located at 479 nm. Large molecular weight units and/or highly conjugated and complex structures contribute largely to the overall fluorescence properties of Suwannee River organic matter and are only slightly affected by photo-oxidation, showing high resistance to photodegradation or to conformational changes. Fraction II is also markedly affected by photo-oxidation, showing a product spectrum very similar to that of the whole sample after the irradiation process and strongly quenched in the intermediate wavelength range (Figure 3c). The most dramatic effect of photo-oxidation is observed in fraction III, which likely reflects the low molecular weight compounds occuring in Suwannee River waters. These molecules appear to be very sensitive to the effects of photoirradiation, showing a noticeable general quenching of the overall fluorescence intensity and a distinct shift to shorter wavelengths in the spectrum. The observed phenomena possibly are ascribed to conformational changes of the dissolved organic compounds after photo-oxidation and apparently reflect molecular fragmentation processes and/or selective disruption of the organic macromolecules.

3.4 Total Luminescence Spectra (TLS)

The TLS are shown in both three-dimensional and contour map distributions in Figures 4 and 5. In TLS, peaks are described by emission-excitation wavelength pairs (EEWP), as indicated in Table 2.

Table 2 *Total Luminescence data of Suwannee River whole sample and its molecular size fractions*

Suwannee	Fraction	EM-EX Wavelength Pairs (EEWP, nm)				
		1st pair	*RFI*	*2nd pair*	*RFI*	*RFI Ratio*
no photox	whole	352, 453	12	252, 450	6.3	1.9
	I	388, 473	4.3	(256,464)	(0.8)	5.4
	II	352, 457	10	250, 455	5.1	2.0
	III	334, 444	11.4	232, 434	18.5	0.6
photox	whole	335, 448	11.7	245, 442	7.5	1.6
	I	358, 464	4.9	256, 456	1.2	4.1
	II	328, 445	10.3	238, 441	8.6	1.2
	III	328, 440	6.9	232, 423	13.7	0.5

Generally speaking, EEWP are characterized by differing fluorescence intensities whose relative contribution is synthetically expressed by the RFI ratio. The TLS of Suwannee River whole sample exhibits two major distinct EEWP at 352, 453 nm and at 252, 450 nm, apparently indicating the occurrence of two main fluorophores in the macromolecular structure of the water dissolved organic material (Figure 4, top). Two main distinct EEWP are also observed in the TLS of the molecular size fractions, and EEWP apparently shifts towards shorter wavelengths as the nominal molecular size decreases (Figure 5, top). In particular, TLS of fraction I is characterized by a major sharp EEWP at 388, 473 nm,[12,13] commonly ascribed to highly conjugated phenolics and

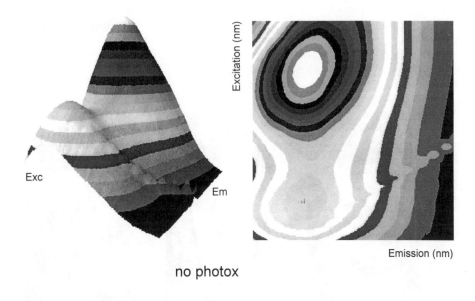

no photox

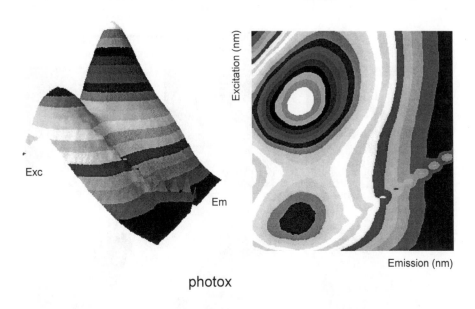

photox

Figure 4 *Total Luminescence Spectra (3-D) of Suwannee River (whole), **(top)** before and **(bottom)** after photo-oxidation*

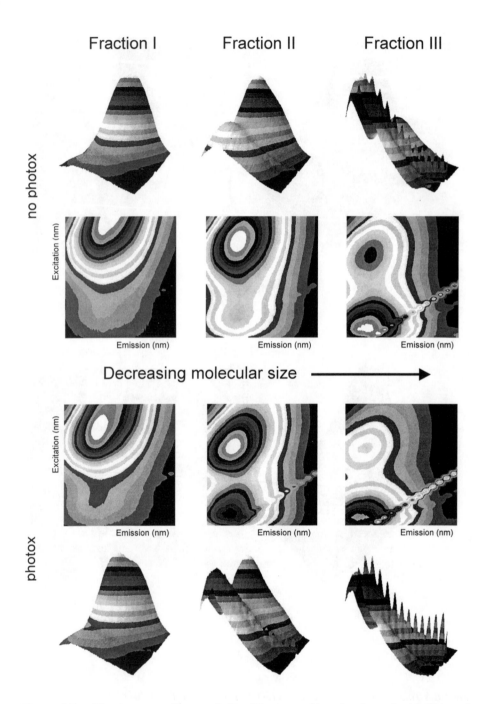

Figure 5 *Total Luminescence Spectra (3-D) of Suwannee River fractions, (top) before and (bottom) after photo-oxidation*

quinoids, which largely dominate the entire luminescence field, and by a secondary weak shoulder located at EEWP 256, 464 nm (Table 2). This result is consistent with the nominally very high molecular weight compounds concentrated in fraction I (MW > 100kDa). The TLS of fraction II, which is dominated by medium to high molecular size substances, shows two EEWP occurring at 352, 457 nm and 250, 455 nm that appear to be very close to the two EEWP of the whole sample.

The two main fluorophores occurring in this group are possibly ascribed to simpler phenolic compounds and other variously substituted aromatic moieties with a high degree of electronic resonance in their molecular configuration.[13] Fraction III is also characterized by two EEWP at 334, 444 nm and 232, 434 nm, the latter being predominant in relative intensity. Small molecular weight aromatic compounds such as aromatic amino acids, hydroxyphenylacetates and highly conjugated aliphatic structures are widely represented in this fraction and could be responsible of the relatively high fluorescence intensity occurring in this region of the TL spectra. As indicated by the RFI ratio, the relative intensities of the three fractions are clearly distinguishable and can be easily related to their different molecular size composition and to major fluorescing areas in the TLS (Table 2).

The effect of photo-oxidation on the whole water sample of the Suwannee River is clearly described in Figure 4 (bottom). It results in a definite narrowing of the two discrete EEWP occurring at 335, 448 nm and 245, 442 nm and a shift towards shorter wavelengths with respect to the TLS of the whole sample. The overall relative fluorescence intensities apparently are not affected by the irradiation process (Table 2). Photo-oxidation is particularly effective for the single fractions of the Suwannee River sample (Figure 5, bottom), mainly causing an overall shift of the fluorescing TLS regions towards shorter wavelengths and forcing the two EEWP closer together.

The RFI ratios generally agree with previous observations. Fraction I shows a main EEWP at 358, 464 nm and a secondary weak EEWP at 256, 456 nm. A prominent shoulder, possibly ascribed to a different fluorophore, is also observed around EEWP 390, 475 nm. This result is consistent with a selective effect of the photo-irradiation, directed either to portions of definite molecular structures or to specific compounds of the water dissolved materials that are weakly bound to the organic macromolecules. Fraction II appears rather distinct from the corresponding non-photo-oxidized fraction, showing a main, sharper EEWP at 328, 445 nm that is strongly shifted to shorter wavelengths and a prominent secondary EEWP at 238, 441 nm of approximately the same fluorescence intensity. Fraction III is also characterized by two EEWP at 328, 440 nm and 232, 423 nm with the latter being of relatively higher intensity. Small aromatic units in the river water appear more sensitive to photo-oxidation, as indicated by the marked reduction in fluorescence intensity and by the definite shift of the EEWP towards much shorter wavelengths.

The effect of photo-oxidation that is particularly evident for the intermediate and the small molecular weight compounds in Suwannee River waters also reflects individual compounds in the mixture and/or structural components that appear to be relatively resistant to molecular changes.

4 CONCLUSIONS

Natural organic matter (NOM) in river waters is a mixture of different compounds of varying molecular sizes and weights, as demonstrated by physical fractionation of

Suwannee River waters. Molecular fluorescence spectroscopy is able to distinguish among fractions and to provide evidence that the overall fluorescence intensity of NOM is produced by different macromolecular entities.

Emission spectra reveal a close direct relationship between emission wavelengths and size and possibly molecular configuration of DOC. Also, high molecular weight compounds apparently show the lowest relative fluorescence intensity. Excitation spectra confirm these results and also show that the medium sized fraction contributes significantly to the overall fluorescence intensity of the whole sample. Synchronous spectra appear more structured and informative than the simple spectra. They reveal greater differences among the different fractions and a definite shift of fluorescence maxima towards shorter wavelengths with decreasing molecular size of the fraction.

TLS exhibit a bi-modal distribution of fluorescence signals, in all cases showing two main EEWP at varying wavelengths and RFI ratios for all fractions. The overall fluorescence intensity of the whole sample clearly consists of several overlapping signals at different EEWP. All EEWPs shift towards shorter wavelengths as molecular size decreases, and TLS reveal an inversion of the relative intensities of the two EEWP peaks.

Photo-oxidation affects fluorescing properties of both the whole sample and fractions of Suwannee River aquatic organic matter. Photo-oxidation led to reduced fluorescence intensity in all spectra, especially for fraction III. It invariably produces a spectral shift towards shorter wavelengths. This is consistent with changes of the molecular configuration in the DOC, including uncoiling phenomena caused by the irradiation energy, rearrangements of the overall structural conformation of the molecules and/or partial or selective disruptions of weakly bound molecular components.

Synchronous Scanning and TLS appear very promising in characterizing natural organic matter and its molecular fractions. They provide further insights regarding macromolecular structures and compositions and may also reveal additional information on aggregation phenomena of organic molecules in aquatic systems.

ACKNOWLEDGEMENT

This research and Dr. T. M. Miano were supported by the Sapelo Foundation through the Visiting Scientist Program at the Marine Institute. This is contribution #790 from the University of Georgia Marine Institute.

References

1.   R. L. Malcolm, D. M. McKnight and R. C. Averett, in 'Humic Substances in the Suwannee River, Georgia: Interactions, Properties and Proposed Structures', R. C. Averett, J. A. Leenheer, D. M. McKnight and K. A. Thorn, (eds.), U. S. Geological Survey Open-File Report 87-557, 1989, 1.

2.   J. J. Alberts, J. R. Ertel and L. Case, *Verh. Internat. Verein. Limnol.*, 1990, **24**, 260.

3.   K. C. Beck, J. H. Reuter and E. M. Perdue, *Geochim. Cosmochim. Acta*, 1974, **38**, 361.

4.   W. R. Stokes, III, R. D. McFarlane and G. R. Buell, 'Water Resources Data For Georgia Water: Years 1974 to 1997', U. S. Geological Survey, Atlanta, GA, 1974-1997.

5.   J. J. Alberts, J. P. Giesy and D. W. Evans, *Environ. Geol. Water Sci.*, 1984, **6**, 91.

6.   T. M. Miano, G. Sposito and J. P. Martin, *Soil Sci. Soc. Am. J.*, 1988, **52**, 1016.

7.   N. Senesi, T. M. Miano, M. R. Provenzano and G. Brunetti, *Sci. Total Environ.*, 1989, **81/82**, 143.

8.   M. R. Provenzano, T. M. Miano and N. Senesi, *Sci. Total Environ.*, 1989, **81/82**, 129.

9.   N. Senesi, T. M. Miano and M R. Provenzano, in 'Lecture Notes in Earth Sciences, 33', B. Allard, H. Boren and A. Grimvall, (eds.), Springer, Berlin, 1991.

10. T. M. Miano, N. Senesi and R. L. Malcolm, *Humus-uutiset* (Finnish Humus News), 1991, **3**, 145.

11. T. M. Miano and N. Senesi, *Sci. Total Environ.*, 1992, **117/118**, 41.

12. N. Senesi, T. M. Miano, M. R. Provenzano and G. Brunetti, *Soil Sci.*, 1991, **152**, 259.

13. J. Lüster, T. Lloyd, G. Sposito and I. V. Fry, *Environ. Sci. Technol.*, 1996, **30**, 1565.

CHANGES IN CHEMICAL COMPOSITION, FTIR AND FLUORESCENCE SPECTRAL CHARACTERISTICS OF HUMIC ACIDS IN PEAT PROFILES

M. Takács and J. J. Alberts

University of Georgia Marine Institute, Sapelo Island, GA 31327, USA

1 INTRODUCTION

Peats develop under anaerobic conditions, which in the case of basin peats is a result of the accumulation of water in a natural depression. Under these conditions organic matter decomposition is retarded and accumulation of the partly decomposed, partly transformed plant and animal materials occurs. Peat formation occurs continuously on those surfaces that remain wet and to which organic matter is constantly being added. However, these fragile ecosystems are very sensitive to environmental changes. In particular, alteration of the water regime, which can cause changes in the vegetation as well as in the microbial activity, also can give rise to other soil formation processes.

In Hungary there are about 10^4 ha of peaty area, which is about the 1.1 % of the whole territory of the country. The peaty areas are located in twelve primary sites and are very different with respect to their size and their physical and chemical characteristics.[1, 2] Also, utilization of these areas and peats is different. In some cases efforts were made to turn them into agricultural fields, while in certain areas they were mined. The peat material can be used as a soil amendment to increase the organic matter content in sandy soils, as a substrate in horticultural production or as a basic component for artificial horticultural soils.

As the major constituent of organic matter, the behaviour of humic substances (HSs) is significant in determining the features of the peats themselves. By studying the characteristics of humic and fulvic acids we may get a better understanding of the characteristics of the peats and the possible processes that can occur in response to environmental changes.

Here we present data on the elemental composition and spectral characteristics of the humic acids (HAs) from peats that have experienced different water regimes during their formation.

2 MATERIALS AND METHODS

The peat samples represent two profiles with two layers each from two sampling sites, Pötréte 0-40 and 40-200 cm, and Zalasztmihály 0-10 cm and 10-170 cm. The deeper layers were light in color, fibrous in structure and looked like preserved plant material. The surface layers had a dark brown color without structure. The organic matter content in the deeper layers was higher that in the surface layers.

The HSs were extracted from the peats under N_2 with 0.1 M $Na_4P_2O_7$-0.1 M NaOH solutions (1:10) for 48 hrs, then acidified to pH 2 with HCl. The HAs were separated by centrifugation, dialyzed against deionized water for 48 hrs and then freeze dried.[3]

Ash contents of the freeze dried samples were determined by loss on ignition (450°C for 5 hrs). Major element (C, N and H) concentrations of the samples were determined with a Perkin Elmer Model 2400 CHN Elemental Analyzer. All analyses were made in triplicate.

FTIR spectra of solid samples were measured with a Perkin Elmer Paragon 1000 FTIR spectrophotometer. The freeze-dried organic matter was weighed (1.0 mg) and mixed with 200 mg KBr. The mixture was then pressed into pellets at 10,000 kg/cm^2 for 30 min. The spectra were acquired between 4000-500 cm^{-1} using 64 scans and 4 cm^{-1} resolution. The spectra were evaluated with the Perkin Elmer Search program and the Savitzky Golay second derivatives of the spectra were determined with GRAMS 32 software (Galactic Industries). We used information from several sources for the peak assignments.[4-9]

Fluorescence spectra were acquired with a Hitachi Model 3010 fluorescence spectrophotometer using about 25 mg/L TOC solutions at pH 6. The total organic carbon concentration (TOC) was determined with a Shimadzu TOC 500 Total Carbon Analyzer. Prior to running the fluorescence spectra, the instrument was calibrated with a standard solution of quinine sulfate (1 ppm in 0.105 M $HClO_4$). Slit widths were 5 nm and scan speed was 60 nm/sec for both the excitation and emission spectral measurements. For the excitation spectra λ_{em} = 450 nm (in the range 350-600 nm), while for the emission spectra λ_{ex} = 335 nm (in the range 350 and 600 nm) were used. The initial wavelengths for the synchronous scan spectra were λ_{em} = 250 nm and λ_{ex} = 268nm, with excitation and emission slit widths of 5 nm, and a scan speed of 120 nm/sec. The spectra were acquired in the 250-600 nm wavelength range. The total luminescence spectra were measured in the excitation wavelength range 220 to 400 nm and the emission wavelength range from 400-600 nm with a scan increment of 6 nm and with 29 cycles. The emission and excitation slit widths were 5 nm and the scan speed was 240 nm/sec. The instrument parameters were set and spectral data collected with LabCalc (Galactic Industries), then the data were transferred to and analyzed with GRAMS 32 (Galactic Industries). For the evaluation of the synchronous scan spectra we also used the Savitzky Golay second derivatives.

The HA sample solutions were exposed to the radiation of a medium pressure Hg Arc lamp in presence of air for 72 hrs in Pyrex glass tubes. After the treatment the fluorescence emission, excitation, synchronous scanning and total luminescence spectra were measured with the same instrument parameters as above.

3 RESULTS AND DISCUSSION

3.1 Elemental Composition

The elemental compositions of the peat HAs (Table 1) from the two sites are similar to each other and also show the same trend in their changes with sampling depth. In the upper layers the carbon contents are lower and the nitrogen contents are higher than in the lower layers the carbon contents are lower and the nitrogen contents are higher than in the lower layers, as is evident in the much smaller C/N ratios of the surface layers.

Table 1 *Elemental composition and COOH content of the humic acids*

Sample	C%	H%	N%	O%	C/N	COOH[a]	
Pötréte 0-40	50.12	4.82	3.33	41.72	17.53	3.96	$C_{35}H_{40}N_2O_{22}$
Pötréte 40-200	52.04	4.87	2.13	40.96	28.51	3.41	$C_{57}H_{64}N_2O_{34}$
Zalasztmihály 0-10	46.50	5.35	3.16	45.00	17.19	3.61	$C_{34}H_{46}N_2O_{25}$
Zalasztmihály-170	52.69	6.66	2.77	37.88	22.21	3.01	$C_{44}H_{67}N_2O_{24}$

[a] Calculated from FTIR spectra;[10] units are meq/g HA.

The surface layers are under semiaerobic conditions, which leads to more extensive microbial decomposition of the organic matter. The microorganisms partially incorporate carbon into their tissue and partially liberate some as CO_2. As nitrogen does not escape in a gaseous form under oxic conditions, the process results in carbon loss and nitrogen enrichment in the HAs.

The calculated basic structural unit formulas also show a trend with the sampling depth. The HAs derived from the upper layers seem to have smaller basic units than their counterparts from the deeper layers. The formulas of the upper layers are more similar to each other and to the composition used as a building block in defining the structure of humic acids.[11] These formula differences suggest that carbon loss occurs in the surface layers of the peat profile and a humification process takes place.

3.2 FTIR Spectra

All the spectra are characterized by a strong band in the 3450-3550 cm^{-1} range that can be assigned to OH stretching in alcohols, phenols and water. The samples were very similar in this respect (Figure 1).

In all the samples we observed aliphatic C-H stretching around 2920 cm^{-1}, also with very similar intensity. The shoulders that appear in the spectra at 2850 cm^{-1} can be assigned to the C-H stretching in O-CH$_3$ ethers. The presence of the methoxyl group is characteristic of lignin and lignin degradation products. The intensity of this peak was highest in the deeper layer from the Zalaszentmihály sample.

The HA sample from the Pötréte surface layer has a peak and the deeper layer a strong shoulder at around 1715 cm^{-1}, which is attributed to the C=O stretching of carboxylic and carbonyl groups. At this wave number the Zalaszentmihály samples have only a very weak shoulder.

The major band around 1620-1600 cm^{-1} in all of the spectra is assigned to aromatic C=C and the asymmetric C=O stretching in COO$^-$ groups. The HA sample from the upper Zalaszentmihály layer had the highest absorbance at 1600 cm^{-1}, while the peaks in the samples from the Pötréte site appeared at higher wave numbers and had lower intensities.

A small, sharp peak at 1515 cm^{-1} assigned to C=C in aromatic rings was found only in the spectra of the two deeper peat layer HAs, suggesting the presence of certain aromatic groups in these HAs that are not present in the more oxygenated surface layers.

Another feature common in all the spectra is the band of medium to strong intensity at 1385-1390 cm^{-1}, which is assigned to OH deformation and C-O stretching in phenols and

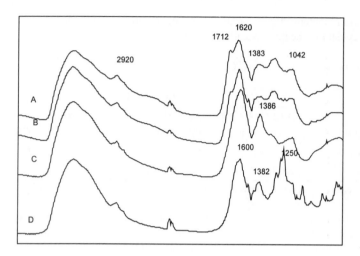

Figure 1 *FTIR spectra of the peat samples. A, Pötréte 0-40 cm; B, Pötréte 40-200 cm, C, Zalaszentmihály 0-10 cm; D, Zalaszentmihály 10-170 cm*

COO⁻ groups. This peak was especially pronounced in the upper layer Zalaszentmihály sample. Coupled with the 1600 cm⁻¹ peak, it suggests that there are carboxylate groups present, presumably because of the ash content of the samples. Although the ash contents of the samples vary about 3-4%, very small changes in the salt quantity and quality can result in carboxylic to carboxylate changes in the material. This hypothesis is supported by our results from lake natural organic matter (NOM) samples,[12] where decrease of the ash content decreased the intensity of the peak around 1600 cm⁻¹ and increased the peak intensity around 1710 cm⁻¹.

All of the spectra have a moderately strong peak between 1200-1280 cm⁻¹, which is attributed to C-O stretch and OH deformation of COOH groups. These peaks appeared at 1232 and 1235 cm⁻¹ with nearly the same intensities in the upper and lower layers of the Pötréte samples, which is in agreement with what we observed with the 1715 cm⁻¹ peak.

The humic acid from the Zalaszentmihály deeper layer sample has a very strong, sharp peak at 1125 cm⁻¹ and a less intense sharp peak at 1157 cm⁻¹ which can be assigned to ethers. With the presence of the 2850 cm⁻¹ peak these are likely to be OCH₃. The Zalasztmihály deeper layer has also a sharp peak at 925 cm⁻¹, which can be assigned to C-O stretching in cyclic ethers and carbohydrates.

An intense band is also found around 1040 cm⁻¹ in all the spectra. This can be attributed to alcoholic and polysaccharide C-O stretching. The samples from Pötréte peats were very similar in this regard.

On the basis of the FTIR spectra we can differentiate the two sites. The HAs from the Pötréte site seem to contain less ethers, OCH₃ and carbohydrates than those in the Zalaszentmihály site. The evaluation of the 1715, 1600, 1390 and 1230 cm⁻¹ bands suggests that the HAs from the Pötréte site are more acidic while those from the Zalaszentmihály site occur in the salt form.

We also observed changes with depth. The peaks assigned to the presence of ethers were more intense in the HAs from the deeper layers, especially for the Zalasztmihály sample. The sharp small peaks at 1515 cm⁻¹ found only in the HAs from the deeper layers suggest the presence of aromatic compounds not occuring in the surface layers.

Assigning the band at 1710 cm^{-1} to the carboxylic groups and the peak at 1600 cm^{-1} to the carboxylate groups the quantity of COOH groups in humic substances can be calculated.[10] We observed higher amounts of carboxylic groups in the HAs from the surface layers (Table 1).

All these trends agree with the changes in the elemental compositions. As the upper layers are not continuously under the influence of water, some susceptible groups may undergo a change either by chemical oxidation or as a carbon source for microorganisms.

3.3 Fluorescence Spectra

The fluorescence emission spectra of all the HA samples are characterized by one broad peak with a maximum centered between 454 and 468 nm. Shoulders are also detectable in all the spectra at around 512 nm. The relative fluorescence intensity (RFI) normalized to carbon was highest in the case of the HA from the deeper layer of the Zalasztmihály sample.

In general, the broad peak in the emission spectra is characteristic of all HSs, both fulvic and humic acids, from different origins. Humic substances with more highly substituted aromatic nuclei with electron donating groups and more conjugated unsaturated systems have their fluorescence emission maxima in the longer wavelength region. However, materials with simpler structural components, a lower degree of aromatic polycondensation and lower levels of conjugated chromophores have their emission maxima at shorter wavelength.

Humic acids usually have broader emission curves, with the maxima at longer wavelength than for FAs. According to the classification system based on the emission and excitation spectra maxima, Leonardite, peat and most soil humic acids belong to Class II (with emission maxima 500-520 nm, excitation peaks at 450 and 465 nm), while soil and peat FAs belong to Class IV (with emission maximum 470-440 nm and excitation maximum 385-395 nm).[13] The emission maxima of the HAs from our peat samples were in the lower wavelength region, which suggests that these organic compounds can be characterized by a lower degree of conjugation.

The main advantage of synchronous scan spectra is that simultaneously scanning the excitation and emission wavelengths with a fixed Δλ should allow better peak resolution.

The main peaks in the spectra of the peat HAs were centered around 466-480 nm. The humic acids from the deeper layers also showed a shoulder around 512 nm. From the second derivatives we can obtain better resolution of the overlapping peaks (Figure 2). This manipulation better reveals the differences among the characteristic features in the two layers. In the deeper layers we can see more nearly equivalent peaks, while the spectra of the HAs from the surface layers tend to be more homogeneous with one dominant peak.

3.4 Photoreaction

We exposed the humic acid solutions to light to study how photoreaction changes their properties, which can be monitored by fluorescence spectroscopy.

In all cases the maximum of the emission spectra shifted towards shorter wavelength (448-452 nm) in response to the photoreaction and the shape of the spectra narrowed. The fluorescence emission spectra of the photooxidized HAs no longer contained shoulders.

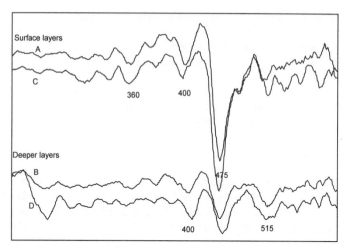

Figure 2 *Second derivatives of the fluorescence synchronous scan spectra of the HAs. A, Pötréte 0-40 cm; B, Pötréte 40-200 cm; C, Zalaszentmihály 0-10 cm; D, Zalaszentmihály 10-170 cm*

The synchronous scan spectra also show HAs dramatic changes due to photo-oxidation in even more detail (Figures 3 and 4). The intensity of the peak around 470-480 nm decreased and the shoulder at around 512 nm eventually became the main peak. Also, a new peak appeared with a maximum around 350 nm. We observed these changes in all of the samples but with different intensities.

The appearance of the peak at around 350 nm suggests that the chemical changes have resulted in the appearance of a smaller component. This can happen by cleavage of a smaller molecular unit from the parent molecule. Simple aromatic systems like vanillic or syringic acids have a peak around 350 nm and are known components of lignin. Thus they represent possible structural models for such compounds. Observations supporting this hypothesis occurred when HAs were photo-oxidized under an oxygen atmosphere, resulting an extensive disappearance of phenolic subunits as well as a substantial synthesis of carboxylic acids.[14]

In the case of HAs from the deeper peat layers the appearance of the 350 nm peak was more pronounced, which suggests that these HAs are more susceptible to cleavage. In the presence of oxygen, the lignin dimers and the lipid fractions were found to be the structural subunits most susceptible to photoreaction in soil HAs.[14] Possibly, the HAs from the deeper peat layers contain more of these compounds. The FTIR data support that hypothesis, as we observed the abundance of OCH_3 groups, especially in the HA from the Zalaszentmihály deeper layer.

In addition to the appearance of the 350 nm peak we observed a shift of the main peak towards longer wavelength in the synchronous scan spectra of the humic acids after photo-reaction. Second derivative analysis reveals a pronounced decrease in the intensity of the peak around 475 nm, which revealed the previously existing peak around 515-520 nm (Figures 3 and 4).

The presence of a component with a fluorescence emission maximum at 350 nm was less pronounced in the HAs from the upper layers. This suggests that the more humified HAs are less susceptible to photoreaction.

Before photoreaction   After photoreaction

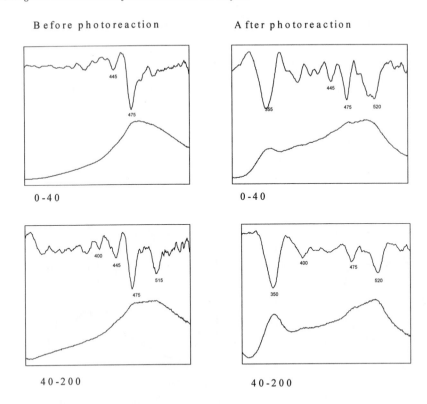

0-40   0-40

40-200   40-200

Figure 3 *Fluorescence synchronous scan spectra and their second derivatives of Pötréte HA samples from different depths before and after the photoreaction*

4 CONCLUSION

The elemental composition, FTIR and fluorescence (emission, excitation synchronous scan and total luminescence) spectra of HAs from two Hungarian peat samples were determined. An experiment was also conducted to study the changes in these characteristic features of HAs due to photoreaction.

Although certain differences were observed among the two sampling sites, the changes related to the depth of the profile were more pronounced than intersite differences.

The HAs from the deeper peat layers showed greater complexity and heterogeneity in their character than those from surface layers. Both FTIR and all the fluorescence spectra had more individual, distinct peaks than their upper layer counterparts. This was more pronounced in the case of the samples from the Zalasztmihály site. The deeper layers tend to have more ether groups and carbohydrates than the HAs from the upper layers. In contrast, the surface layers were richer in carboxylic groups.

From these observations we conclude that these changes are the result of oxidation and humification processes that take place in the upper layers under partially aerobic conditions.

In response to photoreaction we observed a narrowing of the curves of the emission

Before photoreaction   After photoreaction

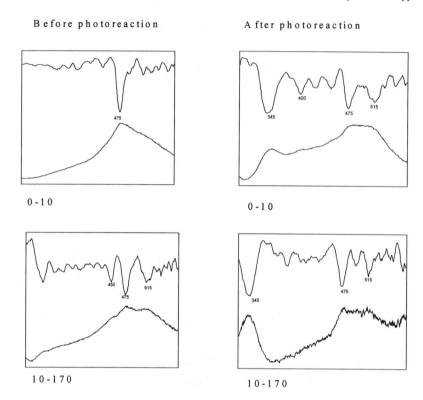

Figure 4 *Fluorescence synchronous scan spectra and their second derivatives of Zalaszentmihály HAs from different depths before and after photoreaction*

spectra and shifting of the absorption maxima towards shorter wavelengths. Evaluation of the synchronous scan spectra and their second derivatives suggests that the processes involved result from probable cleavage of smaller components from the parent material. The fact that the presence of a component with fluorescence emission maximum at 350 nm was less pronounced in the HAs from the upper layers suggests that the more humified HAs are less susceptible to photoreaction.

ACKNOWLEDGEMENTS

This research was supported by the Visiting Scientist Program at the Marine Institute, which was funded in part by a grant from the Sapelo Foundation. This is contribution number 791 of the University of Georgia Marine Institute

References

1.   L. Hargitai, *Kertészeti és Élelmiszeripari Egyetem Közleményei*, 1982, **33**, 223.
2.   J. Dömsödi, 'Talajjavitási útmutató', Mg –i Kiadó, Budapest, 1984.

3.   M. Schnitzer, *J. Soil Sci.,* 1981, **61**, 517.
4.   E. Pretsch, T. Clerc, J. Seibl and W. Simon, 'Spectral Data for Structure Determination of Organic Compounds', Springer, Berlin, 1983.
5.   F. J. Stevenson, "Humus Chemistry" Wiley, New York, 1982.
6.   A. Piccolo and F. J. Stevenson, *Geoderma*, 1982, **27**, 195.
7.   J. Drozd, in 'Humic Substances in the Global Environment and Implications on Human Health', N. Senesi and T. M. Miano, (eds.), Elsevier, Amsterdam, 1994, p. 273.
8.   C. T. Johnston, W. M. Davis, C. Erickson, J. J. Delfino and W. T. Cooper, in 'Humic Substances in the Global Environment and Implications on Human Health' N. Senesi and T. M. Miano, (eds.), Elsevier, Amsterdam, 1994, p. 145.
9.   J. J. Alberts and Z. Filip, *Environ. Technol.*,1998, **19**, 923.
10.   L. Celi, M. Schnitzer and M. Negre, *Soil Sci.*, 1997, **162**, 189.
11.   L. T. Sein, J. M. Varnum and S. A. Jansen, *Environ. Sci. Technol.*, 1999, **33**, 553.
12.   J. J Alberts, M. Takács and D. O. Andersen, 'Proceedings'98 2nd National Symposium on Agricultural Instrumentation (II SIAGRO)', Sao Carlos, Brazil, submitted.
13.   N. Senesi, T. M. Miano, M. R. Provernzano and G. Brunetti, *Soil Sci.,*1991, **152, 259.**
14.   P. Smitt-Koplin, N. Hertkorn, H. R. Shulten and A. Kettrup, *Environ. Sci. Technol.*, 1998, **32**, 2531.

CHROMATOGRAPHIC SEPARATION OF FLUORESCENT SUBSTANCES FROM HUMIC ACIDS

M. Aoyama

Faculty of Agriculture and Life Science, Hirosaki University, Hirosaki 036-8561, Japan

1 INTRODUCTION

Humic substances (HSs) are dark-colored organic polymers with a wide range of molecular weight (MW) distributions. HSs exhibit featureless spectra in the UV and visible regions, with the absorbance decreasing as the wavelength increases.[1] Humic substances also have been considered to possess fluorescent structures.[2] Earlier work demonstrated that the fluorescent intensity of humic acids (HAs) and fulvic acids (FAs) samples increased with decreasing molecular size.[3-5] This phenomenon has been explained by the presence of internal quenching.[5] However, an alternative explanation is that low MW fluorescent substances other than HSs coexist as sample impurities. To examine this hypothesis, high performance size exclusion chromatography (HPSEC) with UV and fluorescence detection was applied to the HAs and FAs extracted from compost and soil samples.[6] For FAs, the peak of the fluorescent substances was eluted slightly later than that of the HSs, but both peaks occurred in practically the same region, suggesting that the fluorescent substances in the FAs were identical to the HSs and that the fluorescence increases with decreasing molecular size. In contrast, the fluorescent substances in the HAs were eluted significantly later than the HSs and were resolved into 10 peaks. This result strongly suggested that the fluorescent substances contained in the HAs are different from HSs and consist of different components.

In this study further attempts were made to separate fluorescent substances from HAs samples by HPSEC using columns with higher resolving power than used in the previous study.[6] In addition, the suitability of Sephadex gel chromatography was investigated for the collection of the fluorescent substances in large quantities needed for identification.

2 MATERIALS AND METHODS

2.1 HA Samples

Eight HA samples were prepared from a rice straw compost and seven soils (Tsukano, Fujisaki, Nishimeya, Takizawa, Kawatabi, Dando and Inogashira). The HAs were extracted twice from compost and soil samples corresponding to 500 mg C with 150 mL of 0.1 M NaOH, separated by acidification to pH 1.0 with H_2SO_4, dialyzed against distilled deionized water and freeze-dried. The Dando and Inogashira HA samples were purchased from the Japanese Humic Substances Society.[7] The origin and total carbon and nitrogen contents of the HA samples are shown in Table 1.

Table 1 *Origin and total carbon and nitrogen contents of HA samples*

Sample	Origin	C %	N %	C/N
Compost	Rice straw compost	54.8	3.63	15.1
Tsukano	Gray lowland paddy soil (*Udifluvent*)	45.0	4.87	9.2
Fujisaki	Brown lowland soil (*Udifluvent*)	47.7	5.40	8.8
Nishimeya	Brown forest soil (*Haplumbrept*)	47.0	4.41	10.7
Takizawa	Allophanic Andosol (*Melanudand*)	52.4	3.77	13.9
Kawatabi	Non-allophanic Andosol (*Melanudand*)	50.3	3.41	14.8
Dando	Brown forest soil (*Haplumbrept*)	53.0	4.49	11.8
Inogashira	Allophanic Andosol (*Melanudand*)	54.8	4.01	13.7

2.2 Measurement of Fluorescence Spectra

For the measurement of the fluorescence spectra, each HA sample corresponding to 5 mg C was dissolved in 5 mL of 0.1 M NaOH. Distilled deionized water was added to bring the total volume to 100 mL, the pH being adjusted to 8.0 with H_2SO_4 before the final filling. Excitation and emission spectra were recorded with a JASCO FP-920 fluorescence spectrophotometer equipped with a 1-cm quartz cell. The emission (λ_{em}) and excitation (λ_{ex}) wavelengths were set at 560 and 360 nm, respectively. The excitation and emission spectra were expressed as difference spectra by subtracting the spectra obtained for a 0.005 M NaOH solution adjusted to pH 8.0.

2.3 HPSEC

The HA solutions used for fluorescence spectra measurements were used for HPSEC. HPSEC was performed with a Hitachi L-6000 high performance liquid chromatograph with Asahipack GS-220HQ and GS-320HQ columns (7.6 mm I. D. x 300 mm) linked in series and a 100 μL injection loop. The nominal MWs at void volume (V_0) of the GS-220HQ and GS-320HQ columns were 3 and 40 kDa, respectively, for pullulans. The columns were kept at 40°C in a Shodex AO-50 column oven. The solvent for HAs consisted of 50 mM phosphate buffer (pH 8.0) containing 300 mM NaCl. CH_3CN in the volume ratio of 4:1 was used as an eluent at a flow rate of 0.5 mL min^{-1}. Each HA sample solution was mixed with an equal volume of 100 mM phosphate buffer (pH 8.0) containing 600 mM NaCl, and with half a volume of CH_3CN, then filtered through a 0.45 μ membrane filter (DISMIC-13CP, Advantec, Tokyo) before injection. The chromatogram was monitored by absorbance at 280 nm using a Hitachi L-4200 UV-VIS detector. The peaks of the fluorescent substances were detected using the JASCO FP-920 fluorescence spectrophotometer equipped with a 16 μL flow cell. The λ_{ex} and λ_{em} were set at 460 and 520 nm, respectively. For the soil HAs, excitation (λ_{em} = 560 nm) and emission (λ_{ex} = 360 nm) spectra were obtained on individual fluorescent peaks and expressed as difference spectra by subtracting the spectra obtained for the eluent. The V_0 and the total effective column volume (V_0+V_i) were determined using Blue Dextran (Sigma) and acetone, respectively. The columns were calibrated with polyethylene glycols as the MW standards.

For the Fujisaki, Nishimeya and Takizawa HAs, the HA solutions adjusted to a concentration of 1 mg mL^{-1} were subjected to HPSEC under the above conditions, and the column effluent was collected over the elution volume ranges of 8-13 mL (Fraction A) and 13-23 mL (Fraction B) using an Advantec SF-2120 fraction collector. The collected fractions were reinjected under identical conditions.

2.4 Sephadex Gel Chromatography

For the preparation of a sample solution, 10 mg of the Tsukano HA was dissolved in 5 mL 0.1 M NaOH and adjusted to pH 8.0 with HCl, then distilled deionized water was added to bring the total volume to 10 mL. A 3 mL volume of the sample solution was applied onto a glass column (37 mm I. D. x 500 mm) packed with Sephadex G-25 (fine grade; Pharmacia Biotech) and eluted with distilled deionized water at a flow rate of 40 mL hr^{-1}. The column effluent was collected as 5 mL fractions using the Advantec SF-2120 fraction collector. All the procedures were carried out at 10°C. The V_0 and V_0+V_i were determined using Blue Dextran and NaCl, respectively. For each fraction, the absorbance at 280 nm was measured in a 1-cm quartz cell with a Hitachi U-1100 spectrophotometer, and the fluorescent substances were estimated by HPSEC using a short column (Asahipack GS-220M, 7.6 mm I. D. x 100 mm) with the same eluent as described above. The peak area when monitored by the fluorescence (λ_{ex} = 460 nm and λ_{em} = 520 nm) was used as an index of the quantity of the fluorescent substances. To obtain detailed chromatograms, selected fractions were subjected to HPSEC using the GS-220HQ and GS-320HQ columns linked in series.

3 RESULTS AND DISCUSSION

3.1 Fluorescence Spectra of HAs

Figure 1 shows the excitation and emission spectra of the HAs. The compost HA showed a broad peak at 380 nm and a shoulder at 450 nm. In contrast, excitation spectra of the soil HAs, except for the HA samples that originated from brown forest soils (Nishimeya and Dando), were characterized by a principal peak at 460-470 nm and a shoulder at 360-380 nm. The brown forest soil HAs showed distinctive peaks at 360-380 nm and 320-330 nm, together with the principal peak at 460-470 nm. All of the emission spectra showed two broad peaks at 460 nm and 510-530 nm. The major peak occurred at 510 nm for the soil HAs, except for the Andosol HAs (Takizawa, Kawatabi and Inogashira). A shift in the emission maximum to 530 nm was observed for the Andosol HAs. In contrast, the compost HA sample had a major peak at 460 nm and a minor peak at 510 nm. Thus, the maximum excitation and emission wavelengths were 460-470 and 510-530 nm, respectively, for the soil HAs, compared to 380 and 460 nm, respectively, for the compost HA. The maximum excitation and emission wavelengths obtained in the present study are within the range of those previously reported for HAs.[8,9] On the basis of these results, the near-optimum excitation and emission wavelengths (λ_{ex} = 460 nm, λ_{em} = 520 nm) of soil HAs were adopted for the detection of the fluorescent substances by HPSEC.

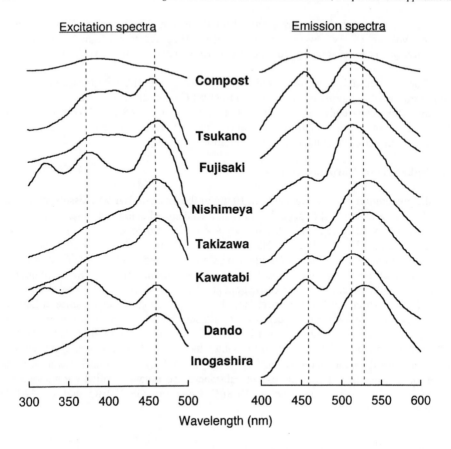

Excitation spectra   **Emission spectra**

Compost

Tsukano

Fujisaki

Nishimeya

Takizawa

Kawatabi

Dando

Inogashira

300 350 400 450 500   400 450 500 550 600

Wavelength (nm)

Figure 1 *Excitation and emission spectra of HA samples*

3.2 HPSEC

Figure 2 shows the HPSEC chromatograms obtained for the HA samples. For all the HAs, a broad peak appeared between V_0 and V_0+V_i, when monitored by the absorbance at 280 nm. The substances responsible for the broad peak can be regarded as the humic substances, as described in the previous study.[6]

Since stationary phases of the columns used in the present study have a small number of negatively charged groups, the presence of anionic groups on the humic substances will result in exclusion effects during HPSEC at a low mobile phase ionic strength.[10] Hence, a neutral salt such as NaCl was generally added to the mobile phase to increase its ionic strength.[11] The increase in ionic strength induces adsorption of the solute onto the stationary phase of the column because of the hydrophobic nature of the solute.[11] The addition of CH_3CN to the mobile phase was reported to be effective in eliminating the adsorption of humic substances onto the stationary phase of the column.[10] The fact that the peak of the humic substances occurred between V_0 and V_0+V_i indicates that the use of the eluent containing 300 mM NaCl and 20% CH_3CN minimized both the charge exclusion and hydrophobic interaction.

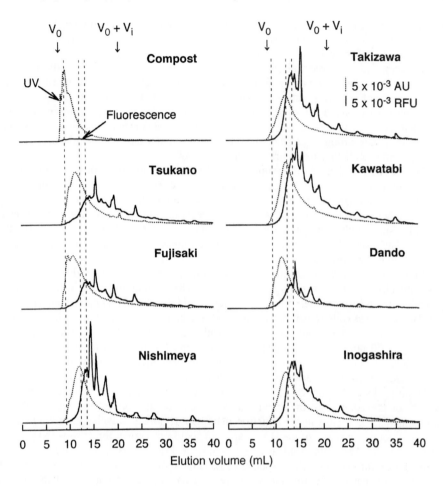

Figure 2 *HPSEC chromatograms of HAs*

Elution of the humic substances was most rapid for the compost HA and least rapid for the Andosol HAs (Takizawa, Kawatabi and Inogashira). The apparent MW of HSs at peak maximum ranged from 1.7 kDa for the Andosol HAs to 10.7 kDa for the compost HA, indicating a 6-fold difference. The other HAs had intermediate MWs. The MW was estimated at 3.8, 4.4, 3.6 and 2.4 kDa at the peak maximum for the Tsukano, Fujisaki, Dando and Nishimeya HAs, respectively.

The MWs of humic substances at peak maximum obtained in the present study were somewhat higher than those in the previous study,[6] except for the Andosol HAs. This is attributed to the fact that in the present experiment the columns with different MW separation ranges were linked in series. Becket et al.,[12] who analyzed the MW distributions of HAs from different sources using flow field-flow fractionation, reported that the MW at peak maximum ranged from 1.75 to 4.05 kDa, in good agreement with the range 1.7-4.4 kDa from the present study with the exception of the compost HA.

When monitored at the near-optimum excitation and emission wavelengths ($\lambda_{ex} = 460$

nm, λ_{em} = 520 nm) of soil HAs, the fluorescent substances were eluted considerably later than the humic substances, and had a wide range of elution volumes, as found in the previous study.[6] However, the peak intensity of the fluorescent substances in the compost HA was very low compared with the soil HAs. The fluorescent substances in the soil HAs were resolved into as many as 15 sharp peaks or more. Each sharp peak appeared at the same elution volume irrespective of the HA samples used. As mentioned before, the humic substances gave a broad peak in HPSEC. This is due to the polydispersity of the humic substances.[13] In contrast, the fact that the fluorescent substances were separated into many sharp peaks strongly suggests that the fluorescent substances in soil HAs consist of different components that are common to the soil HAs.

On the basis of their elution volumes, the fluorescent substances in soil HAs seem to have a lower MW than the humic substances. However, the peaks of the fluorescent substances were observed even at elution volumes larger than V_0+V_i. This fact clearly indicates that the mechanism responsible for the separation of the fluorescent substances is not only size exclusion but also interactions with the stationary phases of the columns. Therefore, the humic and fluorescent substances in soil HAs appeared to differ in their chemical properties.

Figure 3 shows the excitation and emission spectra of the individual fluorescent peaks obtained for three of the soil HA samples. The excitation spectra were similar to each other irrespective of their elution volumes, except for the HA samples that originated from Brown forest soils (Nishimeya and Dando), which were characterized by a gradual decrease in the relative fluorescent intensity with increasing emission wavelength, with a shoulder at 370 nm and a broad peak at 460-470 nm. For the Brown forest soil HAs, a peak at 320-330 nm was observed together with the shoulder at 370 nm and the peak at 460-470 nm, which is similar to the unseparated samples (Figure 1). Emission spectra had two broad peaks at 460 nm and 510-530 nm without exception. The relative fluorescent intensity of the latter peak increased relative to that of the former peak as the elution volume increased. A shift in the emission maximum to longer wavelength was observed for some fluorescent peaks of the HAs that originated from Brown forest soils (Nishimeya and Dando) and Andosols (Takizawa, Kawatabi and Inogashira). The similarities in the fluorescent excitation and emission spectra among the individual peaks and among the HA samples mean that the fluorophore structure of each fluorescent component in the soil HAs is virtually identical.

Figure 4 shows the HPSEC chromatograms of the fractionated samples of Fujisaki HA. When fraction A (elution volume range 8-13 mL) was reinjected under the same conditions, the fluorescent substances were eluted later than the HSs and both peaks appeared at the same elution volumes as those observed for the unfractionated sample (Figure 2). When fraction B (elution volume range 13-23 mL) was reinjected, the fluorescent substances showed the identical elution pattern to that observed for the unfractionated sample. The fluorescent peaks were accompanied by corresponding UV-absorbing peaks, indicating that the fluorescent substances in the HAs exhibit UV absorption. In addition, fraction B contained a small amount of HSs with the same elution volumes as those observed for the unfractionated sample and fraction A. This may be explained in terms of expansion of the contracted humic molecules, as described by Chin and Gschwend.[13] Similar results were obtained for the Nishimeya and Takizawa HA samples. Thus, most of the fluorescent substances in the soil HAs could be separated from the humic substances by HPSEC under the present operating conditions.

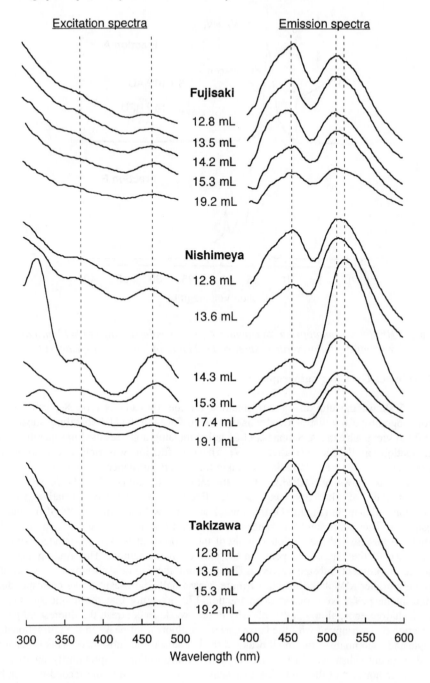

Figure 3 *Excitation and emission spectra of fluorescent peaks obtained for Fujisaki, Nishimeya and Takizawa HAs*

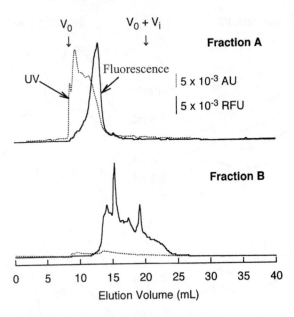

Figure 4 *HPSEC chromatograms of fraction A (elution volume range of 8-13 mL) and fraction B (elution volume range of 13-23 mL) obtained for Fujisaki HA*

3.3 Sephadex Gel Chromatography

Figure 5 shows the elution pattern obtained for the Tsukano HA on Sephadex G-25 chromatography using distilled water as the eluent. For both the UV-absorbing substances and fluorescent substances, Sephadex G-25 gel chromatography resulted in separation into two fractions at V_0 and just after V_0+V_i. The first fraction was rich in UV-absorbing substances, while the second peak was rich in fluorescent substances.

Figure 6 shows the HPSEC chromatograms of fractions 17 and 35, which are representatives of the first and second fractions, respectively. Relative to the unfractionated sample (Figure 2), fraction 17 lacked low MW components of the humic substances. The fluorescent substances in fraction 17 were separately eluted from the humic substances, and the elution volume of the fluorescent peak corresponded with the former part of the fluorescent peaks in the unfractionated sample. This finding suggests that the fluorescent substances are different from the humic substances even in the relatively higher MW fraction. Furthermore, very small fluorescent peaks corresponding to the humic peaks were observed in fraction 17, indicating that the humic substances exhibit fluorescence, although the intensity is very weak. In contrast, the fluorescent peaks in fraction 35 overlapped with the latter part of the fluorescent peaks in the unfractionated sample and accompanied the small peaks of the UV-absorbing substances. This is because the fluorescent substances in the HAs exhibit UV absorption, as previously mentioned. Thus, the major part of the fluorescent substances was recovered in the second fraction by Sephadex gel chromatography.

It has been reported by Posner[14] that HAs are separated into two fractions on Sephadex G-50 by applying a sample containing a relatively high salt concentration and eluting with distilled water. The approximate limiting MW of Sephadex G-25 for

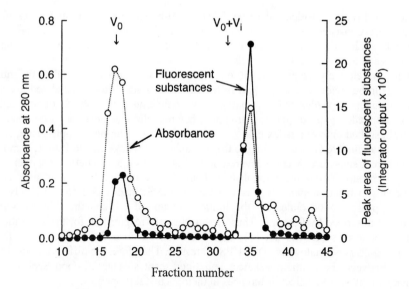

Figure 5 *Elution pattern obtained for Tsukano HA on a Sephadex G-25 column*

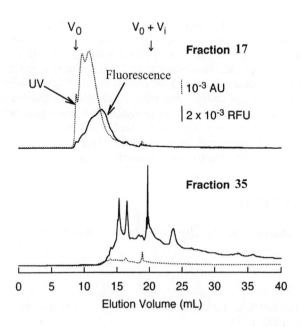

Figure 6 *HPSEC chromatograms of fractions 17 and 35 obtained by Sephadex gel chromatography of Tsukano HA*

complete exclusion was reported to be 5 kDa.[15] Although the humic substances in the Tsukano HA consisted largely of components with MW less than 5 kDa, they were completely excluded. This exclusion effect is the result of repulsion between the humic substances and Sephadex gels.[16]

Only a relatively small amount of fluorescent substances resembled the humic substances in behavior. The major part of the fluorescent substances was eluted just after the V_0+V_i upon Sephadex gel chromatography, indicating the existence of interactions with the Sephadex gels. In general, aromatic, heterocyclic and phenolic compounds are strongly adsorbed onto Sephadex gels.[16,17] When adsorption chromatography using XAD-7 resin was applied to HAs according to the method of Yonebayashi and Hattori,[18] most of the fluorescent substances were recovered in the fraction rich in aromatic components and carboxylic groups (Aoyama, unpublished data). Consequently, the major part of the fluorescent substances apparently was adsorbed on the Sephadex G-25 gel. The adsorbed fraction appeared to be eluted by the limited amount of Cl⁻ contained in the sample solution, as described by Posner.[14] Thus, the mechanism responsible for the separation of the fluorescent substances is not only size exclusion but also the interactions with the Sephadex G-25 gels, analogous to the HPSEC results. Therefore, the differences in elution behavior between the humic substances and fluorescent substances upon Sephadex gel chromatography should reflect differences in their chemical properties.

4 CONCLUSIONS

The fluorescent substances in HAs were mostly separated from the humic substances by HPSEC under the present operating conditions. The fluorescent substances in soil HAs consisted of chemically different components which were common to the soil HAs, and the fluorophore structure of each fluorescent component appears to be virtually identical. These results provide evidence that the fluorescent substances contained in soil HAs are different from the humic substances themselves. Furthermore, the major part of the fluorescent substances in soil HAs was also separated from the humic substances by gel chromatography using Sephadex G-25 and distilled water as the eluent. Accordingly, large-scale column chromatography using Sephadex G-25 will enable collection the fluorescent substances in large quantities for characterization.

ACKNOWLEDGMENTS

I thank Mr. K. Ishizawa and Mr. M. Ohnuma for technical assistance.

References

1.   M. Schnitzer and S. U. Khan, 'Humic Substances in the Environment', Dekker, New York, 1972, p. 55.
2.   W. R. Seitz, *Trends Anal. Chem.*, 1981, **1**, 79.
3.   M. Levesque, *Soil Sci.*, 1972, **113**, 346.
4.   K. Hayase and H. Tsubota, *Geochim. Cosmochim. Acta*, 1985, **49**, 159.
5.   M. Ewald, P. Berger and S. A. Visser, *Geoderma*, 1988, **43**, 11.

6. M. Aoyama, 'Proceedings of the 9th International Meeting of the International Humic Substances Society', Adelaide, in press.
7. S. Kuwatsuka, A. Watanabe, K. Itoh and S. Arai, *Soil Sci. Plant Nutr.*, 1992, **38**, 23.
8. T. M. Miano, G. R. Sposito and J. P. Martin, *Soil Sci. Soc. Am. J.*, 1988, **52**, 1016.
9. N. Senesi, T. M. Miano, M. R. Provenzano and G. Brunetti, *Soil Sci.*, 1991, **152**, 259.
10. A. Katayama, M. Hirai, M. Shoda and H. Kubota, *Soil Sci. Plant Nutr.*, 1986, **32**, 479.
11. S. Mori, 'Size Exclusion Chromatography', Kyoritsu Shuppan, Tokyo, 1991, p. 38.
12. R. Beckett, Z. Jue and J. C. Giddings, *Environ. Sci. Technol.*, 1987, **21**, 289.
13. Y. -P. Chin and P. M. Gschwend, *Geochim. Cosmochim. Acta,* 1991, **55**, 1309.
14. A. M. Posner, *Nature*, 1963, **198**, 1161.
15. E. Gjessing and G. F. Lee, *Environ. Sci. Technol.*, 1967, **1**, 631.
16. H. Gellotte, *J. Chromatog.*, 1960, **3**, 330.
17. J. A. Demetriou, F. M. Macias, M. J. McArthur and J. M. Beattie, *J. Chromatog.*, 1968, **34**, 342.
18. K. Yonebayashi and T. Hattori, *Geoderma*, 1990, **47**, 327.

A STUDY OF NON-UNIFORMITY OF METAL-BINDING SITES IN HUMIC SUBSTANCES BY X-RAY ABSORPTION SPECTROSCOPY

Anatoly I. Frenkel[1] and Gregory V. Korshin[2]

[1] Materials Research Laboratory, University of Illinois at Urbana-Champaign. Mailing address: Brookhaven National Laboratory, Upton, NY 11973
[2] Department of Civil Engineering, University of Washington, Seattle, WA 98195-2700

1 INTRODUCTION

Complexes of heavy metals with humic substances (HSs) are present in the environment at trace levels. This dramatically narrows the range of techniques applicable for structural studies. X-ray absorption spectroscopy (XAS) that includes extended-X-ray absorption fine-structure and X-ray absorption near-edge structure spectroscopies (EXAFS and XANES, respectively) is among the few experimental methods capable of probing metal-HSs complexes in dilute solutions with great structural and chemical sensitivity. These techniques give different but complementary information about the sample.

EXAFS analyzes the X-ray absorption coefficient measured in the region from 40 eV to 1000 eV past the absorption edge and gives information about the local atomic structure surrounding the absorbing atom.[1] Existing analysis methods find the types of the neighbors, coordination numbers, bond lengths and their disorders with the corresponding uncertainties.

The range of XANES comprises the energies between the first symmetry-allowed unoccupied state and the continuum states, that is < ca. 40 eV past the absorption edge. XANES gives information about the electronic structure, density of states and bonding geometry around the absorbing atom.[2] As opposed to the interpretation of EXAFS, where accounting for single-scattering photoelectron processes is often sufficient for nearest neighbor structural analysis and only a limited number of relevant multiple scattering paths should be included in the theory, full multiple scattering calculations are crucially important in the analysis of XANES data. Due to this difficulty, the calculation and interpretation of XANES data for an arbitrary system have been unavailable until recently.[3,4]

Substantial effort has been invested in studies of complexes of Cu^{2+} with HSs and a number of model organic ligands by EXAFS[5-11] and XANES.[3,5,12-14] EXAFS experiments have showed that the first coordination shell of Cu-HS complexes may be represented by an axially distorted CuO_6 octahedron whose axial and equatorial Cu-O distances and their respective disorders have been precisely measured.[5,11] In our previous work we used the theoretical code FEFF8[4] to model and analyze the XANES spectra of Cu(II) complexes in aqueous solution.[14] The axial and equatorial Cu-O distances in Cu-HS complexes derived from these XANES measurements were in good agreement with previous independent EXAFS data. They also supported the conclusion made regarding the tetragonal distortion of the first complexation shell based on the combination of ESR, IR, spectrophotometric and potentiometric data.[5,14-18] The goal of this work is to compare the XANES and EXAFS analyses of Cu-HS complexes at varying Cu/C ratios.

2 MATERIALS AND METHODS

The hydrophobic acids fraction (HPOA) from Suwannee River natural organic matter was used in the experiments. Samples preparation and properties were identical to those described in Korshin et al.[11] HPOA contained 0.7% nitrogen and 0.3% sulfur.[19] The DOC concentration in all HS-containing solutions was 1000 mg/L. The desired amount of metal was added as $Cu(ClO_4)_2$. The Cu/C molar ratios were 0.0005, 0.00125, 0.0025, 0.005, 0.0125 and 0.03. The pH of solutions containing HSs and copper was 4.0 and 12.0. Model systems included the copper(II) aqua-complex $[Cu(H_2O)_6]^{2+}$ and copper complexes with ethylenediamine, salicylic acid, citric acid and glycine (denoted as *Cu-Aqua*, *Cu-EN*, *Cu-Sal*, *Cu-Citric* and *Cu-Glycine*, respectively). The metal and ligand concentrations in solutions containing copper and the model ligands were 0.01 and 0.1 M, respectively. The metal concentration in solutions containing only the copper aqua-complex was 0.1 M. *Cu-Aqua*, *Cu-Citric* and *Cu-Glycine* solution were prepared at pH 4 and *Cu-Sal* and *Cu-EN* were prepared at pH 12.

The X-ray absorption measurements were carried out at the X11-A and X16-C beamlines of the National Synchrotron Light Source at Brookhaven National Laboratory using settings identical to those reported in ref. 11. The X-ray energy was varied from 200 eV below to 400 eV above the absorption K edge of Cu (E_K = 8979 eV) using a Si(111) double crystal monochromator. The data were obtained in the fluorescence mode. The pre-edge and near-edge regions of the data (-30 eV < E-E_K < 40 eV) were acquired with a 0.5 eV energy increment. The EXAFS data in the range 40 eV < E-E_K < 400 eV were acquired with a 2 eV increment. Up to 20 measurements were averaged for the same sample to improve the signal-to-noise ratio. To correct for a small angular drift of the monochromator crystals between the scans, all data sets were aligned vs. their absolute energy and interpolated to the same 0.25 eV-increment grid before the averaging. Cu metal foil was measured in the transmission mode simultaneously with all other samples and was used as the reference for the alignment of energies.

2.1 XANES Data

The edge-step normalized absorption coefficient data for all the Cu-HPOA samples are shown in Figure 1. As discussed by Frenkel et al.,[14] increase of the Cu/C ratios in the Cu-HSs system was accompanied by a pronounced change of a pre-edge feature located between 8976 and 8982 eV. This feature was most intense for the lowest Cu/C ratio (0.0005) and gradually lost intensity with increase of the copper concentration. At the Cu/C ratio of 0.03 it was undetectable. Of all the reference systems (*Cu-Aqua*, *Cu-Sal*, *Cu-EN*, *Cu-Citric* and *Cu-Glycine*), a similar pre-edge feature was found only for the *Cu-EN* complex (Figure 2).

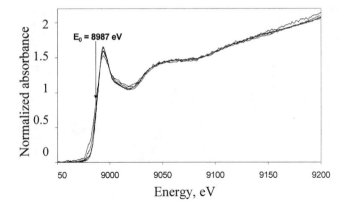

Figure 1 *X-ray absorption coefficients, normalized by the edge step, for the Cu-HPOA complexes (pH 4) with different Cu/C ratios. The energy origin for the background removal is indicated by the arrow*

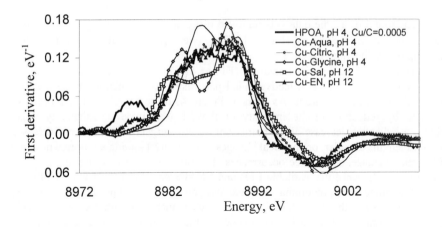

Figure 2 *First derivatives of the X-ray absorption spectra in the XANES region for the Cu-HPOA solution (Cu/C ratio 0.0005) and model complexes*

2.2 EXAFS Data

To evaluate the structural contributions of the atoms surrounding the central radiation-absorbing target atom (Cu in this work), an isolated-atom smooth background function $\mu_0(k)$, where k is the photoelectron wave number, is subtracted from the experimental absorption coefficient $\mu(k)$, and the resultant signal is normalized by the absorption edge step:

$$\chi(k) = \frac{\mu(k) - \mu_0(k)}{\Delta\mu_0(0)} \qquad (1)$$

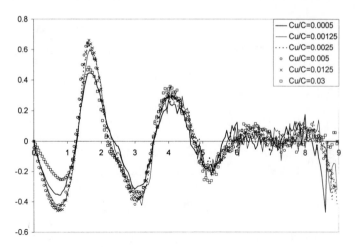

Figure 3 *EXAFS $k\chi(k)$ data in Cu-HPOA solutions at different Cu/C ratios*

The background of all the data was removed using an AUTOBK program[20,21] using the energy reference E_0 = 8987 eV (Figure 1). The k-weighted $\chi(k)$ data obtained for the Cu-HS samples are shown in Figure 3.

The magnitude of the Fourier transform (FT) of $k\chi(k)$ in r space (Figure 4) gives a qualitative representation of the effective radial pair distribution function of the nearest neighbors of the central Cu atom. As seen in Figure 4, decrease of the Cu/C ratio is accompanied by reduction in the intensity of the FT magnitude. Qualitatively, this reduction can indicate a decrease of the number of nearest neighbors to Cu or an increase in their bond lengths disorder, or chemical changes in the nearest atomic environment with the Cu/C ratio decrease, or all three occurrences. The first mechanism is very unlikely to play a role for Cu(II) since the available ESR and EXAFS data for a large variety of Cu-HS complexes indicate the predominance of the tetragonally distorted octahedral structure of the inner shell. Since the strength of Cu-HS complexes increases with decreasing Cu/C ratios,[22] it is very unlikely that this is accompanied by increase of the bond length disorder. This emphasizes the need to ascertain possible changes of the chemical structure around the Cu at low Cu/C ratios. To explore this mechanism, we have attempted to fit the experimental XANES and EXAFS spectra by including the formation of Cu-N bonds at low Cu/C ratios. While the substitution of an oxygen by a nitrogen atom in the inner complexation shell will not per se significantly affect the EXAFS signal, the different bond lengths of the Cu-O and Cu-N bonds resulting from this change of the inner shell composition may cause the changes of the FT magnitude associated with variations of the Cu/C ratios.

3 RESULTS AND DISCUSSION

3.1 EXAFS Data Analysis

In the single-electron single-scattering approximation, the edge-step normalized EXAFS

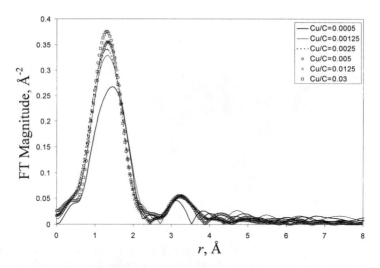

Figure 4 *Fourier transform magnitudes of the Cu K-edge EXAFS kχ(k) data in Cu-HPOA solutions at different Cu/C ratios*

signal $\chi(k)$ generated by a given shell of atoms could be written as:[1]

$$\chi(k) = \frac{NS_0^2}{kR^2} f(k)\exp(-2k^2\sigma^2)\sin(2kr + \delta(k)) \tag{2}$$

where N is the coordination number of the shell of the same atomic species located approximately at the same distance from the central atom, R is the average distance to the shell, σ^2 is the mean square deviation of this distance (i.e., both static and dynamic disorder) and S_0^2 is the many-body factor introduced to account for the shake-up and shake-off effects of the passive electrons. Functions $f(k)$ and $\delta(k)$ are the photoelectron backscattering amplitude and phase shifts, respectively. These two functions can be either calculated theoretically using a model whose structure is expected to be similar to that in the unknown sample or obtained from experimental standards.

We used the *ab initio* FEFF6 code[23] to generate the $f(k)$ and $\delta(k)$ functions for the absorbing Cu atom. The FEFF6 calculations were performed using the $Cu[NH_3]_x[H_2O]_{6-x}$ model octahedron (Figure 5). The equatorial and axial Cu-O distances were set at 1.92 and 2.13 Å, respectively, in accordance with our previous EXAFS results for the Cu-HPOA system with the Cu/C ratio of 0.03.[11] The O-H distances in the water molecules were 0.957 Å and 104.5° was the H-O-H angle.[24] The N-H distances in the NH_3 group were 1.015 Å and the H-N-H angle was 107.3°. The Cu-N distance in the model was 2.00 Å. The theoretical $f(k)$ and $\delta(k)$ functions were calculated for the axial and equatorial Cu-O bonds and for the Cu-N bonds.

The Cu-O bond lengths and their disorders were obtained by fitting the data generated with eq 2 to the experimental data. During the fitting procedure, the Cu-O distances for the equatorial and axial oxygens as well as the mean square disorders of these distances were varied independently until the best fit was achieved.

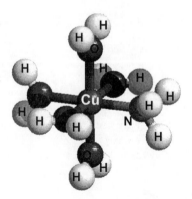

Figure 5 *Model structure of the $[Cu(NH_3)(H_2O)_5]^{2+}$ complex*

To account for the possible Cu-N interactions, the relative number of these bonds in the CuN_xO_{6-x} octahedron and the Cu-N bond lengths and disorders need to be varied as well. In addition, the many-body factor S_o^2 and the correction to the energy origin (ΔE_0) should also be elucidated from the fit. Therefore, the total number of fitting parameters if no constraints are applied is 9. The number of relevant independent data points N_{idp} in the data can be calculated with eq 3;[25]

$$N_{idp} = \frac{2\Delta k \Delta r}{\pi} + 2 \qquad (3)$$

where Δk and Δr are the data ranges in k and r spaces, respectively. For the Cu-HS samples, the Δk (from 2 to 7.5 Å^{-1}) and Δr (from 1 to 2 Å) ranges are rather narrow and the corresponding N_{idp} is ~ 5 points. Therefore, to obtain structural information from the fit it is crucial to circumvent one of the major obstacles in analyzing EXAFS data in disordered samples with a complex chemical and structural environment around the absorbing atom. This obstacle is an excessive number of adjustable parameters compared to the number of relevant independent data points.

To reduce the number of the fit variables, the EXAFS data for the Cu-HPOA system with Cu/C ratios 0.0005, 0.00125, 0.0025, 0.005, 0.0125 and 0.03 were processed using the same constraints for all datasets. These constraints were set as follows. First, the S_o^2 factor was fixed at 0.9, that is, at the median of its most probable variation range (0.8 to 1.0). Second, the equatorial and axial Cu bond lengths were set at 1.92 and 2.13 Å, respectively, in accord with the previous EXAFS results for the Cu-HPOA system with the Cu/C ratio=0.03.[11] It was assumed that, in the first approximation, these Cu-O distances were applicable for all Cu/C ratios used in the experiments. The third constraint presumed that, as found in ref. 11, the σ^2 of the equatorial Cu-O bonds was 0.0032 Å^2. This was applied for all the datasets. The number of the Cu-N bonds was presumed to be zero for the Cu/C ratios 0.005, 0.0125 and 0.03, that is for the Cu concentrations for which the shoulder between 8976 and 8982 eV was close to or below the noise level in the XANES derivatives. The best fit was sought for the set of parameters that included the correction to the energy origin, (ΔE_0), the length and σ^2 of Cu-N bonds, and the σ^2 of the axial Cu-O bonds. The number of the Cu-N bonds in the CuN_xO_{6-x} octahedron for the Cu/C ratios 0.0005, 0.00125 and 0.0025 was also independently varied. Thus, the total number of variables was 7, while the total number of the relevant independent data points was 30. The best fit parameters were established for all the data sets simultaneously.

The experimental EXAFS spectra were fitted with the theoretical EXAFS function (eq 2) generated using the FEFFIT program of the UWXAFS package.[20] This package provides nonlinear least-squares fitting. The estimates of the errors in the parameters were also calculated. The fitting was performed in *r*-space by Fourier transforming both the experimental data and theoretical predictions. The *k* weighting factor and the Hanning window function with margins of 1.8 Å^{-1} were used in Fourier transforms. The fits to all the 6 data sets with different Cu/C ratios are shown in Figure 6. The best fit parameters and predicted number of the N and O atoms in the inner complexation shell of Cu(II) are given in Table 1.

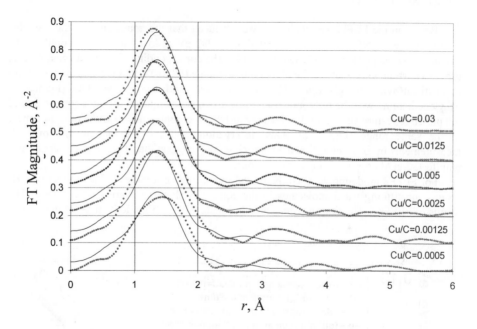

Figure 6 *FT magnitudes of the theoretical fits (solid) and experimental data (symbols) for Cu-HPOA solutions with different Cu/C ratios. The plots are shifted vertically for clarity*

3.2 XANES Data Analysis

In the preceding work[14] we reported the results of the quantitative XANES analysis of the local structure around Cu(II) in aqueous solution. The splitting of the derivative XANES spectrum for $[Cu(H_2O)_6]^{2+}$ between 8985 and 8991 eV (Figure 2) has its origin in the tetragonal distortion of the CuO_6 octahedron caused by the Jahn-Teller effect. For the prototype $[Cu(H_2O)_6]^{2+}$ system, the axial and equatorial Cu-O distances obtained independently from EXAFS and XANES data analysis were in excellent agreement, which attests to the reliability of the *ab initio* full multiple scattering calculations accounted for by the algorithm of the FEFF8 computer code.[4] In the present work we provide further analysis of the manifestations of Cu-N interactions in XANES.

Table 1 *Best fit values of nitrogen and oxygen occupation numbers in the CuN_xO_{6-x} inner coordination shell for different Cu/C ratios obtained by EXAFS data analysis*

Cu/C ratio	Cu-O bonds[a]	Cu-N bonds[a]
0.0005	5.5 ± 0.4(c)	0.5 ± 0.4 (v)
0.00125	5.8 ± 0.3(c)	0.2 ± 0.3 (v)
0.0025	5.9 ± 0.2(c)	0.1 ± 0.2 (v)
0.005	6 (f)	0 (f)
0.0125	6 (f)	0 (f)
0.03	6 (f)	0 (f)

[a] The letter in parentheses indicates whether the corresponding variable was fixed (f), varied (v) or constrained in the fit (c).

Based on the FEFF8 simulations it was deduced that the appearance of the pre-edge feature in the XANES data between 8976 and 8982 eV at low Cu/C ratios could not be explained by the presence of Cu-O bonds only. The corresponding photoelectron scattering processes would contribute to the higher energy range (8980 eV to 8995 eV), as seen from comparison with the *Cu-Aqua, Cu-Sal, Cu-Citric,* and *Cu-Glycine* data (Figure 2). Ethylenediamine was the only model compound whose complexation with copper caused the pre-edge feature similar to that for the Cu-HPOA complex at low Cu/C ratio to appear in the same range (Figure 2). This observation, and the previous results of ESR measurements by Boyd et al.[18] and Mangrich et al.[16] suggest that this pre-edge feature is a manifestation of copper-nitrogen interactions.

For FEFF8 modeling we used the model represented in Figure 5. The number of NH_3 groups substituting water molecules in the equatorial shell was varied from 1 to 4. The results of FEFF8 simulations are presented in Figure 7 as the first derivatives of the

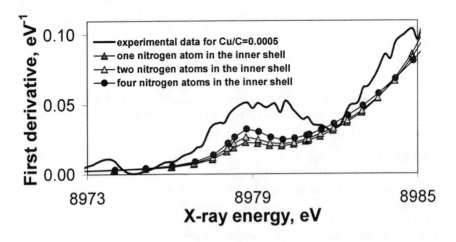

Figure 7 *Influence of nitrogen incorporation in the first coordination shell of Cu(II). Results of FEFF8 simulation. Equatorial and axial Cu-O distances are fixed at 1.92 and 2.13 Å, respectively. The Cu-N distance is fixed at 2.00 Å*

calculated absorption coefficients. One can see that the pre-edge feature between 8976 and 8982 eV was reproduced in these simulations. Its magnitude increased with the number of the N atoms in the complexation shell. The location of the feature associated with the Cu-N interactions was analogous to that observed for Cu/C ratios <0.005 and for the *Cu-EN* model system.

4 CONCLUSIONS

The combined results of our EXAFS and XANES data analysis of a Cu-HS system with varying Cu/C ratios demonstrate the non-uniformity of copper-binding sites in HSs. In the present work this non-uniformity is associated with Cu-N interactions that appear to be particularly important at low Cu/C ratios. The conclusion regarding the incorporation of nitrogen atoms in the inner coordination shell of Cu-HS complexes is supported by both EXAFS and XANES data. However, these complementary methods provide fairly divergent estimates of the number of N atoms in the inner coordination shell at low Cu/C ratios. This is not surprising since the level of sophistication of the models used in the calculations needs to be further refined. The precision of the calculations discussed is also restricted by the fact that changes of the Cu-O distances in the inner shell that may occur upon introduction of nitrogen atoms in the $CuO_{6-x}N_x$ octahedron were not accounted for in either the EXAFS or XANES analyses. Further theoretical *ab initio* XANES calculations of the composite oxygen- and nitrogen-containing inner shell that would account for these phenomena will answer these questions in more detail.

The qualitative agreement between the XANES and EXAFS results demonstrates the possibility to employ XANES to further explore the nature of the metal binding sites in HSs and quantify their non-uniformity. Latest XANES experiments and simulations open yet unexplored possibilities of obtaining accurate structural information for solutions with very low metal concentrations, for which the EXAFS signal is too weak to be reliably processed. In this work the XANES measurements were carried out for solutions with Cu concentrations as low as 4.2×10^{-5} M, which is probably among the lowest target concentrations ever used for XAFS experiments in solutions. It is nevertheless recognized that in order to study the state of heavy metals in non-contaminated water samples, the threshold metal concentration needs to be lowered by several orders of magnitude. This necessitates the use of third generation X-ray synchrotron sources. We have conducted preliminary experiments at the UNI-CAT 33-ID beamline (Advanced Photon Source, Argonne National Laboratory) whose beam intensity is ca. two orders of magnitude higher than that at BNL. Our objective is to obtain the XANES spectra of a Cu-HS system with metal concentrations lower than those used in this work. The relevant data will be described elsewhere.

ACKNOWLEDGEMENTS

We are indebted to Prof. John Rehr and Dr. Alexei Ankudinov (University of Washington) for their help in FEFF8 calculations. The authors express their appreciation to Dr. Jerry Leenheer (US Geological Survey, Denver, CO) and Prof. Jean-Philippe Croué (University de Poitiers, Poitiers, France) for characterization of the HSs samples. A. I. Frenkel acknowledges support from DOE grant DEFG02-96ER45439 through the Materials

Research Laboratory at the University of Illinois at Urbana–Champaign. G. V. Korshin acknowledges support from the U. S. Environmental Protection Agency (Grant #R826645).

References

1   E. A. Stern and S. M. Heald, in 'Handbook on Synchrotron Radiation', E. E. Koch, (ed.), North-Holland, Dordrecht, 1983.
2.   J. Stöhr, 'NEXAFS Spectroscopy', Springer-Verlag, Berlin, 1992.
3.   M. Benfatto, J. A. Solera, J. Chaboy, M. G. Proietti and J. Garcia, *Phys. Rev. B*, 1997, **56**, 2447.
4.   A. L. Ankudinov, B. Ravel, J. J. Rehr and S. D. Conradson, *Phys. Rev. B*, 1998, **58**, 7565.
5.   K. Xia, W. Bleam and P. A. Helmke, *Geochim. Cosmochim. Acta*, 1997, **61**, 2211, 2223.
6.   G. Davies, A. Fataftah, A. Cherkasskiy, E. A. Ghabbour, A. Radwan, S. A. Jansen, S. Kolla, M. D. Paciolla, L. T. Sein, W. Buermann, M. Balasubramanian, J. Budnick and B. Xing, *J. Chem. Soc. Dalton Trans.,* 1997, 4047.
7.   P. D'Angelo, E. Bottari, M. R. Festa, H. -F. Nolting and N. V. Pavel, *J. Chem. Phys.*, 1997, **107**, 2807.
8.   M. Nomura and T. Yamaguchi, *J. Phys. Chem.*, 1988, **92**, 6157.
9.   K. Ozutsumi, Y. Miyata and T. Kawashima, *J. Inorg. Biochem.*, 1991, **44**, 97; K. Ozutsumi and T. Kawashima, *Polyhedron*, 1992, **11**, 169; *Inorg. Chim. Acta*, 1991, **180**, 231.
10.   M. Tabata and K. Ozutsumi, *Bull. Chem. Soc. Jpn.*, 1994, **67**, 1608.
11.   G. V. Korshin, A. I. Frenkel and E. A. Stern, *Environ. Sci. Technol.*, 1998, **32**, 2699.
12.   L. Palladino, S. Della Longa, A. Reale, M. Belli, A. Scafati, G. Onori, A. Santucci, *J. Chem. Phys.*, 1993, **98**, 2720.
13.   J. Garcia, M. Benfatto, C. R. Natoli, A. Bianconi, A. Fontaine and H. Tolentino, *Chem. Phys.*, 1989, **132**, 295.
14.   A. I. Frenkel, G. V. Korshin and A. L. Ankudinov. Submitted to *Environ. Sci. Technol.*
15.   N. Senesi, *Anal. Chim. Acta*, 1990, **232**, 51.
16.   A. S. Mangrich, A. W. Lermen, E. J. Santos, R. C. Gomes, R. R. R. Coelho, L. F. Linhares and N. Senesi, *Biol. Fertil. Soils*, 1998, **26**, 341.
17.   P. R. Bloom and M. B. McBride, *Soil Sci. Soc. Am. J.,* 1979, **43**, 687.
18.   S. A. Boyd, L. E. Sommers, D. W. Nelson and D. X. West, *Soil Sci. Soc. Am. J.,* 1981, **45**, 745; 1983, **47**, 43.
19.   J. -P. Croué, G. V. Korshin, J. A. Leenheer and M. M. Benjamin, *Isolation, Fractionation and Characterization of Natural Organic Matter in Drinking Water*. To be published by AWWA Research Foundation and American Water Works Association, Denver, CO, (1999).
20.   E. A. Stern, M. Newville, B. Ravel, Y. Yacoby and D. Haskel, *Physica B*, 1995, **208/209**, 117.
21.   M. Newville, P. Livins, Y. Yacoby, J. J. Rehr and E. A. Stern, *Phys. Rev. B*, 1993, **47**, 14126.

22. J. H. Ephraim and J. A. Marinsky, *Environ. Sci. Technol.,* 1986, **20**, 367.

23. S. I. Zabinsky, J. J. Rehr, A. L. Ankudinov, R. C. Albers and M. J. Eller, *Phys. Rev. B*, 1995, **52**, 2995.

24. B. Beagley, A. Eriksson, J. Lindgren, I. Persson, L. G. M. Pettersson, M. Sandstrom, U. Wahlgren and E. W. White, *J. Phys.: Condens. Matter*, 1989, **1**, 2395.

25. E. A. Stern, *Phys. Rev. B*, 1993, **48**, 9825.

APPLICATION OF HIGH PERFORMANCE SIZE EXCLUSION CHROMATOGRAPHY (HPSEC) WITH DETECTION BY INDUCTIVELY COUPLED PLASMA-MASS SPECTROMETRY (ICP-MS) FOR THE STUDY OF METAL COMPLEXATION PROPERTIES OF SOIL DERIVED HUMIC ACID MOLECULAR FRACTIONS

Sandeep A. Bhandari,[1] Dula Amarasiriwardena[1] and Baoshan Xing[2]

[1] School of Natural Science, Hampshire College, Amherst, MA 01002, USA
[2] Department of Plant and Soil Sciences, University of Massachusetts, Amherst, MA 01003, USA

1 INTRODUCTION

Humic acids (HAs) are naturally occurring heterogeneous macromolecules in the soil and aquatic environments. They have affinity for inorganic and organic contaminants and play important roles in the uptake and transportation of nutrients and contaminants in surface and subsurface soil environments.[1-3] Carbon-rich HA macromolecules have a strong ability to form metal complexes, but a deep understanding of metal-humate complexes is challenging owing to HAs properties of polydispersity, multi-functionality, interaction with electrolytes and a propensity to form aggregates.[1,4] Various studies have been devoted to understanding the nature of HA-metal complexes, the functional groups involved, reaction mechanisms and conditions involved in the formation of such complexes.[4-7] Metal binding capability of HAs is manifested mostly by carboxylic, phenolic, keto and nitrogen atoms of amide functional groups.[1,8] Carboxylic groups are involved in the formation of metal humate salts (R-COO$^-$ M^{n+}), and humic acids form coordination complexes with polyvalent cations including transition metal ions. Gel filtration uv-visible chromatography,[9] spectrofluorimetry,[10] ion selective electrodes (ISE),[11,12] anodic stripping voltammetry (ASV),[12] atomic absorption spectrophotometry (AAS),[1,13] and more sensitive detectors like inductively coupled plasma-emission (ICP-AES)[4,6] and mass spectrometry (ICP-MS)[13-15] have been used for the investigation of metal complexation by humic acids.

There is growing interest in understanding trace metal binding to dissolved organic matter molecular fractions in water[14,16] and soil HAs.[15] Analytical approaches include the separation of molecular species by high performance size exclusion chromatography (HPSEC),[14-16] capillary electrophoresis (CE),[17] and subsequent determination of trace metals bound to these molecular fractions by uv-visible spectrophotometry,[16] graphite atomic absorption spectrophotometry (GFAAS)[16] or by inductively coupled plasma-mass spectrometric (ICPMS) methods.[14,15,18] By coupling a high performance liquid chromatograph (HPLC) system with an ICP-MS, a very sensitive analytical system is available for the on-line separation and detection of elemental species.[19-21] The HPSEC-ICP-MS system has most often been applied to speciation studies of metalloproteins in biological fluids.[19,22-24] The HPLC system generates molecular size information through a SEC medium, which is necessary to separate different HA molecular fractions according to their molecular sizes, while the ICP-MS generates an ion-chromatogram that gives information on the elements contained within those fractions. HPSEC is widely used for the separation of dissolved organic matter and HAs in water[14,16,25-30] and soil HA molecular species.[15,31] HPSEC coupled to ICP-MS has been successfully applied for the identification of trace metal-bound humic substances (HSs) molecular fractions in water,[14,25-27] soil HA,[15] and halogen species bound to HSs in natural water, sewage and brown water.[26,27]

Information about the physical and chemical nature of trace metals bound to various HA molecular size species has become increasingly significant in characterizing the distribution and environmental implications of toxic and nutritionally important trace metals. The focus of this work is to expand our analytical capabilities for speciation of metal-HA complexes using HPSEC-ICP-MS and to elucidate the distinct affinities and complexation properties of metals towards soil HA molecular fractions.

2 MATERIALS AND METHODS

2.1 Humic Acid Samples and Extraction

HAs extracted from five soils containing moderate to low soil organic matter were investigated. These soils originate from (i) an apple orchard sprayed with a pesticide containing Pb and As (HAAO) in the 1950s at Hampshire College, Amherst, MA; (ii) a control soil from the Hampshire College forest (HAFS); (iii) an agricultural soil from South Deerfield, MA (HASD); (iv) a mineralized soil from East Monroe, Colorado (HACO); and (v) an agricultural soil in Amherst, MA (HAAM). These samples are hereafter referred to by their codes designated in parentheses. The soil types for the extracted HAs are as follows: HAAO (Paxton B sandy loam), HAFS (Sudbury sandy loam), HASD (silt clay loam), HAAM (sandy loam) and HACO (clay loam).

All soil samples were cleaned of debris and plant material. The soil was then dried in an oven at 60°C for about 30 hrs after which it was sifted through a plastic sieve with a pore cross-sectional diameter of 2.5 mm. The sifted soil was then ground with a porcelain mortar and pestle. The ground soil was then tested for free carbonates by mixing approximately 5 g of soil with 10 mL of 0.5 M HCl. No apparent reaction took place in all the soils analyzed except HACO, indicating insignificant amounts of free carbonates in the soil used. Humic acids were extracted from the soil using sodium pyrophosphate extraction methods.[32,33] The extracted HAs were freeze dried in a Labconco Model 775000 freeze-drying unit. The percent ash content of HA samples were determined by ashing pre-weighed HA samples in porcelain crucibles at 730°C for 6 hrs in a Thermoline 6000 muffle furnace. To minimize trace metal contamination, all vessels used in the extraction, separation and purification of HA are made of polypropylene and they were leached with 50% (v/v) nitric acid before use. All solutions were prepared in 18 MΩ cm⁻¹ distilled deionized water.

2.2 Spectroscopic Characterization of Humic Acids

HA samples (2-4 mg) were dissolved in 10 mL of 0.05 M NaHCO$_3$ solution at pH 7.6-8.0. The absorbances were measured in a 1-cm path length cuvette (Fisher Scientific) and scanned from 190 to 820 nm in a diode-array uv-visible spectrophotometer (Hewlett Packard Model 8452A). The E$_4$/E$_6$ ratios (absorbance ratio at 465/665 nm) were determined in triplicate for each HA sample.

A 2 to 4% (w/w) mixture of well dried HA and spectroscopic grade KBr (dried in a oven at 60 °C) was prepared. The mixture was finely ground to powder consistency using an agate mortar and pestle. Dried KBr reference powder was prepared in a similar fashion and used as a blank. The DRIFT-IR spectrometer (Midac Series M2010), equipped with a DRIFT accessory (Spectros Instruments) was used for infrared analysis. Before analysis, the sample compartment was purged for about 10 min with dry N_2 to minimize interference from carbon dioxide and moisture in the air. In addition, a jar of anhydrous $Mg(ClO_4)_2$ was placed in the sample compartment chamber to reduce moisture further. The finely ground HA-KBr was carefully transferred into a stainless steel sample holder and smoothed with a glass microscope slide. Samples were scanned 100 times at a resolution of 4 cm^{-1}. Absorption spectra were obtained, smoothed and converted to Kubelka-Munk functions using the Grams/32™ software package (Galactic Industries).

Metal loading experiments were carried out using the HAFS sample only. This HA sample was shaken separately with individual 0.5 M aqueous solutions of Al^{3+}, Cd^{2+}, Cu^{2+}, Fe^{3+}, Mn^{2+} and Pb^{2+}. Metal solutions were prepared by dissolving $Al(NO_3)_3.9H_2O$, $Cd(NO_3)_2.4H_2O$, $Cu(NO_3)_2.3H_2O$, $Fe(NO_3)_2.9H_2O$, $Mn(SO_4)_2.H_2O$ and $Pb(NO_3)_2$ (Fisher Scientific or Sigma) in distilled deionized water. The initial pH of the metal solutions was 1.2-3.7. Between 198 to 203 mg of HAFS sample was then mixed with 30.0 mL of the respective aqueous metal salt solution in a high speed centrifugation tube (Oak Ridge, Nalgene) and shaken for 37 hrs at 300 rpm. After shaking, the mixture was centrifuged at 10,000 rpm for 10 min. The supernatant was decanted and its pH was 0.9-2.0. The metal loaded (M-HAFS) samples were then re-suspended in 30 mL of a fresh batch of the metal solution in centrifugation tubes, shaken for 30 min at 300 rpm and the mixture was centrifuged again. This entire process was then repeated with a fresh batch of metal solution. The last two steps were carried out to saturate the metal binding sites in HA with the respective cations. Metal loaded HAFS samples were then washed with 30 mL of distilled deionized water for 30 min. The mixture was then centrifuged and the supernatant discarded. The washing step was repeated twice more to remove loosely bound cations from the HAs.

2.3 Separation and Qualitative Analysis of Trace Metals Bound to Humic Acid Molecular Fractions

The chromatographic column used in this work was a Superose 12HR10/30 size exclusion column (Pharmacia Biotech). A Beckman Gold HPLC system was used for chromatography experiments (Table 1). Pumps and the plumbing for the mobile phase are made from PEEK™ (polyetheretherketone), which reduces contamination of analytes and eluents with trace metals. After the column was connected to the HPLC system it was equilibrated with at least three column volumes (75 mL) of the eluent distilled deionized water before samples were injected. Distilled deionized water filtered through a 0.22 μ Millipore filter was used as the eluent. All chromatograms were run in isochratic mode. A flow rate of 0.40 mL/min and an upper pressure limit of <2.0 MPa (410 psi) was maintained during the chromatographic analysis. Each chromatogram was programmed to run for 55-65 min. The uv-visible spectra of separated HA fractions were scanned from 190-350 nm and simultaneously displayed at 254 and 280 nm. The SEC column was calibrated using globular protein markers (Sigma): apoferritin (MW = 443 kDa), human albumin (MW = 65.4 kDa), and cytochrome-c (MW = 12.4 kDa). Known amounts (5 to 10 mg) of each of the standards were separately dissolved in 10.0 mL of distilled deionized water and eluted with a 0.025M phosphate/0.15 M NaCl buffer.

A detailed description of our HPSEC-ICP-MS interface was previously discussed[15] and instrument parameters are summarized in Table 1. Briefly, the effluent containing

Table 1 *HPSEC-ICP-MS instrument operating parameters*

HPSEC:	
HPLC pump	Beckman Gold HPLC system with PEEK™ pumps and plumbing (Model 126 NM)
Detector	Diode-array detector (190-800 nm) with a micro cell (Beckman Model 168NM)
UV detection	254 and 280 nm
Sample injection	Autosampler (Beckman Model 507e) with PEEK™ injection valve and 100 µL loop
Sample volume	50 µL
SEC column	Superose 12HR 10/30 column (Pharmacia/Amersham Biotech)
Mobile phase	Distilled deionized water (18 M cm^{-1})
Flow rate	0.40 mL/min
CP-MS:	Perkin Elmer Sciex/Elan 6000
RF frequency, MHz	40 (free-running)
RF forward power, kW	1
Argon gas flow rates:	L. min^{-1}
Outer	15.0
Intermediate	1.2
Nebulizer	0.83-0.96
Nebulizer	Ryton® Cross flow with GemTips® (Perkin Elmer)
Spray Chamber	Scott type
Scanning Mode	Peak Hop
Detector	Dual
Dwell time, msec	25
Sweeps/Reading	30
Readings/Replicate	420
Number of Replicates	1
Analyte mass/element	^{27}Al, ^{75}As, ^{114}Cd, ^{52}Cr, ^{57}Fe, ^{202}Hg, ^{55}Mn, ^{208}Pb and ^{64}Zn

separated HA fractions was introduced into the ICP-MS spray chamber through a 10-cm long PEEK/polypropylene transfer line. Once the separated sample aerosol was introduced into the hot plasma, the metals associated with HA molecular fractions were dissociated, atomized and ionized. The metal ions were then extracted into the quadrupole mass spectrometer and analyzed according to the mass/charge (m/z) ratios of the analytes. The elemental masses studied were ^{27}Al, ^{75}As, ^{114}Cd, ^{63}Cu, ^{57}Fe, ^{202}Hg, ^{55}Mn, ^{208}Pb and ^{64}Zn. The PE Elan NT mass spectrometer software was applied to obtain respective real-time ion chromatograms (ion counts vs. chromatographic elution time). The same software performed the operation of the ICP-MS and analysis. Ion chromatographic data was subsequently imported via Perkin-Elmer Turbochrome software for further analysis using Grams /32™ with the PE Views™ software facility.

2.4 Total Metal Concentrations in HAs

HA samples (48-56 mg) were digested with 5.0 mL of sub-boiled nitric acid (Fisher Optima Grade), using moderately high pressure vessels (Ultimate Digestion Vessels-UDV, CEM Corp.) in a microwave oven (MSP 1000, CEM Corp.) before the metal analysis. Digestion was carried out for 25 min at a final temperature of 200°C; this temperature is sufficient to decompose the HA organic matrix. Each digestion batch had 5 to 6 vessels containing a blank, control vessel and n = 3 or 4 HA replicate samples plus nitric acid. After each digestion, vessels were allowed to cool to room temperature before depressurization to minimize analyte loss and to facilitate condensation of any volatile elements. The digested samples were transferred into clean polypropylene centrifuge tubes and diluted with distilled deionized water to a final volume of 10-14 mL. The samples were kept in a refrigerator at 4°C until metal analysis was carried out by ICP-MS. Single metal reference standards of 1000 μg/mL (Spex) were used to prepare a 10 μg/mL spike solution containing the metals Al, As, Cd, Cr, Cu, Fe, Mn, Pb and Zn. A 1 mL aliquot of this spike standard was added to HAs and the spike recovery was determined.

Quantitative analyses of total trace metals in digested HA samples were conducted by ICP-MS. For external calibration of the ICP-MS, a set of multi-element standards (Al, As, Cd, Cr, Cu, Fe, Mn, Pb and Zn) covering the appropriate calibration range (1, 10, 100 and 200 ng/mL in 2% v/v sub-boiled nitric acid) were used. The standards were prepared by serially diluting 1000 μg/mL stock standards (Spex). An internal standard solution containing 20 ng/mL ^{9}Be, ^{45}Sc, ^{115}In, and ^{209}Bi was added on-line to both calibration standards and samples before analysis. The metals chosen for the internal standard were representative of major groups of mass (m/z) ranges analyzed across the periodic table. Linear calibration functions were obtained and diluted digested HA samples were analyzed. Further dilution was necessary for Al to ensure that its analytical concentration was within the calibration range.

3 RESULTS AND DISCUSSION

3.1 Spectroscopic Characterization of Soil Humic Acids

Extracted soil HAs were analyzed by uv-visible and infrared spectroscopy to characterize sample maturity, functionality and purity.

3.1.1 UV-visible Spectroscopy, Ash Content and HA solubility. The E_4/E_6 ratios for extracted soil HA samples were between 2.8 and 4.7, indicating a high degree of aromatization, and they were within ranges reported for soil HAs.[1] The ash content of HAFS, HAAM and HASD was less than 3%; however, HAAO and HACO had 17 and 24 % ash contents, respectively. Metal loaded HAFS showed lower solubilities in water and NaHCO₃ than unloaded HAFS. Aluminum loaded HA was most difficult to dissolve while Mn-HAFS was easily soluble in 0.05M sodium bicarbonate solution. This may be due to the loss of hydrogen bonding HA sites in Al loaded HA samples.

3.1.2 DRIFT-IR Analysis. DRIFT-IR spectra were obtained for all extracted soil HAs samples. They were used to assess the gross structural composition and to further investigate the metal-humate interaction sites. Typically, HA spectra (Figures 1a and b) exhibit a large, broad band at 3300-3000 cm⁻¹ resulting from OH and amide groups, a broad yet very diffuse band around 2500 cm⁻¹ (COO-H hydrogen bonded); a peak assigned to C=O ketonic stretch or carboxylic group around 1728-1710 cm⁻¹; 1670-1630 cm⁻¹ bands

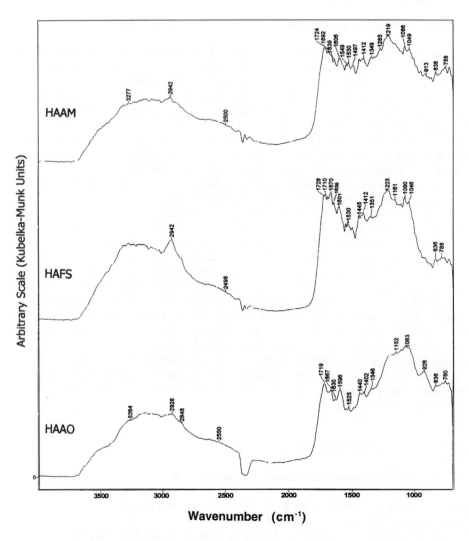

Figure 1a *DRIFT-IR spectra of soil humic acids: HAAM, HAFS, HAAO (see section 2.1*
for codes). Peak markings on the figure indicate wave numbers associated with
each peak

due to aromatic C=C and amide I bend, 1590-1517 cm^{-1} (C=O, symmetric stretching of
COO- and C=N stretching due to an amide II band), medium to weak bands at 1450-1410
cm^{-1} (CH$_3$ asymmetric stretch, CH bend, aromatic ring stretching); a strong band at 1245-
1220 cm^{-1} (C=O stretch, COOH, OH deformation, COC, phenolic OH); a weak to medium
band at 1080-1030 cm^{-1} (aliphatic and aromatic COC, alcohol/ polysacharides and silicate
impurities) and minor mineral bands below 850 cm^{-1}. These bands are akin to IR bands
reported in the literature for soil HAs.[1,33-37] HACO and HAAO show a silicate band at the
1030-1045 cm^{-1} region consistent with the high ash content of these samples. In addition,
the HAFS spectrum had a distinct band at 2942 cm^{-1} (aliphatic CH$_2$ stretching) indicative
of the immature nature of this HA extracted from the A$_1$ horizon of a forest soil.

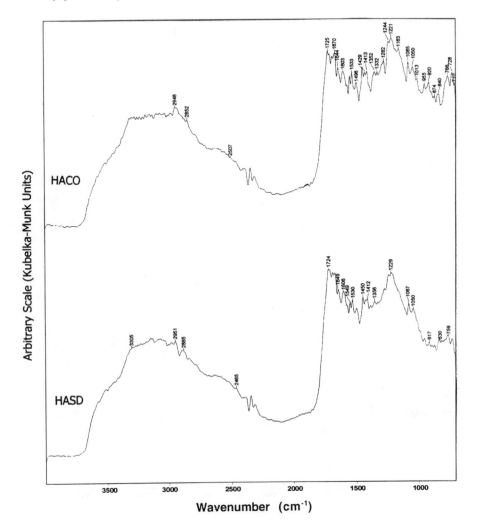

Figure 1b *DRIFT-IR spectra of soil humic acids, HACO and HAAO (see section 2.1 for codes). Peak markings on the figure indicate wave numbers associated with each peak*

The DRIFT-IR spectra of the metal loaded HA were used to identify the metal binding functional groups in metal-HA complexes.[6] These complexes were obtained by separately saturating the HAFS sample with 0.5M aqueous solutions of Al, Fe, Cu, Cd, Mn, Pb cations.

The DRIFT-IR spectra obtained for Fe, Al, Pb and Cd loaded HA (M-HAFS) samples are given in Figures 2a and b, respectively. These spectra were compared with the unloaded HAFS sample spectra to determine any changes observed. The spectra were investigated for the emergence of new peaks, intensity variations of peaks originally present and wavelength shifts in M-HAFS with respect to those in the unloaded HAFS spectra.

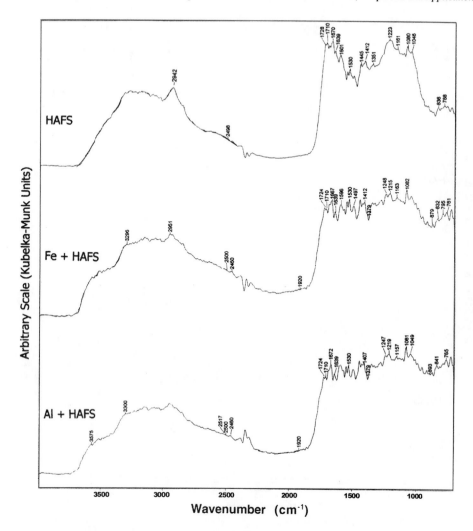

Figure 2a *DRIFT-IR spectra of metal loaded HAFS soil humic acid: Fe + HAFS, and Al + HAFS. Peak markings in the figure indicate wave numbers associated with each peak*

The metal loaded spectra demonstrate a decrease in relative intensity at 1710-1728 cm^{-1} (C=O and COOH) band with respect to unloaded HAFS, which is consistent with KBr pellet IR spectral information reported in the literature.[3-6] This is also true for Mn and Zn bound M-HAFS samples (spectra not shown). Senesi et al.[6] reported the disappearance of the 1710 cm^{-1} shoulder in HA loaded with Cu^{2+}, Mn^{2+} and Fe^{3+}. Furthermore, the emergence of a well-defined peak at 1639 cm^{-1} (except for Pb where the band is at 1644 cm^{-1}) in M-HAFS spectra is indicative of COO-M^{n+} salts, amide I and C = O stretch. The increase in relative intensity of doublet peaks at 1549 and 1530 cm^{-1} in all metal loaded spectra is possibly due to the metal interaction with amide II functional sites. A strong but broad absorption band present in the 1249-1223 cm^{-1} region (amide, OH deformation of phenols and carboxylic groups) in unloaded HAFS spectra decreases and splits into two

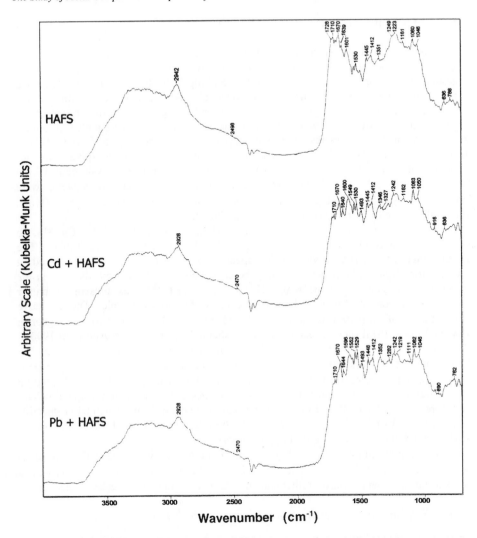

Figure 2b *DRIFT-IR spectra of metal loaded HAFS soil humic acid: Cd + HAFS, and Pb + HAFS. Peak markings on the figure indicate wave numbers associated with each peak*

sub-peaks in metal loaded spectra. As reported in the literature[1,33] peaks in the 1170-950 cm^{-1} region are representative of polysaccharide or polysaccharide-like substances. In this respect, a well-defined peak at 1081 cm^{-1} (C-O stretch of aliphatic alcohol) had increased intensity in metal loaded HAFS spectra, indicating the participation of these groups in metal complexation. These observations confirm the participation of carboxylic, phenol, and amide groups in metal-humate complex formation as discussed in other studies.[3-7] A multitude of bands emerge below the 900 cm^{-1} frequency region, particularly in Al (845 cm^{-1}) and Fe (795 cm^{-1}) loaded DRIFT-IR spectra. The nature of binding here is believed to be of the covalent M-OH type (Al-OH) or associated with mineral residues of HA in the form of Fe-O-Si or Al-O-Si groups.[37]

3.2 Total Elemental Analysis

The microwave assisted nitric acid (5 mL) digestion at 200°C and elevated pressures (170-270 psi) is capable of digesting 48-56 mg of soil HA. The appearance of the humic acid digestions was clear and pale green in color, resulting from dissolved nitrous acids and some brown fumes owing to dissolved nitrogen dioxide. Upon dilution with distilled deionized water, these solutions turned colorless, indicating complete dissolution of the organic matrix. However, a few digestions yielded pale yellow colored but clear solutions. A trace amount of siliceous residue was observed in some samples (i. e., HAAO, HACO with high ash content) that had not dissolved in nitric acid alone. Average spike recoveries obtained for ^{27}Al, ^{75}As, ^{114}Cd, ^{63}Cu, ^{55}Mn and ^{64}Zn from ICP-MS analysis ranged from 114 ± 5 % to 91 ± 1% (n = 4); however, lead (70 ± 6 %, n = 4) and iron recoveries were not satisfactory. The spike recovery of ^{52}Cr was somewhat high, 117 ± 4% (n = 4), and this could be due to ^{40}Ar^{12}C polyatomic ion interference.

The average concentrations and standard deviations for Al, As, Cd, Cr, Cu, Mn, Pb and Zn in digested HA samples determined by ICP-MS are shown in the Table 2. The metal content of soil HAs is very much dependent on its natural environment including contributions from anthropogenic origin. The concentrations of most metals were in the lower ppm, except for Al (HAAO, HACO), in which Al was present in higher concentrations consistent with the higher ash content of these two samples. The respective concentration of Pb (619 ± 16 µg/g, n = 6) and As (41 ± 1 µg/g, n = 6) was higher in HAAO. This soil still showed residues of the pesticide lead arsenate sprayed from 1900 to the early 1950's for the control of leaf rollers, moths and chewing insects.[38,39] The control humic acid (HAFS) sample extracted from uncontaminated soil in near-by woods demonstrated that the concentration of Pb and As was 94 ± 1 µg/g (n = 4) and 7.1 ± 0 µg/g (n = 4), respectively. The reported concentration ranges of Pb and As in the apple orchard soil were 411 to 35 and 85 to 13 µg/g as opposed to the control soil, which had respective concentrations of 67 to 36 and 5 to 2 µg/g.[38] These findings further illustrate the widely studied metal binding and sequestering properties of soil HAs.

Another point of interest is the higher concentration of Cu in HASD (1035 ± 16 µg/g) with respect to other HAs (67 ± 1 to 248 ± 7 µg/g). Information about the history of this soil was not sufficient to make further conclusions. Furthermore, Cd was below detection

Table 2 *Total metals concentrations in HAs measured by ICP-MS*

HA Sample	HAFS	HAAO	HAAM	HASD	HACO
Metal			Concentration (µg/g)		
Pb	94 ± 1	619 ± 16	131 ± 2	6.5 ± 0.2	47 ± 1
As	7 ± 0	41 ± 1	12.0 ± 0.4	2.1 ± 0.1	14.0 ± 0.4
Cu	100 ± 2	67 ± 1	248 ± 7	1035 ± 16	243 ± 6
Zn	11 ± 1	25 ± 16	23 ± 2	< 2.0	51 ± 1
Al	605 ± 2	1.13 ± 0.04[a]	800 ± 29	97 ± 2	4.5 ± 0.1[a]
Cr	34.9 ± 1.3	58.2 ± 2.4	86.5 ± 1.4	48.0 ± 1.9	69.2 ± 0.2
Mn	21.2 ± 0.8	51.4 ± 1.4	15.6 ± 0.5	< 0.01	75.6 ± 1.4
n[b]	4	6	6	5	4

[a] Percentage; [b] number of digestion replicates.

limits (< 4μg/g) for all the HA samples. This might indicate either that Cd is not a naturally present element in these HA samples or at least is not at levels detectable with direct ICP-MS analysis. However, studies indicate that Cd binds to soil HAs.[7,39]

3.3 Trace Metal Bound Humic Acid Molecular Fractions

Size exclusion chromatograms were obtained for the extracted and metal loaded HAs with distilled deionized water as eluent. Their uv absorbance was simultaneously monitored at 254 and 280 nm. All soil HAs had a wide range of molecular sizes. Early retention times represented large HA molecular sizes while delayed retention times represent smaller HA molecular sizes. Typical chromatograms for the unloaded soil HAs (HAAM, HACO and HASD) are illustrated in Figure 3. Chromatograms for all extracted HA samples exhibited a well-defined monomodal peak (labeled M_1) with a retention time ranging from 1081 to 1136 sec. In the case of HAAM, a small, medium molecular weight M_2 peak was observed (t_r = 1708 sec). Other HA samples showed a chromatographic peak slowly emerging around 1500 sec and culminating into a sharp peak (M_3) at a retention time of 2264-2280 sec. The shape of the major peak (M_1) is consistent with findings reported by Ruiz-Haas et al.[15] The M_1 peak is close to the exclusion limit (1050 sec), and this could be partly due to the large hydrodynamic radius of HAs in water or negative charges on the HA molecule.[16,30] The later peak M_3 demonstrated slight polydispersity of HA molecular fractions ranging from medium to larger amounts of smaller fractions. In addition to those two major peaks, very tiny sub-peaks can be observed at the tailing end of the M_1 peak for

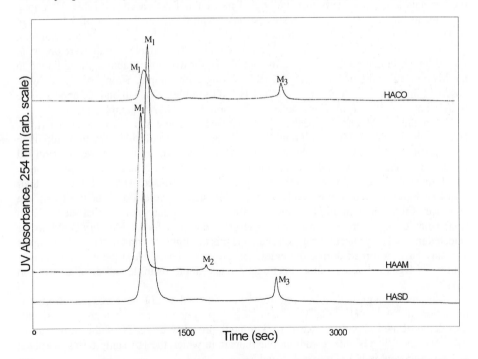

Figure 3 *Size exclusion chromatograms (uv absorbance at 254 nm) of HACO, HAAM and HASD soil derived humic acids. M_1, M_2 and M_3 are separated molecular fractions (M_1 > M_2 > M_3)*

the HASD and HACO samples at high resolution. In general, chromatograms had similar features to those reported by a number of other groups who have worked with aquatic[14,18,25,26,28-30,32] and soil HAs.[15]

The resolution of SEC separation can be further improved with other eluents like sodium acetate[16,40] and phosphate buffers.[29,40] However, in this study water was used as a more natural medium to which soil HA is exposed and also to minimize the potential matrix interferences in on-line ICP-MS analysis.

Determination of molecular weights of HA fractions by SEC is experimentally challenging.[1,16,29] It is known that HAs uncoil in different conformations in the mobile phase, depending on the ionic strength and pH of the eluent. Although a linear relationship is observed for retention volume vs. log MW of protein standards, a number of studies have argued that calibration with globular protein standards tends to over-predict the MW of humic substances.[16] For this reason, apparent MW observed here are over-predicted and not emphasized. Other workers have suggested random coil polymers such as polystyrene-sulfonates (PSS) as alternative molecular markers for the determination of molecular weights of HAs. However, even PSS do not mimic the more cross-linked and branched nature of soil HAs.

Trace metals tightly bound to molecular fractions of extracted HAs were determined by ICP-MS after separation with HPSEC.[15] Ion chromatogram of HAAM obtained with ICP-MS correspond with the peaks observed in uv-chromatograms of ^{27}Al, ^{114}Cd, ^{63}Cu, ^{64}Zn and ^{208}Pb (Figure 4). Individual metal ion chromatographic profiles (^{27}Al, ^{75}As, ^{63}Cu, ^{114}Cd, ^{202}Hg, ^{55}Mn, ^{208}Pb and ^{64}Zn) of all unloaded HAs were qualitatively examined and trace metals associated with each individual molecular fraction are in Table 3. Among those metals Al, Cu, Zn, Cd and Pb are associated with the M_1 major molecular fraction in all HAs, which was consistent with those reported.[15] Although the total Cd concentration was below detectable levels (< 4 µg/g) in HA samples, discernable signals were observed for Cd ion chromatographic profiles. In contrast, as shown in the Table 3, metal bound to the M_3 molecular fraction is highly variable among each sample. Strong Al signals were observed in the M_3 fraction for HAFS and HAAO samples. Cu, Mn and Zn appear to be weakly bound to the M_2 medium molecular fraction in the HAAM sample only.

In addition, Al and Cu peaks were observed in some ion chromatograms (i. e., HAAM, Figure 4) prior to eluting the M_1 major HA molecular fraction, although corresponding uv peaks were absent. This could be due to uv-insensitive, unidentified molecular species eluting early. These kinds of peaks are worth detailed consideration in the future. Although As and Mn are present in extracted HAs (Table 2), their signal/background (s/b) ratios in chromatographic signals were too weak to discern the presence of those ions in all HA molecular fractions. This may indicate that they exist in weak complexes present in either free or inorganic complex forms. Mercury could not be detected in any HA molecular fraction. Perhaps it is not tightly bound to these HA samples and may have desorbed during the extraction process. Nevertheless there is evidence of binding of Hg^{II} to reduced sulfur functional groups (thiol, disulfide and disulfane) in soil HAs.[41]

Elevated backgrounds observed for ^{57}Fe and ^{27}Al could have resulted from the formation of ^{40}Ar^{16}O^{1}H$^+$, ^{40}Ca^{16}O^{1}H$^+$ and ^{13}C^{14}N$^+$ polyatomic ions. ^{63}Cu and ^{64}Zn signals also demonstrated slightly elevated backgrounds probably as result of the formation of ^{40}Ar^{23}Na$^+$ and ^{32}S^{16}O$_2{}^+$ due to sodium bicarbonate in which the HA sample was dissolved and sulfur present in HA (Figures 4, 5 and 6).

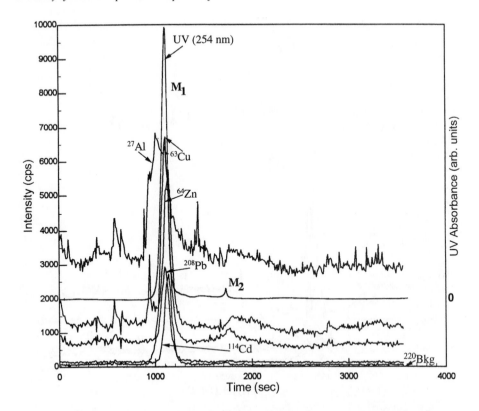

Figure 4 *ICP-MS ion chromatogram and size exclusion chromatogram (uv absorbance at 254 nm) of HAAM soil derived HA. M₁, M₂ are separated molecular fractions*

Table 3 *Metals tightly bound to soil HA molecular weight fractions*

Soil HA Sample	M₁ (Large)	M₂ (Medium)	M₃ (Small)
HAFS	Al, Cu, Zn, Cd, Pb Mn (w), As (vw)	NA	Al, Cu, Cd, Zn, Mn (w)
HAAO	Al, Cu, Zn, Cd, Pb Mn (w), As (vw)	NA	Al, Cu, Zn, Mn (w), Pb (vw)
HAAM	Al, Cu, Zn, Cd, Pb Mn (w), As (vw)	Cu (w), Zn (w), Mn (w)	NA
HASD	Al, Cu, Zn, Cd, Pb Mn (w), As (vw)	NA	Cu, Cd (m), Zn(m), Al (w), Pb(w), Mn (w)
HACO	Al, Cu, Zn, Cd, Pb	NA	Cu, Zn, Cd, Al (w), Pb(w), Mn(w)

NA - no corresponding molecular fraction; m - moderate signal (s)/background (b) ratio peak; w - low s/b peak; vw - very low s/b peak

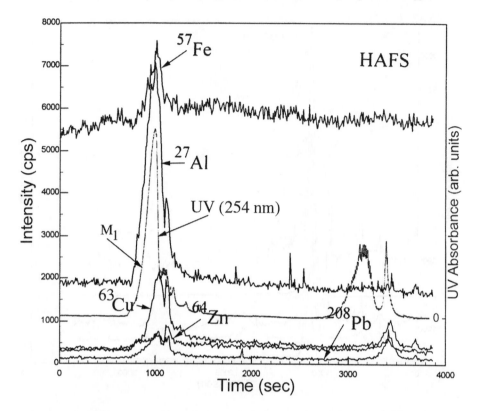

Figure 5 *ICP-MS ion chromatogram and size exclusion chromatogram (uv absorbance at 254 nm) of HAFS soil HA*

3.4 Extent of Metal Binding in HA Molecular Species Using HPSEC-ICP-MS

A feasibility study of the application of SEC-ICP-MS for the determination of the extent of metal binding to molecular fractions of HAs was made. The size exclusion chromatograms of the metal–loaded HAs consisted of three peaks at 960, 3180 and 3360 sec. The uv and ion chromatograms of HAFS and Fe-HAFS are illustrated in Figures 5 and 6. Their ion chromatograms indicate peaks at about 960 and 3360 sec and are associated with metal binding, while there is no corresponding ion chromatographic peak at 3180 sec. The broad peak at 3180 sec and the emergence of a sharp peak around 3360 sec indicate the presence of small molecular weight fractions, and are close to the permeation volume (V_p). Whether the broader peak eluting at 3180 sec is an organic residue or an artifact resulting from the adsorption of some HA on the column medium[1,30] is presently unknown. Thus, the degree of metal-binding properties of the larger M_1 molecular fraction was the focus of this study.

The intensities of the different background corrected metal peaks in the ion chromatograms were normalized against the intensities of the uv-chromatogram (254 nm) of M_1 molecular fractions for the metal-humate (M-HAFS) and unloaded HAFS samples. This was done to correct for the slightly different HA sample concentration of the HAFS and M-HAFS samples injected into the SEC column. The elemental signals observed on the ion chromatogram are dependent on the concentration of the respective cations detected by the ICP-MS. These normalized intensities were used to distinguish the extent

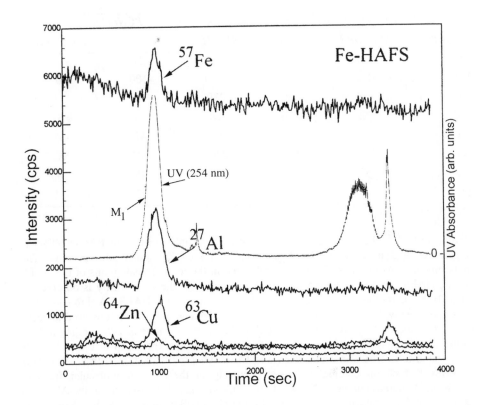

Figure 6 *ICP-MS ion chromatogram and size exclusion chromatogram (uv absorbance at 254 nm) of Fe loaded HAFS soil HA*

of metal binding by the M_1 molecular fraction. The percentage change in normalized peak intensities, PCNI = [(normalized intensity of metal loaded M-HAFS - normalized intensity of unloaded HAFS)/normalized intensity of unloaded HAFS] * 100 of the metal loaded chromatograms were also computed to evaluate the use of peak intensities as a qualitative measure of the relative concentration of metals bound to this HA molecular fraction.

The intensities of the different metals bound to the M_1 fraction were first compared in Table 4, using the PCNI of the respective metals after loading of the HA with respect to unloaded HAFS. A large positive PCNI indicates an increase in concentration of the metal investigated in the M-HAFS sample. A negative PCNI is characteristic of a decrease in concentration. The general tendency observed (Table 4) is an increase in the PCNI index of the metal (except for Pb) that the HA was saturated with, indicating that these metals are binding to the HAs M_1 molecular fraction. This is consistent with the metal binding properties of HAs,[42] and demonstrates the application of HPSEC-ICP-MS to describe metal binding on various molecular species in HAs. The observed order of decrease in PCNI was found to be Fe (788%) > Al (132%) > Cu (120%) > Cd (57%) > Mn (20%).

In other words, PCNI for the respective metals indicates the degree of binding with HA. Interestingly, much higher binding capacity of Fe relative to Al, Cd, Cu and Mn was observed, and the binding sequence of those metals follows the literature metal-HA

Table 4 *PCNI of metal peaks (M_1 fraction) in ion chromatograms for Cd-HAFS, Cu-HAFS, Al-HAFS, Fe-HAFS and Mn-HAFS*

Element	The percent change in normalized peak intensities - PCNI (%)				
	Cd-HAFS	*Cu-HAFS*	*Al-HAFS*	*Fe-HAFS*	*Mn-HAFS*
Al	-23	45	132	-21	-38
Mn	-100	-100	-100	-100	20
Fe	-58	-58	230	788	-37
Cu	22	120	78	45	0.52
Zn	-66	-44	249	47	-67
As	NC[a]	NC	NC	NC	NC
Cd	57	-100	272	45	-100
Pb	-44	-51	176	47	-54

[a] NC: No signal change

binding orders.[42] The PCNI data for the Fe-HAFS sample indicate that Fe (+788%) is sequestered and it replaces Al (-21.0%) and Mn (-100%) bound sites in HAs. The replacement of Al by Fe is also very clearly seen in the ion chromatograms of the HAFS and Fe-HAFS samples (Figure 5 and 6). A large drop in the Al intensity is observed at about 1000 sec on the time axis of the Fe-HAFS relative to the HAFS ion chromatogram. This coincides with the retention time of the M_1 molecular fraction. An observed increase of 132% in the PCNI of Al for the Al-HAFS sample is indicative of Al^{3+} binding to HA. In the case of the HA samples loaded with Cu, Cd and Mn, the PCNI of these elements were found to decrease in that order (i. e., Cu (120%) > Cd (57%) > Mn (20%)). This confirms that cation binding is indeed occurring in the M_1 HA molecular fraction. However, an increase in the PCNI of Cd, Cu, Pb, and Zn is also observed for the Al-HAFS sample. Similar trends are also observed for the Fe-HAFS sample where only Mn and Al were replaced and is accompanied by an increase in all other metals analyzed.

This increase perhaps could be due to the presence of these metals as trace impurities in the reagent grade $Al(NO_3)_3$ and $Fe(NO_3)_3$ aqueous solution used in the metal loading process. With the exception of the Al-HAFS and Fe-HAFS samples, the PCNI for Cd, Mn, Fe, Pb, and Zn decreased in all metal-humate samples except the ones where the metal under question is being loaded. Manganese, one of the weakly binding metals,[42] is completely replaced. This further demonstrates the weak binding between a cation like Mn^{2+} and HAs, which is consistent with its lower position in the binding series.[4,42] The findings of the PCNI approach are promising for semi-quantitative characterization of metal binding signals in HA molecular species.

Although this preliminary experiment demonstrates the potential capabilities of this method to unravel important metal binding information on HA molecular fractions, it is stressed that a more quantitative approach by HPSEC-ICP-MS is essential. This could be achieved by calibration of the HPSEC-ICP-MS using appropriate external standards. Another possibility is to use on-line isotope dilution calibration of HPSEC-ICP-MS as described by Rottman and Heumann.[25]

4 CONCLUSIONS

Five different HAs extracted for this work have uv and DRIFT-IR spectra that are typical of soil HAs. Metal loaded DRIFT-IR spectra of HAs clearly indicate the involvement of O (carboxylic and phenolic) and N (amide) containing functional groups in metal binding. A microwave assisted nitric acid method for the digestion of HAs was developed using moderately high-pressure microwave vessels. Approximately 50 mg of HA samples were digested in 5 mL of nitric acid at a temperature of $200^{\circ}C$ and pressure between 170-270 psi. Satisfactory metal spike recoveries for a number of metals (Al, As, Cd, Cu, Cr, Mn and Zn) were obtained by ICP-MS except for lead. Elevated Al levels were found to be associated with HACO and HAAO samples, consistent with their high ash contents. The increased concentrations of Pb and As in the apple orchard humic acid (HAAO) sample further illustrate the well known metal sorbing and binding capacities of SOM, particularly in HAs. SEC-ICP-MS chromatograms also support this observation, indicating that lead is bound to the largest molecular fraction M_1.

The on-line coupling of HPSEC-ICP-MS[15] has been successfully applied and verified for the qualitative determination of trace metals bound to soil HA molecular fractions. The results demonstrate that Zn, Cu, Pb, Al, Fe and Cd are preferentially and tightly bound to the largest molecular species (M_1) in extracted soil HAs. Results also show that HA bound Mn is present at low levels while As and Hg are loosely bound and desorbed during the soil HA extraction process. However, as determined by total metal analysis, the presence of Mn and As in HAs suggests that these metals could be present in inorganic forms. This observation needs to be further investigated. Metal complexation of HA with Al, Fe, Cu, Cd and Mn was also semi-quantitatively investigated using HPSEC-ICP-MS, and it is observed that Fe^{3+} competitively replaces Al^{3+} during the loading of HA with Fe. Furthermore, Al, Cu, Cd and Mn bind to the M_1 molecular species of HAs after saturation with these metals. PCNI data indicate that the binding affinities are in the order Fe > Al > Cu > Cd > Mn. This work shows that the HPSEC-ICP-MS system can be used for the separation and identification of trace metal bound soil HA molecular species and is a sensitive technique to study metal-HA complexing phenomena in environmental applications.

ACKNOWLEDGEMENTS

Dula Amarasiriwardena would like to thank the National Science Foundation (Award No: BIR 9512370), the Howard Hughes Medical Institute (HHMI), the Kresge Foundation, and the Keck Foundation for instrument and financial support for this project. Thanks are extended to the CEM Corporation for providing high-pressure digestion vessels and technical assistance for this project. The undergraduate research grant awarded by the HHMI to Sandeep A. Bhandari is gratefully acknowledged. Baoshan Xing greatly appreciates the support from the Federal Hatch Program (Project No. MAS00773) and a Faculty Research Grant from the University of Massachusetts at Amherst.

References

1.   F. J. Stevenson, 'Humus Chemistry: Genesis, Composition, and Reactions', 2nd edn., Wiley, New York, 1994.

2.   P. MacCarthy, C. E. Clapp, R. L. Malcolm and R. R. Bloom, 'Humic Substances in Soil and Crop Sciences: Selected Readings', American Society of Agronomy, Madison, WI, 1990.
3.   N. Senesi, *Trans. World Congr. Soil,* 1994, **39**, 384.
4.   G. Davies, A. Fataftah, A. Cherkasskiy, E. A. Ghabbour, A. Radwan, S. A. Jansen, S. Kolla, M. D. Paciolla, L. T. Sein, Jr., W. Buermann, M. Balasubramanian, J. Budnick and B. Xing, *J. Chem. Soc., Dalton Trans.,* 1997, 4047.
5.   N. Senesi, G. Sposito, K. M. Holtzclaw and G. R. Bradford, *J. Environ. Qual.,*1989, **18**, 186.
6.   N. Senesi, G. Sposito, G. R. Bradford and K. M. Holtzclaw, *Water, Air and Soil Pollution,* 1991, **55**, 409.
7.   K. M. Spark, J. D. Wells and B. B. Johnson, *Aust. J. Soil. Res.,* 1997, **35**, 89.
8.   Y. Chen and F. J. Stevenson, in 'The Role of Organic Matter in Modern Agriculture', Y. Chen and Y. Avnimelch, (eds.), Martinus Nijhoff, Boston, 1986, Ch. 5, p. 73.
9.   R. F. C. Mantoura and J. P. Riley, *Anal. Chim. Acta,* 1975, **78**, 193.
10.   R. A. Saar and J. H. Weber, *Anal. Chem.,* 1980, **52**, 2095.
11.   R. A. Saar and J. H. Weber, *Environ. Sci. Technol.,* 1980, **14**, 877.
12.   T. A. O'Shea and K. H. Mancy, *Anal. Chem.,* 1976, **48**, 1603.
13.   R. F. Breault, J. A. Colman, G. R. Aiken and D. McKnight, *Environ. Sci. Technol.,* 1996, **30**, 3477.
14.   L. Rottmann and K. G. Heumann, *Anal. Chem.,* 1994, **66**, 3709.
15.   P. Ruiz-Haas, D. Amarasiriwardena and B. Xing, in 'Humic Substances: Structures, Properties and Uses', G. Davies and E. A. Ghabbour, (eds.), Royal Society of Chemistry, Cambridge, 1998, p. 147.
16.   L. Zernichow and W. Lund, *Anal. Chim. Acta,* 1995, **300**, 167.
17.   P. Schmitt, A. Kettrup, D. Freitag and A. W. Garrison, *Fresenius J. Anal.Chem.,* 1996, **354**, 915.
18.   J. Vogl and K. G. Heumann, *Fresenius J. Anal. Chem.,* 1997, **359**, 438.
19.   K. Sutton, R. M. C. Sutton and J. A. Caruso, *J. Chromatogr. A,* 1997, **789**, 85.
20.   A. Seubert, *Fresenius J. Anal. Chem.,* 1994, **350**, 210.
21.   P. C. Uden, *J. Chromatogr. A,* 1995, **703**, 393.
22.   B. Gercken and R. M. Barnes, *Anal. Chem.,* 1991, **63**, 283.
23.   S. C. K. Shum and R. S. Houk, *Anal. Chem.,* 1993, **65**, 2972.
24.   I. Leopold and D. Günther, *Fresenius J. Anal. Chem.,* 1997, **359**, 364.
25.   L. Rottmann and K. G. Heumann, *Fresenius J. Anal. Chem.,* 1994, **350**, 221.
26.   G. Rädlinger and K. G. Heumann, *Fresenius J. Anal. Chem.,* 1997, **359**, 430.
27.   K. G. Heumann, L. Rottman and J. Vogl, *J. Anal. At. Spectrom.,* 1994, **9**, 1351.
28.   J. Knuutinen, L. Virkki, P. Mannila, P. Mikkelson, J. Paasivirta and S. Herve, *Wat. Res.,* 1988, **22**, 985.
29.   Y. P. Chin, G. R. Aiken and E. O'Loughlin, *Environ. Sci. Technol.,* 1994, **28**, 1853.
30.   Y. P. Chin and P. M. Gschwend, *Geochim. Cosmochim. Acta,* 1991, **55**, 1309.
31.   H. K. J. Powell and E. Fenton, *Anal. Chim. Acta,* 1996, **334**, 27.
32.   M. Schnitzer, in 'Methods of Soil Analysis, Part 2: Chemical and Microbiological Properties', A. L. Page, R. H. Miller and D. R. Keenly, (eds.), 2nd edn., Am. Soc. of Agronomy, Madison, WI, 1982, Ch. 30, p. 581.
33.   Z. Chen and S. Pawluk, *Geoderma,* 1995, **65**, 173.
34.   M. Schnitzer, in 'Soil Organic Matter', M. Schnitzer and S. U. Khan, (eds.), Elsevier, Amsterdam, 1978, Ch. 1, p. 1.

35.   A. U. Baes and P. R. Bloom, *Soil Sci. Soc. Am. J.,* 1989, **53**, 695.
36.   J. Niemeyer, Y. Chen and J. M. Bollag, *Soil Sci. Soc. Am. J.,* 1992, **56**, 135.
37.   M. M. Wander and S. J. Traina, *Soil Sci. Soc. Am. J.,* 1996, **60**, 1087.
38.   M. N. J. Hochella, Senior Undergraduate Thesis, Hampshire College, 1996.
39.   S. -Z. Lee, H. E. Allen, C. P. Huang, D. L. Sparks, P. F. Sanders and W. J. G. M. Peijnenburg, *Environ. Sci. Technol.*, 1996, **30**, 3418.
40.   J. Peuravuori and K. Pihlaja, *Anal. Chim. Acta,* 1997, **337**, 133.
41.   K. Xia, U. L. Skyllberg, W. F. Bleam, P. R. Bloom, E. A. Nater and P. A. Helmke, *Environ. Sci. Technol.*, 1999, **33**, 257.
42.   M. B. McBride, 'Environmental Chemistry of Soil', Oxford University Press, New York, 1994, p. 144.

INTERACTION OF ORGANIC CHEMICALS (PAH, PCB, TRIAZINES, NITROAROMATICS AND ORGANOTIN COMPOUNDS) WITH DISSOLVED HUMIC ORGANIC MATTER

Juergen Poerschmann, Frank-Dieter Kopinke, Joerg Plugge and Anett Georgi

Center for Environmental Research Leipzig-Halle, 04318 Leipzig, Germany

1 INTRODUCTION

1.1 Background

Widely distributed humic organic matter (HOM) is a high proportion of our world's organic carbon (OC). Soil HOM accounts for more than two-thirds of the terrestrial carbon reserve (about 2×10^{15} kg). However, HOM structures and functions are the least understood of all natural materials.

Particulate and dissolved HOM (POM and DOM, respectively) strongly influence the transport, chemistry and biological activity of hydrophobic organic compounds (HOCs) in the environment. Generally, HOC sorption on POM leads to a retardation of solute diffusion, whereas sorption on DOM results in enhanced mobility. Sorption of HOCs occurs by partitioning/dissolution into a phase such as the amorphous, polymeric HOM or by physical adsorption on surfaces. The intermolecular interactions available to neutral HOCs are similar for both mechanisms. They include hydrophobic (dispersion) or specific (for example, charge-transfer) forces. In contrast to surface adsorption, partitioning is non-competitive and results in linear sorption isotherms. Hydrophobic partitioning can be considered the result of van der Waals binding forces and a substantial entropic term that drives the hydrophobic molecules out of the water phase.[1] In this latter sense, HOM represents an organophilic medium into which the HOCs can escape from water.

The partition coefficient referred to HOM (K_{OM} or K_{DOM}) or organic carbon (K_{OC} or K_{DOC}) is a common measure of sorption. In the case of DOM we can write eqs 1 and 2,

$$K_{DOM,i} = \frac{C_{bound,i}}{C_{free,i}} \cdot \frac{m_{water}}{m_{DOM}} \tag{1}$$

$$C_{total,i} = C_{free,i} + C_{bound,i} = c_{free,i} + K_{DOM,i} \cdot C_{free,i} \frac{m_{DOM}}{m_{water}} \tag{2}$$

where $c_{free,i}$, $c_{bound,i}$ and $c_{total,i}$ are the concentrations of the analyte i in the freely-dissolved state, reversibly bound on the DOM phase and in total, respectively, all in relation to the amount of water in the sample. $K_{DOM,i}$ is the partition coefficient for the analyte i between DOM and water, and m_{DOM} and m_{water} are the masses of DOM and water in the sample.

A simple physical phase separation is a convenient method of measuring the freely dissolved portion of HOC interacting with particulate sorbents. It allows the calculation of sorption coefficients (eqs 1 and 2). In cases where no mass balance assumption is made (that is, the analyte is measured both in the solution and in the sorbed phase), attention should be paid to analyte losses on inner glass walls, septa, stirrers and so on. These losses are often mistakenly assigned to the humic-bound fraction. This is particularly important when investigating highly hydrophobic compounds such as 5-ring PAHs or high chlorine content PCB congeners.

The phase separation method is valuable in determining freely dissolved sorbate concentrations in DOM studies. Alternative ways to investigate sorption on DOM include fluorescence quenching (FQ), rapid solid phase extraction (SPE), flocculation, solubility enhancement, dialysis and the headspace partitioning method.[2] None of these techniques fully satisfy the demands (i) not to disturb the sorption equilibrium, (ii) not to be laborious and involve many manipulation steps, (iii) to be applicable to multicomponent systems, and (iv) also to be applicable to systems that include POM. The latter criterion allows the environmental impact of DOM and POM to be more reliably compared. The approach being developed to investigate sorption on DOM may also be extended to other multicomponent aqueous systems. It is well known that dose-response relationships in pharmacology depend on the freely available drug concentration. This makes it necessary to distinguish the total from the freely dissolved fraction. This also applies to pollutants in a multicomponent system.

1.2 Solid Phase Microextraction to Determine Sorption Coefficients

The Solid Phase Microextraction (SPME) method introduced and pioneered by Pawliszyn and co-workers at the University of Waterloo in the early '90s has become an established extraction/preconcentration method for the solvent-free extraction of non-polar and polar, volatile and semi-volatile organic analytes.[3] SPME is an equilibrium rather than an exhaustive extraction technique. It utilises polymer-coated fused silica fibres for analyte extraction from aqueous or gaseous matrices or the headspace over liquids and solids (Figure 1).

The fibre is mounted in a syringe-like device for protection and ease of handling. Analytes partition between the aqueous phase or headspace and the fibre polymeric phase until equilibrium is reached. Widely used fibre coatings include 100μ non-polar polydimethylsiloxane (PDMS) and 85μ polar polyacrylate (PA) coatings to extract preferentially non-polar and polar organic analytes, respectively. Similar to pesticides sorption on HOM, the analyte mass extracted by the fibre depends on the partition coefficient between the fibre and the sample. The fibre distribution coefficient K_F in the conventional solution SPME can be calculated from eq 3 provided that the fibre volume is negligible compared to the sample volume. Here, n is the mass of analyte extracted by the

$$\mathbf{K}_F = \frac{\mathbf{n}}{\mathbf{C}_0 \bullet \mathbf{V}_F} \tag{3}$$

fibre coating at equilibrium, C_0 is the initial concentration of the analyte in the aqueous sample and V_F is the fibre volume (e.g. $V_F = 2.57 \times 10^{-5}$ mL with 7µ PDMS, $V_F = 6.6 \times 10^{-4}$ mL for 100µ PDMS)

The analyte extracted by the fibre can be desorbed in a hot split/splitless GC injector (the dominant current approach) or in the eluent flow of an HPLC device. SPME allows extraction and analysis to be conducted in a single step. It is simple, fast and prevents depletion of the sample if sufficiently large sample volumes are used. The extraction mechanisms in SPME and liquid-liquid extraction (LLE) are identical, but the volumes of the extracting phases are quite different. Another important difference is that LLE measures analyte concentrations whereas SPME measures analyte activities.[2] However, in many cases SPME data can be taken as analyte concentrations because the calibration procedure is conducted with aqueous solutions of known analyte concentrations and equal activity coefficients.

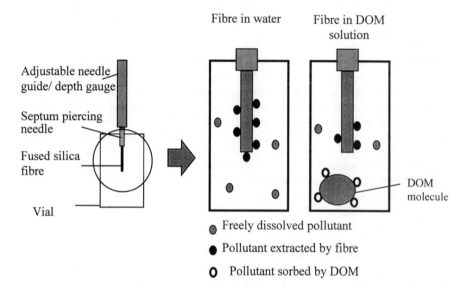

Figure 1 *Schematic of conventional SPME*

The basic assumption of SPME to study sorption on DOM is that the fibre samples only the freely dissolved analyte fraction (Figure 1).[4] Therefore, partition coefficients can be calculated from SPME data based on external calibration ($c_{SPME,external}$). An alternative approach is particularly straightforward when investigating sorbates of unknown total concentration in solutions rich in DOM. It is based on the assumptions that (i) the partitioning behaviour of a non-deuterated and a deuterated analyte is identical and (ii) the sorption equilibrium is established very quickly (see section 3.6). The internal calibration using deuterated surrogates and GC/MS analysis ($C_{SPME,internal}$) gives the total concentration of the analyte in the sample and the sorption coefficient from eq 4.[5]

$$K_{DOM,i} = \frac{C_{SPME,internal,i} - C_{SPME,external,i}}{C_{SPME,external,i}} \cdot \frac{m_{water}}{m_{DOM}} \qquad (4)$$

This paper highlights the usefulness of the SPME technique to determine the thermodynamics and kinetics of HOC sorption on DOM. The theoretical fundamentals are discussed in a previous paper.[2] The present paper applies the SPME technique to sorbate classes that include PAHs, PCBs, nitroaromatics, triazines and organotin compounds, all of which are subjects of environmental concern.

2 MATERIALS AND METHODS

2.1 Isolation of Humic and Fulvic Acids

Humic and fulvic acids (HAs and FAs) were isolated from heavily polluted coal wastewater obtained from a pond in East Germany. They are called "anthropogenic" HOM due to their anthropogenic origin. Another source was a water plant in Fuhrberg, near Hannover. The isolation procedure is detailed elsewhere.[6] The humic acids from the sediment at the bottom of the pond were isolated by extraction with 1 M NaOH without using XAD resins.

2.2 Standards

Deuterated and non-deuterated PAHs were obtained from Promochem (Wesel, Germany) and Supelco. Nitroaromatics, PCBs and triazines were from Supelco and Sigma-Aldrich. ^{13}C-labelled PCBs were obtained from Cambridge Isotope Laboratories. Analytical reagent grade solvents (Merck) were used throughout. All glassware and glass-coated stirrers were thoroughly washed and silanised prior to use. Drops of sodium azide solution added to the DOM solutions gave a 0.05% solution of this microbial inhibitor. Octanol-water coefficients were taken from the Daylight Chemical Information Systems database CLOPG-4.34 (Irvine, CA).

2.3 Instruments

The following instruments were employed. GC-MS: Model 5973 GC-MS (Hewlett Packard) equipped with electron impact and chemical ionisation, data acquisition in SIM mode; GC-AED: Model G 2350A (Hewlett Packard), tin emission line at 326 nm, chlorine emission line at 479 nm. Both devices were run with HP-5 columns 30m × 0.25mm × 0.25μ (Hewlett-Packard). The splitless injector (pressure pulse mode) was at 300°C for the 7μ PDMS and the PA fibre, at 280°C for the 100μ PDMS fibre, and at 260°C for the CW fibre.

2.4 SPME

SPME syringe and fibres with 7μ and 100μ PDMS, 85μ polar polyacrylate (PA) and Carbowax-divinylbenzene (CW) coatings were purchased from Supelco. Fibres were preconditioned at 270°C under helium overnight prior to use (for CW the temperature was 250°C).

SPME study of pesticide sorption on DOM with external calibration was performed in a 40 mL amber vial (PCBs: 250 mL vial) using deionised water containing known concentrations of analytes: PAHs (2 ppb), PCBs (50 ppt), nitroaromatics (100 to 500 ppb, depending on the analyte hydrophobicity), triazines (5 ppb), organotin compounds (50 ppb) along with sodium azide to prevent microbial activity. The sampling time in each experiment was sufficient to ensure equilibrium conditions. For example, an overnight procedure was applied for measuring PAHs and PCBs using the PDMS fibres. Afterwards, DOM solutions with concentrations from 20 ppm to 800 ppm (depending on analyte hydrophobicity) were spiked with the analytes at the same concentration as with external water calibration. The SPME sampling time was identical to that of the external calibration procedure, thus ensuring equilibrium conditions. In order to determine the absolute masses of fibre uptakes, the calibration was done using cold on-column injection with standard solutions in non-polar to medium polarity organic solvents.

SPME study of PAH sorption kinetics on DOM was conducted by spiking a solution of 100 ppm DOM (aquatic FA isolated from the coal wastewater pond) in a 40 mL vial to give 5 ppb each of naphthalene, anthracene and pyrene; the system was then allowed to equilibrate overnight. After spiking again with 5 ppb deuterated PAH surrogates in acetone, a 100μ PDMS fibre was dipped into the solution for 10 sec under vigorous stirring and then exposed to desorption for 10 sec in a hot GC injector connected to a 3m x 0.1mm fused silica capillary deactivated with phenylmethylpolysiloxane (Hewlett-Packard) at 270°C. GC-MS data acquisition was restricted to monitor only the ion traces of deuterated and non-labelled PAHs (e.g. $m/z = 128$ and 136 for naphthalene). This procedure was repeated 20 times within 30 min to monitor the time-dependent ion trace ratio of deuterated and non-labelled PAHs.

3 RESULTS AND DISCUSSION

3.1 Polycyclic Aromatic Hydrocarbons

Figure 2 shows the sorption of PAHs on a dissolved HA of anthropogenic origin. These data were obtained with the conventional (solution phase) SPME approach. Headspace SPME-based results are almost identical.[2] However, the time to establish equilibrium in this three-phase system is quite long at room temperature, especially for 4- and 5-ring PAHs. As can be seen from Figure 2, the larger the K_{DOM} partition coefficients the larger the differences in the analyte's fibre uptake when comparing data for water and the DOM solution. For naphthalene, the small peak area difference in water and DOM solution indicates a small partition coefficient on the DOM. By contrast, chrysene is overwhelmingly sorbed by the humic acid, as indicated by the very different peak areas in Figure 2. Sorption coefficients were calculated from eq 1. The strong linear correlation ($r^2 > 0.96$) of partition coefficients on the DOM under study with the analyte hydrophobicity as expressed by the octanol-water coefficient (K_{OW}) indicates the dominance of hydrophobic partitioning (Figure 3, dashed line and Table 1). The same applies to the other DOM studied, the sample isolated from a water plant (the straight line designated with "FA" in Figure 3). This FA is more hydrophilic and of lower average molecular weight as indicated by acid-base titration, elemental analysis and ESI-MS (data not given here).

The experiment summarized in Figure 2 was conducted with a DOM concentration of 50 mg/L. Very similar results were obtained with higher DOM concentrations up to 800 mg/L. This suggests that increasing aggregation/micellization with increasing DOM concentration has little impact on the overall sorption at neutral pH in the absence of multivalent cations. DOM is assumed to form cage-like, pseudomicellar structures as a

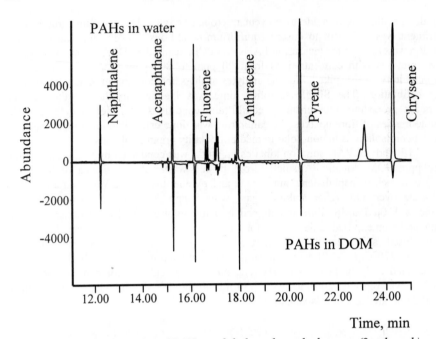

Figure 2 *MSD chromatogram of PAHs naphthalene through chrysene (2 ppb each) sampled by SPME from water (top) and DOM solution. SPME: 7 μ PDMS, sampling time 4 hrs, stirring rate 800 rpm; DOM: humic acid isolated from coal wastewater, C_{DOM} = 50 mg/L; oven: 50°C (3') to 280°C with linear temperature program 8°C/min; Data acquisition: SIM, target ions corresponding to molecular ions of the PAHs*

Table 1 *Sorption coefficients on DOM, PDMS fibre distribution coefficients and octanol-water partition coefficients of PAHs*

PAH	Log K_{OW}[a]	Log K_{DOM} (HA)	Log K_{DOM} (FA)	Log K_F
Naphthalene	3.30	3.05	2.62	3.09
Acenaphthene	3.92	3.38	2.95	
Fluorene	4.18	3.55	3.08	3.69
Phenanthrene	4.44	3.85	3.26	4.01
Anthracene	4.47	3.97	3.31	4.13
Fluoranthene	5.15	4.35	3.66	4.47
Pyrene	5.19	4.46	3.81	4.62
Chrysene	5.50	4.75	4.06	4.84

[a] Data from ref. 2.

result of intra- and intermolecular aggregation.[7] The former DOM-concentration-independent contribution entails coiling and folding. Intermolecular aggregation, which is assumed to influence sorption by sequestering in hydrophobic domains,[8] may be more significant at DOM concentrations above 800 mg/L. However, this is beyond the scope of this paper and should be studied by headspace-SPME.

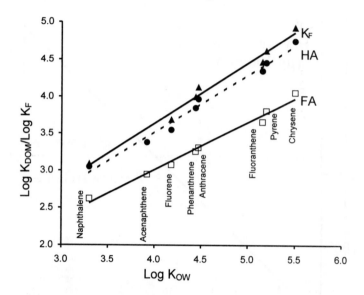

Figure 3 *Sorption coefficients and (7 μ PDMS) fibre distribution coefficients vs. hydrophobicity for PAHs naphthalene through chrysene*

In experimental work with conventional SPME (as above with PAHs), the partition coefficients for 4- and 5-ring PAHs have high inaccuracies at high DOM concentrations because the freely available PAH portion is very small. Furthermore, fibre fouling occurs at high DOM concentrations when using the conventional SPME mode.

As expected from the marginal impact of DOM concentration on sorption coefficients, HOC loading of the DOM was insignificant within the HOC concentration range studied (Figure 4), which is limited by HOC water solubilities (for phenanthrene, 1.1 ppm) and the detector linearity. The linear sorption isotherm for phenanthrene has a regression coefficient close to 1.000, which indicates a non-competitive partitioning mechanism in the range 10 to 7500 ppm phenanthrene loading on DOM. Linear and non-competitive sorption also prevails for POM in a "rubbery" state, whereas sorption by "glassy" geosolids can be accounted for by a hole-filling mechanism resulting in nonlinear sorption uptake isotherms, site-specific competition and desorption hysteresis.[1]

The literature contains many empirical relationships between partition coefficients and octanol-water coefficients of the type $\log K_{OM} = a \log K_{OW} + b$, which is basically a free energy relationship. However, the coefficients a and b are quite empirical and they may be different for different classes of sorbents. This lowers the value of the concept.

A modified concept describes the partitioning behaviour of a given HOM relative to that of octanol.[9] The solubility parameters (δ) of any HOM, which are a measure of polarity, can be calculated from the literature solubility parameters of the sorbates and their octanol-water coefficients. The δ-values for the hydrophilic FA and the hydrophobic HA used here were calculated to be $\delta = 12.85 \pm 0.4$ (cal cm^{-3})$^{0.5}$ and $\delta = 11.85 \pm 0.55$ (cal cm^{-3})$^{0.5}$, respectively. They are within the range 12.5 ± 1.0 (cal cm^{-3})$^{0.5}$ that is characteristic of many DOM and POM investigated by our laboratory and by other

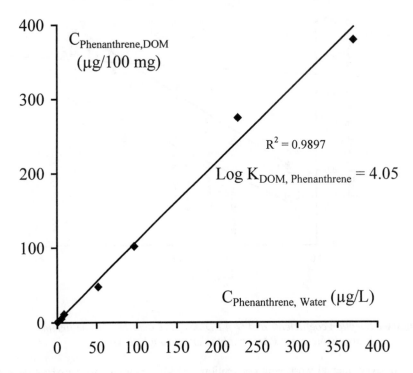

Figure 4 *Phenanthrene sorption isotherm on an anthropogenic humic acid (C_{DOM} = 100 ppm)*

authors.[9] According to the Flory-Huggins theory, similar solubility parameters of the sorbate and the amorphous, polymeric sorbent favour the partitioning process. Considering the low solubility parameters of PAHs as sorbates (naphthalene and anthracene have δ = 9.9 (cal cm^{-3})$^{0.5}$), their better partitioning with the anthropogenic HA compared to the hydrophilic FA is as expected.

As mentioned above, there are some similarities between the sorption of analytes on HOM and SPME sampling. The strong correlation between the PDMS-water distribution coefficients K_F for PAHs under study and their hydrophobicities (r^2 > 0.97) supports the partitioning mechanism. Although the PDMS coating is a very hydrophobic polymer, its PAH-partition coefficients are only slightly larger than those of the HA. The average solubility parameter of PDMS estimated from the measured K_F values is δ_{PDMS} = 7.6 ± 1.0 (cal cm^{-3})$^{0.5}$, which is close to the literature value[28] (7.1 (cal cm^{-3})$^{0.5}$). In other words, PDMS is too non-polar for optimal PAH partitioning whereas HOM is too polar.

3.2 Polychlorinated Biphenyls

Manufacture of PCBs was halted about 20 years ago. Nonetheless, due to their persistence, their formation in combustion processes and their introduction in freshwater systems via the atmosphere, PCBs still are pollutants.[10] The toxicity of the 209 congeners depends on their degree of chlorination and spatial substitution.[11] As is typical for hydrophobic pollutants, PCBs tend to adsorb on non-polar surfaces and to accumulate in lipophilic materials along the aquatic and terrestrial food chain.

High resolution GC combined with selective and sensitive detection methods including electron impact MS (EI-MS), negative chemical ionisation MS (NCI-MS) or atomic emission detection (AED, using the chlorine emission line at 479 nm) are highly sophisticated tools for PCB congeners analysis. Very selective second generation AED allows the detection of 14 pg/sec and the MS sensitivity in the SIM acquisition mode is below 1 pg/sec. However, the soft NCI-ionisation technique is not very convenient for detecting PCB congeners with low chlorine content Cl_1-Cl_3. Substantial target analyte enrichment is necessary to reach ppt-(and sub-ppt) PCB levels in aqueous matrices. The very high hydrophobicity of many PCB congeners (log K_{OW} in Table 2) results in high fibre distribution coefficients with non-polar coatings, and PDMS fibres are the media of choice. Therefore, using SPME to measure PCB sorption on HOM allows analysis at challenging trace levels.[1] Because of the large fibre distribution coefficients, the volume of the sample vials should be at leat 250 mL; otherwise, extensive fibre uptake disturbs the sorption equilibrium.

Figure 5 shows selected PCB congeners sampled by SPME in pure water (top) and in a DOM solution (bottom). Detection was by NCI-MS using methane as the moderating gas. As shown for selected PCB congeners in Europe's regulation measures (Table 2), a strong linear correlation between sorption coefficients on DOM and octanol-water coefficients is evident. However, the slope of the regression line is about 0.5. Obviously, the increasing hydrophobicity of PCB congeners with increasing numbers of chlorine atoms is not adequately reflected in their sorption coefficients. Because of the lack of solubility parameters for highly chlorinated PCB congeners, we cannot apply the Flory-Huggins concept to this class of HOC. However, our data agree well with PCB sorption data measured on sediments.[13] As with PAHs, partition coefficients remain constant over a wide range of solid-to-solution ratios. This allows sorption to be viewed as a phase partitioning process that is quantifiable with a single K_{OC} or K_{OM} equilibrium constant.

Use of the analytical SPME to determine aqueous PCB congener concentrations or the "physico-chemical" SPME to determine sorption coefficients has problems because of high PCB hydrophobicity. Unfortunately, the literature has many misleading PCB fibre distribution coefficients. In these reports, the PCB concentration is claimed to be the concentration spiked via a PCB stock solution into the SPME vial filled to capacity with water. In reality, the concentration of dissolved PCB is much lower due to sorption on glass walls, the stirrer, the vial cap and so on. As a result, the denominator of eq 3 is much smaller than assumed. Findings that have been explained by physical adsorption of PCB congeners on the fibre (i.e. a surface effect) can in fact often be ascribed to adsorption effects in the system.

To circumvent these problems we applied a recently introduced dynamic SPME approach that will be detailed elsewhere.[14] Briefly, a generator column is loaded with PCB congeners under study and flushed with water at high flow rates (about 3 mL/min). After equilibrium is reached the concentration of any PCB congener is equal to its water solubility. Undesired PCB losses are compensated for by redelivering analytes from the generator column. Moreover, no organic solvent (e.g. from the PCB stock solution) that may change the PCB congener activity and falsify the fibre distribution coefficient is introduced into the system. With these precautions the fibre distribution coefficients correlate well with PCB hydrophobicities (data not shown here).

When measuring PCB sorption coefficients for DOM solutions, the fibre extracts the freely dissolved PCB fraction from the circulating DOM solution. The total PCB congener concentration (see denominator of eq 3) can be determined by adding the isotopically labelled congener into the SPME vial (at a concentration about 1/5 of the theoretical water solubility so as not to exceed it dramatically), followed by SPME and GC/MS monitoring

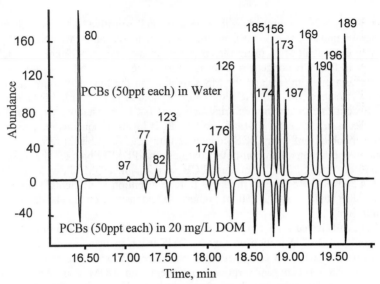

Figure 5 *NCI-MS chromatogram of selected PCB congeners with Cl_4 to Cl_8. Sampled by SPME from water (top) and DOM solution. SPME: 7 μ PDMS, 250 mL vial with polyethylene cap (no silicone or teflon septum), sampling time 15 hrs ("overnight" procedure), stirring rate 500 rpm; DOM: humic acid isolated from coal wastewater, C_{DOM} = 20 mg/L; oven: 50°C (3') to 300°C with linear temperature program 12°C/min; Data acquisition: SIM*

Table 2 *Sorption coefficients and hydrophobicity of selected PCB congeners*

PCB congener (Cl number)	Log K_{OW}	Log K_{DOM}^a (HA)
28 (3)	5.62	4.3
52 (4)	6.26	4.6 (4.46[b])
118 (5)	7.12	4.9(4.9[c])
153 (6)	7.44	5.3
180 (7)	8.26	5.6

[a] Average of 3 measurements with standard deviations from 0.20-0.35; [b] Data from ref. 12 for Aldrich HA; [c] data from ref. 5.

of traces of the labelled and non-labelled PCB congener. This strategy is based on the valid assumptions that the non-labelled and labelled congener are sampled by the fibre in the same manner and that equilibrium on DOM establishes quickly. The data in Table 2 were measured in this way.

3.3 Nitroaromatic Compounds

Nitroaromatic compounds and their transformation products, including nitrobenzenes/ toluenes, nitrophenols and nitroanilines are toxic. They can be mobilised in soil and flushed into the groundwater because of their high water solubility. Nitroaromatics mainly have been used in explosives and pesticides applications. It is important to control their mobility and to assess their fate in the subsurface. Also important is to study the sorption behaviour and sorption mechanisms with particulate and dissolved matrices.

The sorption of nitroaromatic compounds on dissolved and particulate HOM was investigated with the SPME approach using polar polyacrylate (PA) fibres instead of non-polar PDMS fibres. The sorption coefficients of nitroaromatics on the anthropogenic, sediment-originated HA are shown in Figure 6 with the interpolated log K_{DOC}-log K_{OW} line for PAHs. Data for BTX aromatics with hydrophobicities similar to nitroaromatics fall below this line. A plausible explanation within the framework of our modified Flory-Huggins concept is given by lower BTX solubility parameters than for PAHs (for example, δ = 8.9 (cal cm^{-3})$^{0.5}$ for toluene vs. δ = 9.9 (cal cm^{-3})$^{0.5}$ for naphthalene and anthracene).[28]

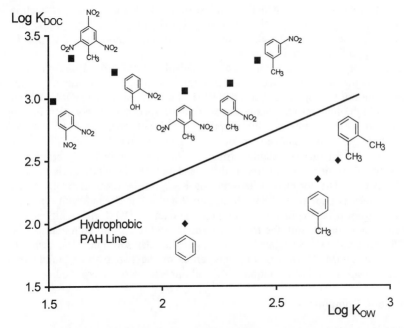

Figure 6 *Sorption of nitroaromatic compounds on DOM vs. hydrophobicity (DOM, isolated from the associated coal wastewater sediment)*

The solubility parameters of BTX aromatics are more distant from those of the HOM under study (δ = 11.3 (cal cm^{-3})$^{0.5}$). Thus, the solubility parameters of PAHs and HOM are more compatible than BTX and HOM.

Sorption of nitroaromatic compounds with electron-withdrawing and electron-delocalising substituents is much stronger than expected from non-polar hydrophobic interactions (Figure 6). There appears to be a significant contribution to overall sorption from specific polar interactions. Both contributions are important. Surface electron donor-acceptor complexation or ion-exchange reactions with the charged inorganic surface must be considered for HOM matrices coated on mineral surfaces.[15,16]

As mentioned above, the SPME approach can be applied to compare dissolved and particulate HOM by eliminating method-related biases. Problems with some nitroaromatic compounds that are difficult to analyze by GC (for example, nitrophenols, dinitrophenols and nitroanilines) may be avoided by SPME-HPLC coupling or by in-fibre analyte derivatization. These approaches were outlined in ref. 17 and will be detailed elsewhere.

3.4 s-Triazines

The sorption of widely used s-triazine herbicides on soil HOM has been well documented.[18,19] However, interactions with DOM and DOM in soil porewater that influence triazine bioavailability and cause groundwater hazards have not been subjects of intensive research.

Figure 7 shows the GC analysis of selected triazines with positive CI detection. This selective and sensitive detection method was very useful for investigating triazines in highly contaminated environmental samples. Use of SPME as an extracting and preconcentrating technique in combination with PCI-MS allows entry to the ppb- and sub-ppb level (Figure 8, top) and meets groundwater regulatory guidelines (0.1 ppb for one pesticide, 0.5 ppb for the sum of triazines).[21]

Triazine sorption on a commercially available HOM (Fa. Roth, Germany) is illustrated in Figure 8. As expected, triazines sorption coefficients are orders of magnitude lower than those of hydrophobic PAHs and PCBs. For this reason the DOM concentration was set very high (400 ppm) to obtain pronounced differences in the polar PA fibre uptake in water and DOM solution. Our sorption coefficients measured on DOM were significantly larger than available sorption coefficients measured on soils (Table 3). This is most noticeable for sulfur-containing triazines (ametryne, prometryne, terbutryne) with differences of about one log unit (recall that our data refer to DOM, while the literature data refer to OC). The correlation between log K_{DOC} on the commercial DOM and log K_{OW} of triazines is poor ($r^2 = 0.83$; see Figure 9 and Table 3), indicating a significant contribution from polar interactions. Although the data on DOM are significantly larger compared to those on the soil, the trend of sorption coefficients on both matrices is quite similar for all triazines investigated. This suggests similar sorption mechanisms for the DOM and soil HOM. However, some properties of the two matrices might be quite different and a substantial contribution of the soil mineral matrix is expected.

3.5 Organotin Compounds

Organotin compounds and their breakdown products, including mono-, di-, tri- and tetrabutyltin and triphenyltin, have harmful environmental effects.[22] Partitioning of ionogenic organotin species (e.g. mono-, di- and tributyltin) between particulate matter and water and their sorption onto sediment and soil matrices is complicated by the existence of cationic and neutral species. This has been a subject of intensive research.[23]

Sorption of non-ionogenic tetraalkylated organotin species (e.g. tetrabutyltin) on DOM can easily by measured with the SPME approach. The partition coefficient of tetrabutyltin on DOM is similar to that of pyrene.[24]

Sorption of organotin on particles can be measured by the batch approach with a centrifugation step to separate the phases. After phase separation, sodium tetraethylborate (STEB) is added to the aqueous buffered phase at pH = 4.5 to allow the ionogenic organotins (for example, monobutyl through tributyl tin) to be ethylated, thus making them amenable to GC. The freely dissolved organotin fraction can be extracted by SPME sampling. This provides the basis for K_{OM} calculation from eq 1 with external calibration.

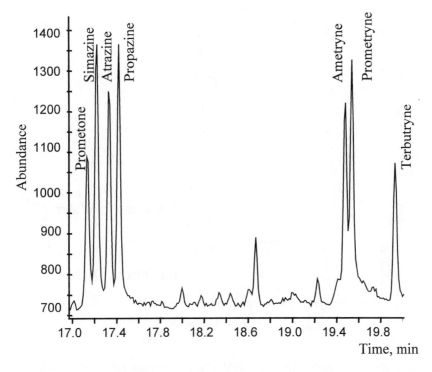

Figure 7 *Detection of symmetrical triazines (each 50 ppb, standard solution in acetone) on HP-5; injection: 1 μL splitless; detection: positive CI using methane as moderating gas, SIM acquisition mode*

Table 3 *Sorption coefficients and hydrophobicity of s-triazines on DOM*

Triazine	Log K_{OW}	Log K_{DOM}[a] (HA)
Prometone	2.97	2.72 (2.19[b])
Simazine	2.18	2.15 (2.02[b])
Atrazine	2.61	2.68 (2.19[b], 2.33[c])
Propazine	2.93	2.78 (2.14[b])
Ametryne	2.98	3.35 (2.50[b])
Prometryne	3.51	3.54 (2.66[b])
Terbutryne	3.38	3.84 (3.17[b])

[a] Average of 3 measurements with standard deviations from 0.15 – 0.30; [b] data from ref. 20 for soil HOM; [c] data from ref. 19 for soil HOM.

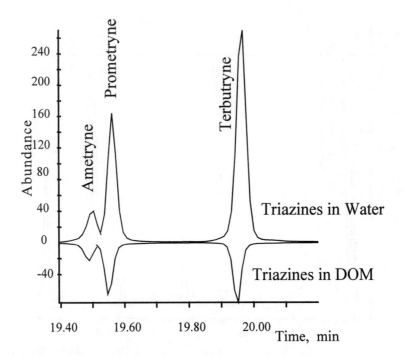

Figure 8 *Sorption of selected triazines on DOM (C_{DOM} = 400 ppm, pH = 6.0). SPME:*
polar polyacrylate fibre, sampling time 3 hrs, 1 ppb each triazine in a 40 mL
vial, stirring rate 500 rpm; GC: see legend of Figure 7; detection by PCI in SIM
acquisition mode, target ions m/z = 242 for both prometryne and terbutryne, m/z
= 228 for ametryne

Interactions between organotin cations and deprotonated carboxyl and phenol groups of
HOM contribute significantly to the overall interaction.

The same applies to DOM under natural water conditions (for example, the acidity
constant pK_a of tributyltin is 6.3).[25] Partition coefficients are governed mainly by
hydrophobic interactions in soils and sediments that are rich in organic carbon (OC $\cong$ 25)
and have low polar groups content.[24] The influence of mineral surfaces becomes more
important with decreasing OC content, for instance via exchange of organotin cation
species.[23]

Conversion to GC-amenable derivatives is necessary in studies of sorption of
ionogenic organotins on DOM by SPME with GC analysis. However, an *in situ*
derivatization would affect the sorption equilibrium (for example, sorption of ethylated
organotin would be measured when using STEB for derivatization). Therefore,
derivatization should occur on the fibre or on desorption in the hot GC injector.

Our experiments indicate that sodium tetraphenylborate is an appropriate reagent for
in-fibre derivatization: Firstly, a Carbowax fibre was used to sample the freely dissolved
ionogenic and nonionogenic organotin species from the DOM solution. The fibre was then
loaded with a 0.5 % sodium tetraphenylborate solution for some minutes. Lastly the fibre
was inserted into the hot GC injector to finish the reaction. The determination of sorption
coefficients was conducted using an external calibration as described above.

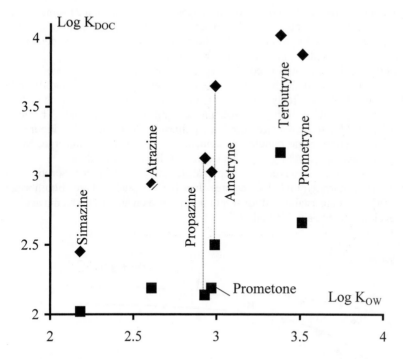

Figure 9 *Sorption of triazines on DOM (♦ our results) and on soil (■ literature) vs. hydrophobicity*

Although this method of "in-fibre derivatization" works in principle, the results scatter widely. The log K_{DOM} value of tributyltin on the dissolved anthropogenic HA was estimated to be 4.7 ± 0.4 at pH = 5.5. This is similar to the value measured on Suwannee River Reference HA (log K_{DOM} = 5.1 at pH = 5.8) measured by the dialysis method.[25]

In contrast to neutral HOC sorption, studies of sorption of ionic (e.g. tetraalkylammonium) or ionizable (e.g. trialkylorganotin and quinolines) species on DOM is much more difficult because of the large impact of pH and ionic strength. Therefore, sorption data should be compared with literature data measured under different conditions.

3.6 Sorption Kinetics on DOM

Kinetic data for sorption on DOM are very scarce in comparison to soil and sediment data. Sorption on porous geosolids is usually characterised by an initial rapid step followed by slow approach towards equilibrium, for example with organotin species on a sediment very rich in HOM.[24] The interaction of HOCs with dissolved polymers is supposed to be fast because physisorption does not require high activation energies and long diffusion pathways.[26]

The relatively slow SPME approach had to be modified to study the sorption kinetics (see section 2). To circumvent the poor reproducibility of a fast, non-equilibrium SPME, the ion traces of the target analytes (non-deuterated naphthalene, phenanthrene and pyrene) were referred to the corresponding deuterated surrogate. The non-deuterated sorbates were added to the DOM solution (anthropogenic FA, see section 2) some 15 hrs before adding the deuterated standards. This ensured that the sorption equilibrium was already established based on the assumption that the target peak intensity of the deuterated and non-deuterated PAH is identical and that no analyte losses occur.

Figure 10 shows the fast sorption equilibrium with phenanthrene and pyrene. An ion trace ratio (signal intensity of the molecular ions of the two isotopomers) significantly below 1.00 shows that the freely dissolved concentration of the deuterated PAH immediately after its addition is higher than under equilibrium conditions. The non-deuterated PAH have definitely reached sorption equilibrium. As illustrated in Figure 10, sorption is completed after some minutes, as evidenced by the constant ion trace ratios $SI_{m/z = 178}/SI_{m/z = 188}$ and $SI_{m/z = 202}/SI_{m/z = 212}$. This agrees well with earlier data.[27]

Similar kinetics were observed for naphthalene and phenanthrene sorption on dissolved HAs with more hydrophobic features than the FA studied here. Equilibrium needs about 1 hr to become established for sorption of pyrene on an aquatic humic acid. A model was developed to explain this finding.[2]

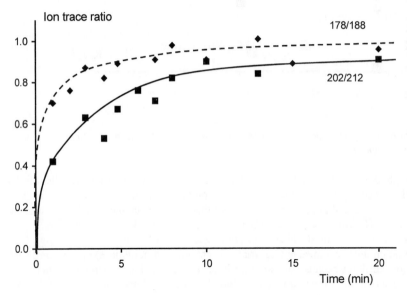

Figure 10 *Kinetics of phenanthrene and pyrene sorption on DOM*

4 CONCLUSIONS

SPME is a very useful method for determining coefficients and kinetics of sorption of organic compounds with a wide range of hydrophobicity and polarity on DOM. Advantages of SPME over traditional approaches include ease of handling, applicability to multicomponent analysis, the possibility of studying sorption with particulate and dissolved HOM matrices and preservation of the integrity of sorption processes.

Sorption of PAHs and PCBs is clearly dictated by hydrophobic partitioning. Strong correlations between sorption coefficients and sorbate hydrophobicity were observed. By contrast, specific interactions also have to be considered in the sorption of triazines, nitroaromatics and organotin compounds on DOM.

The SPME approach may be extended to further classes of compounds of environmental concern that cannot be directly subjected to GC. Alternatives include SPME-HPLC coupling and in-fibre derivatization.

References

1.   R. G. Luthy, G. R. Aiken, M. L. Brusseau, S. D. Cunningham, P. D. Gschwend, J. J. Pignatello, M. Reinhard, S. J. Traina, W. J. Weber, Jr. and J. C. Westall, *Environ. Sci. Technol.*, 1997, **31**, 3341.
2.   F. -D. Kopinke, J. Poerschmann and A. Georgi; in 'Applications in Solid Phase Microextraction', J. Pawliszyn, (ed.), Royal Society of Chemistry, Cambridge, Ch. IX, in press.
3.   J. Pawliszyn, 'Solid Phase Microextraction – Theory and Practice', Wiley-VCH, New York, 1998.
4.   F. -D. Kopinke, J. Pörschmann and M. Remmler, *Naturwiss.*, 1995, **82**, 28.
5.   J. Poerschmann, Zh. Zhang, F. -D. Kopinke and J. Pawliszyn, *Anal. Chem.*, 1997, **69**, 597.
6.   C. E. Clapp and M. H. B. Hayes, in 'Humic Substances and Organic Matter in Soil and Water Environments', C. E. Clapp, M. H. B. Hayes, N. Senesi and S. M. Griffiths, (eds.), International Humic Substances Society, St. Paul, MN, 1996, p. 3.
7.   R. E. Engebretson and R. von Wandruszka, *Environ. Sci. Technol.*, 1998, **32**, 488.
8.   R. E. Engebretson, T. Amos and R. von Wandruszka, *Environ. Sci. Technol.*, 1996, **30**, 990.
9.   F. -D. Kopinke, J. Poerschmann and U. Stottmeister, *Environ. Sci. Technol.*, 1995, **29**, 941.
10.   H. Fiedler and C. Lau, in: 'Ecotoxicology', G. Schüürmann and B. Markert, (eds.), Wiley, Heidelberg, 1997, Ch. 11, p. 317.
11.   R. Looser and K. Ballschmiter, *Fres. J. Anal. Chem.*, 1998, **360**, 816.
12.   R. M. Burgess and S. A. Ryba, *Chemosphere*, 1998, **36**, 2549.
13.   P. M. Gschwend and S. C. Wu, *Environ. Sci. Technol.*, 1985, **19**, 90.
14.   J. Poerschmann and M. Hoyer, 'Ultratrace Detection of PCBs using SPME and NCI-Mass Spectrometry', Pittcon '99, Orlando, FL, March 1999, Contribution 1452.
15.   S. K. Xue, I. K. Iskandar, and H. M. Selim, *Soil Sci.*, 1995, **160**, 317.
16.   S. B. Haderlein, K. W. Weissmahr and R. P. Schwarzenbach, *Environ. Sci. Technol.*, 1996, **30**, 612.
17.   J. Poerschmann and C. Maier, 'Determination of Nitroaromatic Compounds in Aqueous Solutions by Means of SPME', Pittcon '98, New Orleans, LA, March 1998, Contribution 257.
18.   C. G. Zambonin, F. Catucci and F. Palmisano, *Analyst*, 1998, **123**, 2825.
19.   A. B. Paya, P. A. Cortes, M. N. Sala and B. Larsen, *Chemosphere*, 1992, **25**, 887.
20.   A. Kaune, R. Brüggemann, M. Sharma and A. Kettrup, *J. Agric. Food Chem.* 1998, **46**, 335.
21.   U. Dörfler, E. A. Feicht and I. Scheunert, *Chemosphere*, 1997, **35**, 99.
22.   K. Fent, *Crit. Rev. Toxicol.*, 1996, **26**, 1.
23.   A. Weidenhaupt, C. Arnold, S. R. Müller, S. B. Haderlein and R. P. Schwarzenbach, *Environ. Sci. Technol.*, 1997, **31**, 2603.

24.   J. Poerschmann, F. -D. Kopinke and J. Pawliszyn, *Environ. Sci. Technol.*, 1997, **31**, 3629.
25.   C. Arnold, A. Ciani, S. R. Müller, A. Amirbahman and R. P. Schwarzenbach, *Environ. Sci. Technol.*, 1998, **32**, 2976.
26.   J. J. Pignatello, F. J. Ferrandino and L. Q. Huang, *Environ. Sci. Technol.*, 1993, **27**, 1573.
27.   M. A. Schlautman and J. J. Morgan, *Environ. Sci. Technol.*, 1993, **27**, 961.
28.   A. F. M. Barton, 'CRC Handbook of Solubility Parameters and other Cohesive Parameters', CRC Press, Boca Raton, FL, 1985, p. 257.

HUMASORB-CS™: A HUMIC ACID-BASED ADSORBENT TO REMOVE ORGANIC AND INORGANIC CONTAMINANTS

H. G. Sanjay, A. K. Fataftah, D. S. Walia and K. C. Srivastava

Arctech, Inc., Chantilly, VA 20151, USA

1 INTRODUCTION

Contamination of water resources is a serious world wide problem. Toxic organic and inorganic chemicals such as metals, radionuclides, oxoanions and hydrocarbons are present in water. The contamination is primarily due to industrial wastes disposal. The presence of toxic heavy metals, volatile organic compounds (VOCs) and pesticides affects the safety of drinking water. Decontamination of surface and groundwater can be achieved with treatments such as acid precipitation, ion exchange, microbial digestion, membrane separation and activated carbon adsorption. These approaches treat only one class of compounds at a time, which results in complex and costly processing steps. A typical approach is to remove organics with activated carbon and then to remove metals by ion exchange. Typical remediation thus requires the use of two or more stepwise processes.

Activated carbon adsorbs organic compounds onto its highly porous surface. Powdered and granular activated carbon is used in these applications. The organic compound removal capacity of the carbon depends on its available surface area. Contaminated water may be introduced at the top (fixed bed) or at the bottom (expanded bed) of the holding vessel. The activated carbon columns require regeneration to minimize the cost. Activated carbon cannot remove metal ions and is seldom used as a stand-alone technology in treating groundwater. For example, in a recent field test at Rocky Flats, the Department of Energy (DOE) concluded that use of activated carbon to capture radionuclides and chlorinated solvents would be difficult and pointless.

Ion exchange resin systems depend on chemical reactions in which ions contained in the water are exchanged with ions in the resin. There are four major types of ion exchange resins: strong acid, weak acid, strong base and weak base. Strong acid resins are effective in removing heavy metals and are employed in treating wastewater. Weak acid resins are mainly used in special chemical processes, for example to isolate alkaloids and antibiotics. Strong base resins will remove all mineral acids. Weak base resins will selectively remove strong mineral acids (hydrochloric, sulfuric and nitric) and allow weak mineral acids (carbonic and silicic) to pass through. No single ion-exchange resin will remove all metals and radionuclides. The ion-exchange resin must therefore be selected based on the specific ions in the waste water.

HUMASORB-CS™ was developed by Arctech as a single step process to remediate water that contains mixed waste contaminants. The new material is based on humic acid (HA) derived from coal. It has the following desirable properties: high cation exchange capacity, ability to chelate metals,[1,2] reduction of metal species[3-5] and ability to sorb organics.[6-8] These natural mechanisms work simultaneously, not sequentially. HAs are complex macromolecules containing a number of functional groups, including carboxylic acid, phenol, aliphatic, enol-hydroxyl and carbonyl structures. However, a typical HA is soluble in water above pH 2. HUMASORB-CS™ was developed by cross-linking and immobilizing a coal-derived humic acid. HUMASORB-CS™ is a stable and water insoluble adsorbent that retains the properties of HAs to bind metals, radionuclides and organic contaminants from aqueous waste streams in a single processing step.[9]

2 MATERIALS AND METHODS

The starting material for the development of solid HUMASORB-CS™ is a dissolved HA extracted from coal and termed HUMASORB-L™. Purified humic acid was isolated by precipitation of HUMASORB-L™ by addition of HCl. It is labeled HUMASORB-S™. HUMASORB-CS™ was produced by cross-linking and immobilizing HUMASORB-L™ with proprietary methods.[9]

The solubility of HUMASORB-CS™ was determined at pH 3-12 and compared with that of HUMASORB-S™. In these tests, HUMASORB-CS™ (0.5 grams) was mixed with water and the pH was adjusted to the various pHs with sodium hydroxide or hydrochloric acid. The mixture was shaken at 300 rpm and 25°C for two hrs. The mixture then was centrifuged. The liquid phase pH was measured and analyzed for humic acid by reducing the pH to < 2 with HCl, washing the precipitated HA with water and weighing the dried solid. The solubility tests also were conducted in water in the presence of 100 ppm Na_2SO_4 or 100 ppm Na_2CO_3 (1-day, 5 months), 100 ppm $CaSO_4$ or 100 ppm $CaCO_3$ (1-day, 5 months), and 10,000 ppm Na_2SO_4 or 10,000 ppm Na_2CO_3 (3 months).

Stability tests were performed by taking approximately two grams of HUMASORB-CS™ in 50-mL centrifuge tubes. 10 mL of tap water was added. The mixture was then subjected to different temperatures for different periods of time. At the end of the test period the mixture was centrifuged. The liquid phase was analyzed for humic acid as described above. The amount of dry humic acid obtained was used to estimate the solubility of HUMASORB-CS™. The solid phase was then dried in an oven at 50°C and used in batch tests to evaluate removal of a target contaminant (Cr(III)). The stability tests conducted in water at different temperatures include ambient conditions (1, 4 and 6 months), 4°C (3, 5.5 and 6.5 months) and 50°C (1, 2, 3, and 4 weeks).

Simulated waste solutions were prepared by dissolving the desired contaminants in water with no pH adjustment. The batch tests designed to measure contaminant removal were conducted by shaking the simulated waste solution (25 mL) and HUMASORB-CS™ or HUMASORB-S™ (one gram) at 300 rpm and 25°C for the desired contact time. The mixture was then centrifuged. The liquid phase was analyzed for the metals by inductively coupled plasma (ICP) or atomic absorption (AA) spectroscopy. Gas chromatography with electron capture detection (GC-ECD) and gas chromatography-mass spectrometry (GC-MS) were used to analyze for organics.

Column tests were conducted in glass columns with an internal diameter of 22 mm and an approximate bed height of 20 cm. The columns were packed with 80% sand and

20% HUMASORB-CS™ on a weight basis. The sand and HUMASORB-CS™ were uniformly mixed and wet-packed into the column. The packed column was visually inspected for uniform distribution of HUMASORB-CS™. Simulated waste streams were passed through the columns in downflow mode via gravity flow. The flow rate was adjusted using a valve at the column outlet. Column tests were conducted with simulated waste streams at relatively similar rates defined as Empty Bed Contact Time (EBCT). EBCT is the time required for the fluid to pass through the volume occupied by the adsorbent bed. EBCT and the bed volumes used in these tests are based on the volume occupied by dry HUMASORB-CS™ in the column. The amount of HUMASORB-CS™ in the column and the bulk density (~1 gm/mL) were used to estimate the volume to calculate EBCT.

Zero-valent iron was incorporated into the HUMASORB-CS™ by proprietary methods at 2.5 % and 7.5 % on a weight basis. A simulated waste stream containing tetrachloroethylene (TCE) and trichloroethylene (PCE) was contacted in 35-mL zero-head space vials with 1.5 gms of the solid material (either HUMASORB-CS™ (H-CS) or HUMASORB-CS™ with 2.5 % zero-valent iron (2.5%ZVI+H-CS) or HUMASORB-CS™ with 7.5 % zero-valent iron (7.5%ZVI+H-CS)). After contact times of 12 and 24 hrs the vials were centrifuged at 2000 rpm. The liquid phase was then analyzed for TCE and PCE using GC-ECD. Controls containing TCE and PCE were also analyzed.

3 RESULTS AND DISCUSSION

HUMASORB-S™ is insoluble in water at lower pH but will dissolve at higher pH in the presence of monovalent metal ions such as sodium and potassium. Cross-linked humic acid polymer HUMASORB-CS™ was produced to overcome this limitation and lower the solubility at higher pH values. The dissolved humic acid HUMASORB-L™ and the purified humic acid HUMASORB-S™ were evaluated for metal removal to provide baseline data for comparison with data for the water insoluble, cross-linked product HUMASORB-CS™.

3.1 Solubility and Stability

The solubility of HUMASORB-CS™ at various pH values was evaluated. Figure 1 indicates that the solubility of HUMASORB-CS™ determined under the conditions of this study is significantly lower than that of HUMASORB-S™. The pH quoted in Figure 1 is that of the liquid phase measured after the stated contact time.

The stability of HUMASORB-CS™ was evaluated with solubility under various test conditions as a criterion. In addition, HUMASORB-CS™ was tested in batch mode to evaluate its effectiveness for removal of a target contaminant (chromium(III)) after being subjected to the stability test.

The stability test results show that the solubility of HUMASORB-CS™ was less than 0.5 %, indicating that HUMASORB-CS™ is stable under the conditions used in this study. However, HUMASORB-CS™ was soluble in the tests with 10,000 ppm sodium carbonate. HUMASORB-CS™ from the stability tests was used to evaluate chromium(III) removal from a simulated waste stream. More than 90% of chromium(III) was removed as shown in Figures 2-4. The tests clearly show that HUMASORB-CS™ is not only stable under the conditions evaluated in this study, but it also retains ability to remove

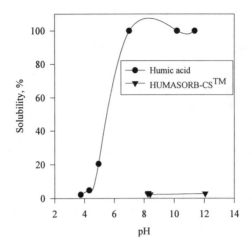

Figure 1 *HUMASORB-CS™ insolubility over a wide pH range*

contaminants (as shown with chromium(III)) from contaminated waste streams.

3.2 Removal of Inorganic Contaminants

Metals are bound to the carbon skeleton of a humic acid through heteroatoms such as nitrogen, oxygen or sulfur. The most common metal binding occurs via carboxylic and phenolic oxygen, but nitrogen and sulfur also have a positive effect on metal binding. The cation exchange capacity (CEC) of humic acid derived from Leonardite is 200-500 meq/100 grams, whereas the CEC of Leonardite is only 50 meq/100 grams.[1,2] The removal of metals by HUMASORB-CS™ was evaluated by contact with simulated waste streams. Similar tests were conducted using the purified humic acid HUMASORB-S™. The results of testing both materials were compared and used to show that the cross-linked material (HUMASORB-CS™) retains the properties of humic acid. The HUMASORB-CS™ material was tested for metal removal in both batch and column modes.

Batch tests were conducted with HUMASORB-CS™ and with HUMASORB-S™ at pH 2-2.5. The pH was not adjusted in these tests to avoid any competition for binding sites by metals such as sodium (in sodium hydroxide) or calcium (in calcium hydroxide) used to adjust the pH. The tests were conducted using 100 ppm of metal (As, Cd, Ce, Cs, Cr, Cu, Pb, Hg, Ni, Sr, U, Zn) in the waste stream. Experiments were conducted with only one metal in solution and with solutions containing all 12 metals at 100 ppm each.

The results from the batch tests are presented in terms of percent removal of metals (Figures 5 and 6). A comparison of results from tests with waste streams containing a single metal using the two forms of HUMASORB™ clearly shows that metal removal is higher with HUMASORB-CS™ for most metals (Figure 5). However, with a waste stream containing multiple metals (Figure 6), HUMASORB-S™ is more effective for a few of the metals (copper, lead and mercury). It is clear from the batch test results that the proprietary methods used to produce HUMASORB-CS™ improve the solubility characteristics while at the same time retaining the ability of HUMASORB-L™ and HUMASORB-S™ to remove metals from aqueous solutions.

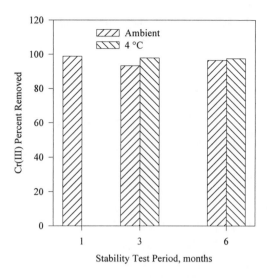

Figure 2 *Chromium(III) removal after storage at ambient temperature and 4°C*

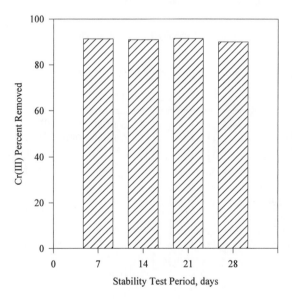

Figure 3 *Chromium(III) removal after storage at 50°C*

HUMASORB-CS™ was also evaluated in batch mode for removal of contaminants such as lead and chromium in the presence of high concentrations of background metals such as calcium. In these tests, simulated waste streams were prepared by spiking water with approximately 10,000 ppm calcium and nearly 200 ppm of either lead or chromium(III). The results of these tests (Table 1) clearly show the effectiveness of HUMASORB-CS™ for toxic metal removal even in the presence of high background concentrations of metals such as calcium. The increase in calcium concentration is due to the presence of calcium in the HUMASORB-CS™ matrix.

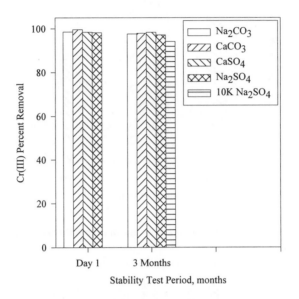

Figure 4 *Chromium(III) removal after storage in the presence of anions*

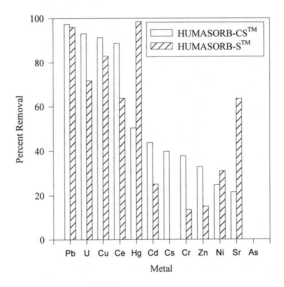

Figure 5 *Removal of individual metals using HUMASORB-S™ and HUMASORB-CS™*

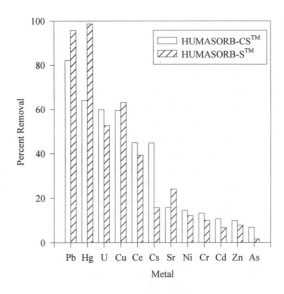

Figure 6 *Removal of multiple metals using HUMASORB-S™ and HUMASORB-CS™*

Table 1 *Metal removal by HUMASORB-CS™ in the presence of high background concentration of calcium*

Contaminant	Simulated waste concentration, ppm		
Waste stream containing Cr, Ca	Initial pH 3.56	Final pH 5.18	Removal %
Chromium(III)	196	28.5	85.5
Calcium	11,300	13,500	-
Waste stream containing Pb, Ca	Initial pH 5.70	Final pH 7.58	Removal %
Lead	169	4.78	97.2
Calcium	9,700	11,100	-

Most adsorption/ion-exchange operations involve continuous processes with the media packed in columns. The objective of our tests was to evaluate HUMASORB-CS™ in bench scale columns and to develop preliminary data that could lead to design of pilot and commercial scale applications.

The results from the column tests were used to develop breakthrough curves. In the breakthrough curves, the ratio of the column output to column input concentration is plotted against number of bed volumes passed through the column. The column was assumed to be saturated when the output concentration was nearly 95% of the input concentration. The breakthrough point was assumed when the output concentration was between 2-5% of the input concentration. The flow rates (and thus EBCT) selected were designed to allow for relatively quick breakthrough and saturation of the column for logistical reasons. The number of bed volumes passed at breakthrough and at saturation was used to estimate the breakthrough and saturation capacity of HUMASORB-CS™.

The breakthrough curve for removal of lead from a simulated waste stream is shown in Figure 7. In this experiment, approximately 2700 bed volumes of water contaminated with lead (20 ppm) was passed through the column. The contact time based on

HUMASORB-CS™ in the column was less than four minutes. As shown in Figure 7, there was no breakthrough of the contaminant after 2700 bed volumes. To obtain breakthrough of lead and to estimate the breakthrough and saturation capacity, the same column was used again with the input concentration of lead increased to 200 ppm. Lead breakthrough was observed after an additional 500-600 bed volumes were passed through the column. The column was approximately 70% saturated after 3600 bed volumes.

A column test was also conducted using a simulated waste stream containing cerium, a surrogate for radioactive plutonium. The input concentration of cerium in this test was 200 ppm. The breakthrough curve shown in Figure 8 indicates no breakthrough for at least 600 bed volumes; the column was saturated after nearly 1700 bed volumes.

Column tests with a simulated waste stream containing chromium(VI) were also conducted. The empty bed contact time (EBCT) based on the HUMASORB-CS™ in the column was approximately 20 min and the waste stream was passed through the column for approximately 50 hrs (approximately 180 bed volumes). Chromium(VI) breakthrough occurred after 20 bed volumes (Figure 9). The concentration profile of chromium(VI) and total chromium is identical and the concentration difference remains constant. This indicates the rapid removal of chromium(III) formed by reduction of chromium(VI) by HUMASORB-CS™. The number of bed volumes treated at breakthrough should increase if the contact time in the column is increased.

3.3 Removal of Organic Contaminants

Adsorption of organic chemicals by humic substances such as humic acids (HAs) has been studied extensively.[6-8] The studies include adsorption of non-ionic organics such as benzene, halobenzenes, chlorinated hydrocarbons such as tetra- and trichloroethylene (TCE and PCE), nitrogen compounds such as urea and anilines, polychlorinated biphenyls (PCBs), pesticides and herbicides. The mechanisms postulated for the adsorption of

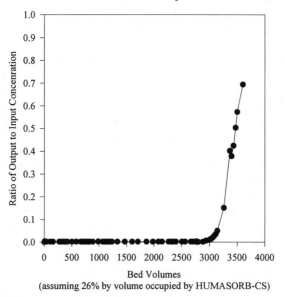

Figure 7 *Column breakthrough curve for lead*

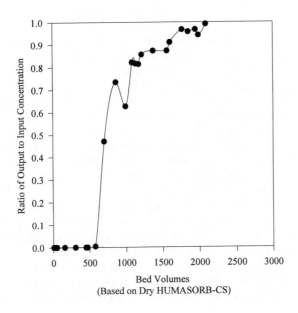

Figure 8 *Column breakthrough curve for cerium*

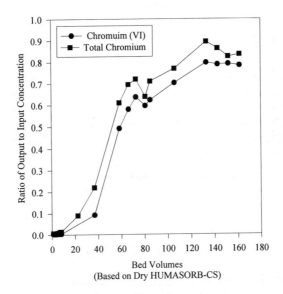

Figure 9 *Column breakthrough curve for chromium(VI)*

organic compounds include van der Waals attractions, hydrophobic bonding, hydrogen bonding, charge transfer, ion-exchange and ligand exchange.

Many batch and column tests performed using HUMASORB-CS™ to remove organic contaminants confirm the effectiveness of this material. Most of the tests were performed with simulated waste streams containing chlorinated hydrocarbons like TCE and PCE. Column studies were conducted with a simulated waste stream containing PCE. In that study, two tests were conducted with one of the columns pretreated with 1 N sulfuric acid before the test and the other washed with water. The tests were conducted for 48 hrs (approximately 80 bed volumes) and the samples collected were analyzed. There was no breakthrough of the contaminant PCE, as shown in Figure 10. However, since there was a time lag between the collection of samples and the analyses, the column could not be used again to continue the test. The results from the two columns indicate that HUMASORB-CS™ effectively removes organic contaminants and that there is no significant effect of column pretreatment under the conditions of this test.

Zero-valent iron (ZVI) is a new approach for removal of chlorinated organic compounds from contaminated groundwater. This new technology has been evaluated at sites contaminated with TCE and Cr(VI). Removal is believed to depend on the ability of of ZVI to reduce chlorinated hydrocarbons such as TCE and PCE. The performance of HUMASORB-CS™ in removing organic contaminants was evaluated after incorporation of ZVI into its matrix. The objective was to find the organic removal capacity after incorporation of ZVI and compare it with data for the original HUMASORB-CS™.

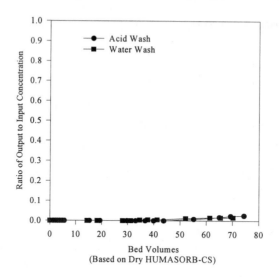

Figure 10 *Column breakthrough curve for PCE*

The test results show that TCE and PCE removal was improved by the incorporation of zero-valent iron into the HUMASORB-CS™ matrix (Figure 11) from about 60 % (H-CS) to about 90 % (2.5%ZVI+H-CS) and to about 93 % (7.5%ZVI+H-CS) when the contact time was 12 hours. At 24 hours contact time, however, the TCE and PCE removal was the same for all three materials (Figure 11). The results indicate that the effectiveness

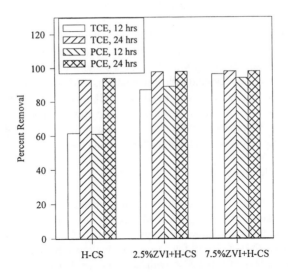

Figure 11 *TCE and PCE removal by HUMASORC-CS™ after incorporation of zero-valent iron after 12 and 24 hrs contact time*

of HUMASORB-CS™ is comparable to derivatives containing zero-valent iron if given adequate contact time.

Humic acids can function as reducing agents and influence the oxidation-reduction state of metal species such as chromium(VI), mercury, vanadium and plutonium.[5-7] This fact led us to investigate the ability of HUMASORB-CS™ to reduce chlorinated organic compounds. A preliminary study in collaboration with Temple University shows that HUMASORB-CS™ sorbs chlorinated organic contaminants and apparently reduces them in the same way as ZVI. These studies show that the reaction of TCE with HUMASORB-CS™ may follow a similar mechanism to that with ZVI (reductive dehalogenation). In these tests, HUMASORB-CS™ was reacted with a simulated stream containing TCE. The reaction mixtures were analyzed with GC-MS, NMR and ion chromatography. The GC-MS results confirmed the disappearance of TCE and ethylene was detected in the headspace. The ion chromatography data accounted for 67-80% (eight trials) of chloride ion produced from reduction of TCE, assuming a one electron reduction reaction mechanism. The NMR studies show that the vinylic proton (characteristic of TCE) disappears completely from the reaction mixture after two hrs of contact time. The NMR spectra of the solid (dissolved in NaOD/D$_2$O) after exposure to TCE show new peaks in the alkyl halides region that may assigned to the sorbed degradation products. Also, these spectra do not contain vinylic proton features. This result indicates that HUMASORB-CS™ removes TCE by sorbing and then degrading it into non-chlorinated compounds such as ethylene.

4 APPLICATIONS OF HUMASORB-CS™

HUMASORB-CS™ can be deployed for remediation and for treatment of industrial waste streams to minimize pollution. HUMASORB-CS™ also can be used as part of a process for recovery of valuable resources such as metals and micronutrients from contaminated sites (see section 4.2). For groundwater remediation, HUMASORB-CS™ can be deployed in an existing pump and treat system. In addition, HUMASORB-CS™ can be deployed for in situ groundwater remediation in shallow and deep applications. In a shallow application it can be deployed in trenches or in funnel and gate systems. For deep applications it can be installed as a barrier wall with commercial drilling equipment.

4.1 Barriers Applications

HUMASORB-CS™ has been evaluated for contaminant removal under simulated barrier conditions at pressures of 10 psig and 100 psig. These conditions were used to simulate barrier installation depths of approximately 10 feet and 100 feet. The tests were conducted using only HUMASORB-CS™ in the barrier for shallow applications (pressure of 10 psig) and by mixing HUMASORB-CS™ with sand in the barrier used to simulate deeper applications (pressure of 100 psig). A simulated mixed waste stream containing chromium(VI), lead, copper, cerium and the two chlorinated hydrocarbons TCE and PCE was passed through the barriers.

The results show no breakthrough in the barriers for any of the contaminants, even after passing more than 200 bed volumes of the mixed waste stream through the shallow barrier (8 months) and 160 bed volumes through the deep barrier (7 months). Thus, HUMASORB-CS™ is effective in removing contaminants (chlorinated organics, metals and a radionuclide surrogate) to non-detectable levels under simulated barrier conditions at different depths in a single treatment step.

4.2 Berkeley Pit Demonstration

Berkeley Pit is a deep, highly polluted waste water deposit in Butte, Montana. The demonstration was designed to establish the applicability of the HUMASORB™ process to remove toxic heavy metals such as cadmium from Berkeley Pit water samples while producing a chelated micronutrient-enriched fertilizer product for agriculture applications. The process combines two diverse properties of coal-derived humic acid as an ion-exchange chelating material and as a soil amendment product to make a process that is both effective and yet simple to implement. In Stage 1, the Berkeley Pit water is treated with HUMASORB-L™ to remove iron and other agricultural micronutrients by formation of humates, which are precipitated as flocs. The precipitated complex is easily separated in a solid/liquid separation unit such as a centrifuge. Metals remaining in the water from Stage 1 and all other toxic metals are removed in Stage 2 using HUMASORB-CS™. Scoping, optimization and confirmation tests were conducted at Arctech and a Process Demonstration Unit (PDU) was designed based on the results. The PDU was used for a demonstration in Butte, Montana. The results from the demonstration were similar to those from the batch mode optimization and confirmation tests. The optimum combination of identified parameters was effective in a continuous process that meets the demonstration goals and objectives.

4.3 Treatment of Waste Brines

A HUMASORB-CS™ based treatment system was used recently at the Johnston Atoll facility of the Department of Defense (DOD) for treatment of spent decontamination solution (SDS) to remove lead, mercury and arsenic. Arctech mobilized a system on-site at Johnston Atoll and successfully treated more than 30,000 gallons of SDS to meet the levels set by the US Environmental Protection Agency (EPA).

5 COST BENEFITS OF HUMASORB-CS™

As indicated earlier, activated carbon typically is used to remove organics and ion exchange resins typically are used to remove metals in remediation efforts. The projected cost of HUMASORB-CS™ is $ 0.5-1.0 per pound. Arctech recently developed a life cycle cost savings analysis for potential remediation of contaminated groundwater at an Air Force site. The costs of state-of-the-art pump and treat systems versus deploying HUMASORB-CS™ technology as a passive treatment system or as a subsurface barrier were compared (Figure 12). It is clear from this cost comparison that HUMASORB-CS™ technology can be deployed not only at low cost but also at a substantial life cycle cost savings.

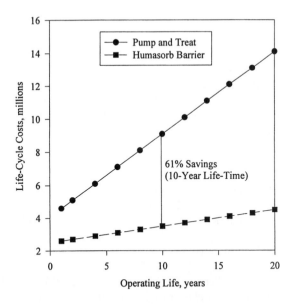

Figure 12 *Comparison of life cycle costs of pump and treat and HUMASORB™ technologies*

6 CONCLUSIONS

The objective in this work was to develop a sorbent/ion-exchange polymer based on the unique properties of HAs for binding and reacting with various types of contaminants. The goal was to obtain a material that is insoluble in water under basic conditions and retains the ability of HAs to remove multiple contaminants in a single process step.

HUMASORB-CS™ was developed using proprietary cross-linking and immobilization methods. The results demonstrate the improved solubility characteristics of HUMASORB-CS™ compared to the solid humic acid HUMASORB-S™. The contaminant removal ability was either retained or enhanced by crosslinking in HUMASORB-CS™. Column tests were used to develop breakthrough curves for different types of contaminants including metals, radionuclide surrogates, oxoanions and organics. In addition, the ability to regenerate and re-use HUMASORB-CS™ was evaluated. The test results demonstrate that HUMASORB-CS™ can be used for removal of multiple types of contaminants (metals, radionuclides and organics) in a single-step process and that it can be regenerated for re-use.

The results with organics indicate that the effectiveness of HUMASORB-CS™ is comparable to derivatives containing zero-valent iron (ZVI) if given adequate contact time. Additional test results confirm that HUMASORB-CS™ sorbs chlorinated organic contaminants and also reduces them in the same way as ZVI.

HUMASORB-CS™ installed in many modes of applications provides life cycle cost saving over other technologies. The results from barrier tests indicate that HUMASORB-CS™ is effective in removing contaminants (metals, organics and a radionuclide surrogate) under simulated barrier conditions at different pressure values in a single treatment step.

ACKNOWLEDGEMENT

The authors acknowledge, with thanks, the support and technical guidance provided by the Department of Energy over the course of this study. We also acknowledge Dr. Susan Jansen and her group at Temple University for testing the chlorinated organics destruction with HUMASORB-CS™ and the NMR, GC-MS and ion chromatography measurements.

References

1. L. Musani, P. Valenta, H. W. Nurnberg, Z. Konard and M. Branica, *Estuarine and Coastal Marine Science*, 1980, **II**, pp. 639-649.
2. J. E. Pahlman, and S. E. Khalafalia, 'Bureau of Mines report of investigations', 1988
3. K. Nash, S. Fried, A. M. Friedman and J. C. Sullivan, *Environ. Sci. Technol.,* 1981, **15,** 834.
4. J. J. Alberts, J. E. Schindler and D. E. Nutter, *Science*, 1974, **184**, 895.
5. A. Szalay, M. Szilagyi, *Geochim. Cosmochim. Acta*, 1967, **1,** 31.
6. G. G. Choudhary, *Toxicol. Environ. Chem.*, 1983, **6**, 127.
7. C. T. Chiou, in 'Humic Substances in Soil and Crop Sciences: Selected Readings', P. MacCarthy, C. E. Clapp, R. L. Malcolm and P. R. Bloom, (eds.), American Society of Agronomy, Madison, Wisconsin, 1990.
8. P. K. Ghosh, G. P. Gupta and P. K. Pal, *Technology* (Sindri), 1966, **3**, 156.
9. H. G. Sanjay, K. C. Srivastava and D. S. Walia, 'Adsorbent', US Patent 5, 906, 960 (1999).

STIMULATION OF PLANT GROWTH BY HUMIC SUBSTANCES: EFFECTS ON IRON AVAILABILITY

Y. Chen,[1] C. E. Clapp,[2] H. Magen[1] and V. W. Cline[3]

[1]Department of Soil and Water, Faculty of Agricultural, Food and Environmental Quality
 Sciences, The Hebrew University of Jerusalem, POB 12, Rehovot 76100, Israel
[2]USDA-ARS and Department of Soil, Water and Climate, College of Agricultural, Food
 and Environmental Sciences, University of Minnesota, St. Paul, MN 55108, USA
[3]The Toro Company, Bloomington, MN 55420-1196, USA

1 INTRODUCTION

Studies of the effects of humic substances (HSs) on plant growth under conditions of adequate mineral nutrition often have shown positive effects on plant biomass. Increases in root length and stimulation of the development of secondary roots have been observed for HSs in nutrient solutions (NS). The typical response curve shows enhanced growth with increasing HSs concentrations in the NS, followed by a decrease in growth at high concentrations of HSs. Shoots generally show similar trends in growth response to HSs but the magnitude of the growth response is smaller. Foliar sprays can also enhance both root and shoot growth. Humic substances can complex transition metal cations, which can sometimes result in enhanced uptake or competition with the roots for the metal and a decreased uptake.

Hormone-like activity was attributed to HSs in a number of studies.[1-3] Effects on various enzymes activities and on membrane permeability have also been suggested.[1,3-6] Work carried out by Chen et al.[7] utilizing precise and modern techniques showed that HSs do not contain plant growth regulators such as indoleacetic acid, cytokinin or abscisic acid.

Addition of organic matter (OM) to soils can stimulate growth beyond the enhancement provided by mineral nutrients, presumably because of the effects of HSs. Addition of Fe-enriched organic materials can alleviate lime-induced chlorosis. Application of HSs to soils in the quantities needed to significantly affect their concentration in the soil solution is not economic, but the response to foliar sprays has the potential to gain the interest of farmers since much smaller quantities are needed. These issues have been reviewed extensively.[1,2,4,5]

The increasing interest in HSs and the fact that mechanisms of plant growth stimulation need further corroboration have led us to conduct the research described here. We hypothesized that the growth enhancement of plants by HSs is the result of an improved Fe and possibly Zn nutrition. Thus a major aim of this research was to provide evidence supporting our hypothesis from both NS chemical studies and plant growth experiments.

2 MATERIALS AND METHODS

2.1 Trace Metal Solubility - A Nutrient Solution Study

We conducted studies on NS containing 0 or 50 mg L^{-1} of a Leonardite (IHSS reference) HA. The NS used to test various trace elements solubilities at equilibrium contained: $Ca(NO_3)_2$, 2 mM; KCl, 0.1 mM; $MgSO_4$, 0.5 mM; KH_2PO_4, 0.2 mM; K_2SO_4, 0.7 mM; H_3BO_3, 0.01 mM; $CuCl_2$, 5 x 10^{-5} mM; $MnCl_2.4H_2O$, 0.001 mM; $ZnSO_4.7H_2O$, 0.005 mM; $FeSO_4.7H_2O$, 0.01 mM; $(NH_4)_6Mo_7O_{24}.4H_2O$, 5x10^{-5} mM.

The same NS, with slight modifications, was used for plant growth experiments. It is based on plant growth physiology studies carried out by Romheld and Marschner.[8] Calculations using Geochem-PC showed that all Fe would be precipitated out of the NS in the form of $FePO_4$ (pH < 6.8) or $Fe(OH)_3$ (pH > 6.8) at equilibrium.

Following the Geochem-PC calculations we conducted experiments to verify the results and provide information on the stability of Fe, Mn and Zn in this solution. Highly purified HA extracted from Leonardite was added to the solution at a concentration of 0 (control) or 50 mg L^{-1}, and at pH values of 5, 6, 7 and 7.5. The pH was buffered with MES for pH 5 and 6 and HEPES for pH 7 and 7.5 with buffer concentration 0.02 M and pH adjustment with metal free NaOH. Aliquots (10 mL) of each solution were prepared in a scintillation vial (25 mL). Twenty μL of a ^{59}Fe solution (0.25 μCi) and 20 μL of ^{54}Mn (0.4 μCi) were added to one set of solutions and 10 μL of ^{65}Zn (2.0 μCi) to a second set. From each solution an aliquot (1.5 mL) was added to a small centrifuge tube and shaken for 7 days at 25 ± 1°C to achieve equilibrium. Following the shaking process an RP-18 chromatography column (Supelco) was loaded with 0.4 mL of the tested NS. The column was pretreated with chloroform to increase its hydrophobicity. The column then was eluated with 1 mL chloroform to provide the first leachate referred to as containing the 'free metal' (not bound to the HA). The 'organically-bound metal' was obtained by elution of the column with 2 mL of 0.1 M acetonitrile/NaOH solution. Both eluates of each tested NS were transferred to polypropylene vials for scintillation counting (Automatic Gamma Counting System MR 480). ^{59}Fe was counted at 800-900, ^{54}Mn at 600-700 and ^{65}Zn at 800-900 kEv. Separate calibration for ^{59}Fe and ^{54}Mn allowed us to calculate the actual reading for ^{54}Mn using the equation: $Mn_t = Mn_a - 0.35 Fe_a$, where Mn_t = calculated Mn and Mn_a and Fe_a are the scintillation readings for these elements. Since Zn was read separately, we avoided the need to correct the readings. These tests were carried out in 4 replicates. The maximum standard error was ± 1.8%.

2.2 Plant Growth Experiments

Two plant growth experiments were designed to provide evidence that plant growth stimulation by HSs results from effects on trace element availability. These experiments were bioassays that tested whether chemical data of the NS would explain plant growth in the same solutions.

Iron is the main element of concern. Due to the complexity of its uptake mechanisms we conducted tests on two major plant species that differ in their Fe uptake strategies.[8,9]

1. Strategy I: dicotyledonous plants represented by melon (*Cucumis melo* L. var. Ein Dor) in Experiment 1;
2. Strategy II: monocotyledonous plants represented by creeping bentgrass (*Agrostis palustris* Huds., cv. Providence) in Experiment 2.

2.2.1 Strategy I - Experiment 1. This experiment tested the response to various sources of Fe of melon plants starved for Fe of a total of 14 days. After this period of starvation the plants were transferred for 7 days to solutions containing different Fe sources and then their chlorophyll contents and dry weights were measured.

Melon seeds were germinated on a filter paper soaked in saturated $CaSO_4$ solution until germination was completed (4 days). Ten seedlings were transferred to 3 L containers filled with the same NS as above but excluding Fe. The seedlings were placed in holes made in the container lid and supported with a polyurethane sponge. Fresh air was continuously provided to the plants via polypropylene tubes. The growth experiment was conducted in a growth chamber (16 hrs light, 8 hrs dark, temp. $25 \pm 2°C$). After 4 days the plants were transferred to a fresh NS saturated and buffered with $CaCO_3$ to pH 7.5 for 5 days. This treatment was employed to obtain Fe-starved plants. The plants then were transferred to 0.5 L containers filled with the same solutions as in the NS study above except for the elimination of the $FeSO_4.7H_2O$ component (-Fe and $CaCO_3$ buffered) for an additional 5 days. One plant was grown per container. At this stage each of the plants had developed 2 mature yet very chlorotic leaves. The plants were, therefore, ready for a remedy test using different Fe sources. The plants were transferred to the same NS as above (excluding the listed Fe treatments) buffered to pH 7.5 in 0.5 L containers. The treatments were as follows: 1. Control (NS, no Fe, no HA); 2. NS + 10 μM $Fe(NO_3)_3$; 3. NS + 10 μM $Fe(NO_3)_3$ + 50 mg L^{-1} HA; 4. NS + 10 μM FeEDDHA. Three replicates were employed. Chlorophyll was determined with a Spad-502 Chlorophyll Meter (Minolta) and a calibration curve of the meter's readings vs. chlorophyll concentration in the leaves (extracted with 80:20 acetone:water). Since 6 days are insufficient to create significant yield differences (which would develop when chlorophyll concentration varies), we measured the diameter of the youngest mature leaf on each plant.

2.2.2 Strategy II - Experiment 2. Based on our earlier experience with plants used in the turfgrass industry, we selected creeping bentgrass as a test plant representing the strategy II plants, which release phytosiderophores to mobilize Fe.[9]

The plants were grown in microtiter plates (Fisher) with 12 wells, each of about 5 mL. Four holes (3 mm I. D.) were drilled at the bottom of each well. One of the holes held a filter paper wick extending about 5 mm above and below the bottom of the hole to supply the NS. Earlier experiments showed that four holes were optimal for solution exchange and root growth. Coarse sand (1 to 2 mm), and then fine sand (<1 mm) was layered into each well to provide a desired level of water-holding capacity. Six creeping bentgrass seeds were placed in the sand and kept moist using a bottom microplate containing water. After 1 week of germination and growth in a 20°C constant temperature chamber, the upper plate was placed in a plastic tray arranged to hold three microtiter plates, and containing NS as in the trace metals study (excluding $FeSO_4.7H_2O$ and $ZnSO_4.7H_2O$) or NS+ HSs + Fe and/or Zn solutions. Each treatment tray contained a total of 36 wells. Seedlings were thinned to three plants per well. A 'constant-head' bottle was used to provide additional NS to the plastic tray. Plants remained in the plates for three additional weeks before harvesting. Further details of the method are given by Clapp et al.[10] Chlorophyll was determined as in Experiment 1.

For harvest, 3 plants from each well were washed with water to remove the sand. After separating shoots from roots, shoots or roots from five wells were combined into one sample; six replicates were collected from each of the treatments. Root and shoot samples were placed into vials, dried at 65°C to constant weight (about 2 days) and weighed for tissue mass.

The NS used in this experiment was the same as above except for the contents of Fe and Zn. The control NS was free of these two elements. Treatments to which Fe and/or Zn were added [(+Fe) or (+Zn)] were supplemented with 50 μM $FeSO_4.7H_2O$ and/or 2.5 μM $ZnSO_4.7H_2O$. The HSs tested in this experiment were IHSS reference preparations. The HA was a Leonardite HA (as in experiment 1) and the fulvic acid (FA) was a peat-derived product. The HSs were added to the NS at a level of 50 mg L^{-1} (the same concentration as in the chemical equilibrium experiment and in the experiment conducted with melons). All NS were buffered to saturation with CaCO$_3$ and a pH of 7.5 was maintained throughout the growth period. The following nine treatments were tested: 1, NS; 2, NS + FA; 3, NS + HA; 4, NS + FA + Zn; 5, NS + HA + Zn; 6, NS + FA + Fe; 7, NS + HA + Fe; 8, NS + FA + Fe + Zn; 9, NS + HA + Fe + Zn.

3 RESULTS

3.1 Trace Metal Solubility - A Nutrient Solution Study

The relative concentrations of 'organically-bound metal' and 'free metal' for ^{59}Fe, ^{65}Mn and ^{65}Zn in the NS at concentrations of 0 or 50 mg L^{-1} HA and pH values of 5, 6, 7 and 7.5 are presented in Table 1. Concentrations in the table are give as a relative percentage of the total added metal. At pH values of 5.0 to 7.0 in the absence of HA, most of the Fe was found as free metal (75 to 83%) and the rest had precipitated. At pH 7.5 only 20% of the free metal was present in solution and 80% had precipitated. In the presence of HA, however, a major fraction was maintained in solution in the organically bound form. In the presence of HAs, 46, 59, 57 and 39% of the total Fe added remained in solution in the form of organically-bound Fe (after 7 days of interaction in solution) at pH values of 5, 6, 7 and 7.5, respectively. In the absence of HA, only 6% of total Fe was still measurable as free Fe in the solution at highest pH 7.5.

Table 1 *Relative concentrations of 'free' and 'organically-bound' metal in nutrient solutions amended with a Leonardite humic acid*

Metal and its chemical form in solution	Solution pH							
	5		6		7		8	
	Humic acid concentration (mg L^{-1})							
	0	50	0	50	0	50	0	50
	(% of total added)							
^{59}Fe-free	83	23	73	9	71	10	20	6
^{59}Fe-bound	0	46	0	59	0	57	0	39
^{65}Zn-free	80	74	81	71	15	25	16	6
^{65}Zn-bound	0	15	0	12	0	19	0	25
^{54}Mn-free	82	76	84	74	79	73	83	66
^{54}Mn-bound	0	18	0	16	0	15	0	15

The distribution of Zn between the three fractions (precipitated, free or organically bound) was different from that of Fe. At pH values of 5 and 6 most of the Zn remained soluble in the free form, but at pH 7.0 and 7.5 most of it (~ 85%) precipitated as Zn(OH)$_2$. At pH 7 and 7.5, 19 and 25%, respectively, of the total Zn was found as organically bound Zn. At pH 7.5 only 6% of the Zn was present as free Zn and 25% was in solution as organic complexes. This suggests that in calcareous soils or in NS buffered with CaCO$_3$ (pH 7.4 - 7.9), Zn and Fe could only be provided to plants in sufficient quantities from organically bound metal forms.

The distribution of Mn between the various forms is different from those of Fe or Zn. In NS devoid of HA, most of the Mn (79 to 84%) was found as a free metal ion at all pH levels, and only 16 to 21% had precipitated. Addition of HAs to the NS resulted in binding of 15 to 18% of the Mn to the HA while 66 to 76% existed as free metal in the nutrient solutions at pH from 5 to 7.5.

The conclusion drawn for Fe and Zn, stressing the difficulty imposed on plants grown in NS at pH 7 and 7.5 with regard to the uptake of these essential elements (Fe in particular), does not hold for Mn, which exists in NS solution in its free form. Humic acids or synthetic chelates can provide the ligands necessary to bind and solubilize Fe and Zn. Our observations strongly support the hypothesis that improved Fe and possibly Zn nutrition are the major mechanism of plant growth stimulation by HSs. This hypothesis was further tested in the plant growth experiments described below.

3.2 Melon Plants Growth - Experiment 1

During the 6 days of treatment we observed that plants that contained Fe in their NS were recovering and regreening. Differences between treatments became distinct. Chlorophyll concentration in the leaves (young yet mature leaves) differed according to the treatments (Table 2). The measured levels were 0.2, 1.3, 1.7 and 2.1 mg g^{-1} (fresh weight) for the NS, NS + Fe(NO$_3$)$_3$, NS + Fe(NO$_3$)$_3$ + HA and NS + FeEDDHA, respectively. A remedy for Fe-deficiency induced chlorosis was only partly achieved with a mineral form of Fe, whereas the presence of either FeHA or FeEDDHA greatly improved chlorophyll synthesis. FeEDDHA seemed to be more effective than the FeHA in enhancing chlorophyll formation. A very similar trend related to leaf size was observed after the 6 days treatment period (Table 2). This experiment clearly indicates that HA stimulates plant growth and improves plant health by enhancing Fe availability at neutral to slightly basic pH. This agrees with other reports suggesting that HSs are important to the Fe nutrition of dicotyledonous plants.

Table 2 *Chlorophyll concentration and relative diameter of the youngest mature leaf of the melon plants*

Nutrient solution	Chlorophyll (mg g^{-1})	Relative leaf diameter (%)[a]
NS	0.2 ± 0.01	0
NS + Fe(NO$_3$)$_3$	1.3 ± 0.08	100
NS + Fe(NO$_3$)$_3$ + HA	1.7 ± 0.09	155
NS + FeEDDHA	2.1 ± 0.04	189

[a] % relative to the control.

3.3 Creeping Bentgrass Growth - Experiment 2

We chose creeping bentgrass to represent strategy II plants.[8] It is a very common plant on golf courses and is therefore important to the turfgrass industry. Growing these plants under optimal conditions for chlorophyll production ensures a healthy green appearance, a major criterion for successful management.

Seedlings were allowed to germinate in microtiter plates. After 7 days they were transferred to the plastic trays supplied with the individual NS. Chlorophyll concentration was measured at harvest, 3 weeks later. Concentrations of leaf chlorophyll are given in Table 3. Significant differences were observed between the various treatments: control NS exhibited

Table 3 *Chlorophyll concentration in leaves of creeping bentgrass plants grown in microtiter plates and supplied with different solutions*

Nutrient solution	Chlorophyll concentration(mg g^{-1})
NS	0.61 f[a]
NS + FA	0.95 e
NS + HA	1.45 d
NS + FA + Zn	2.26 c
NS + HA + Zn	2.41 b
NS + FA + Fe	2.81 b
NS + HA + Fe	3.39 a
NS + FA + Fe + Zn	3.26 a
NS + HA + Fe + Zn	3.31 a

[a]Statistical analysis: Tukey-Kramer test ($p < 0.05$).

the lowest chlorophyll concentration and the highest were measured on leaves of plants treated with NS to which either FA or HA and Fe and Zn were added. Iron addition along with HA enhanced chlorophyll concentration to the highest level obtained, whereas the effect of the FA + Fe combination was slightly less. Nutrient solutions to which only HA or FA were added slightly improved plant chlorophyll levels, but the levels fell far short of those of healthy plants. This was also the case for the NS + HA + Zn or NS + FA + Zn treatments, although these treatments affected the plants in a more favorable manner than the HSs alone. These observations clearly show that highly purified HSs do not exhibit any stimulation of chlorophyll production due to the presence of plant regulators, as suggested in a number of publications.[1-3] The addition of Zn and HSs does not seem to be as effective as that of the combination of Fe and HSs. The individual roles of Fe and Zn in plant growth enhancement by HSs needs further research underway in our laboratory.

The results obtained for leaf chlorophyll concentration are in line with shoot and root weight data (Table 4). The statistical differences between the different treatments are smaller because of the short-term of the experiment. Turfgrasses generally produce low levels of biomass, especially when grown in NS in a growth chamber. Since differences in chlorophyll

Table 4 *Root and shoot dry weight of creeping bentgrass plants grown in microtiter plates and supplied with different nutrient solutions*

Nutrient solution	% yield increase[a]	
	Root	Shoot
NS	0 bc[b]	0 c[b]
NS + FA	17.4 bc	25.4 bc
NS + HA	39.6 bc	11.6 c
NS + FA + Zn	75.2 ab	78.0 a
NS + HA + Zn	74.6 ab	90.1 a
NS + FA + Fe	79.4 ab	79.7 a
NS + HA + Fe	81.7 ab	87.9 a
NS + FA + Fe + Zn	87.3 a	87.8 a
NS + HA + Fe + Zn	89.6 a	63.6 ab

[a]Yield increase % indicated are relative to the control; [b]Statistical analysis: Tukey-Kramer test ($p < 0.05$).

synthesis are exhibited only after about 10 to 12 days, inhibitory effects on biomass formation become effective mostly during the later part of the growth period and differences in growth are therefore smaller than those for chlorophyll concentrations.

The data in Table 4 are presented as a relative yield compared to that for the control plants. The yields of both shoots and roots were slightly enhanced by the addition of either FA or HA to the NS. A much larger increase in yield was achieved when Zn or Fe or both were added to the NS. As observed for the chlorophyll concentrations, the highest root weight was observed in NS to which HSs were added along with both Fe and Zn.

We have no doubt that larger differences in yield would be obtained in a long-term experiment in which plants in some treatments develop chlorosis while others exhibit a healthy and green appearance. However, we believe that experiments in which distinct differences in chlorophyll synthesis can be determined provide the evidence needed to understand plant growth stimulation mechanisms.

4 DISCUSSION

Complexation of transition metals such as Cu, Zn, Fe and Mn by HSs has been the focus of a large number of publications. This topic was reviewed in relation to plant growth[2,11] and will not be discussed in great detail here. Solubilization of micronutrients from their inorganic forms may be the major factor in the promotion of plant growth in soils by HSs. The same situation may apply to NS in which the solubility of most micronutrients is limited. The presence of HSs either in a nutrient or soil solution may contribute to improved availability of elements.

Iron more than any other microelement has drawn the attention of researchers. De Kock and Strmecki[12] concluded that lignite-derived HSs maintained Fe in solution in both NS and plant tissues even at high phosphate concentrations. Untreated chlorotic plants contained high concentrations of Fe in their roots, probably due to precipitation as ferric phosphate. The HSs not only increased the solubility of Fe in solution but also affected Fe translocation from roots to shoots.[12] Dyakonova and Maksimova[13] reported that soluble FeHA complexes occurred in natural peat and prevented chlorosis in plants. Linehan and Shepherd[14] compared the effects of FAs with those of polymaleic acid and other polycarboxylates. Addition of FA to NS at concentrations up to 25 mg L^{-1} enhanced Fe uptake by shoots of wheat seedlings. Polymaleic acid and other polycarboxylates showed similar effects.

Chen and collaborators have shown that Fe-enriched organic materials such as peat or manure serve as a remedy for lime-induced chlorosis in soils.[15-20] The corrective effect was attributed to complexation of Fe by HSs in the organic materials.

The effect of HSs on the uptake of Zn and Cu was investigated with intact plants or plant tissues. Vaughan and McDonald[21] studied Zn uptake by beet root tissue cut into discs and found that addition of HAs slightly inhibited Zn uptake by aged discs when concentrations exceeded 25 mg HA L^{-1}. Lower concentrations did not affect Zn uptake. On intact plants, Jalali and Takkar[22] reported that Fe, Cu, and Zn uptake by rice plants was enhanced by increased levels of OM.

In contrast to the substantial literature on the activity of HSs in the stimulation of plant growth in nutrient cultures, information on the activity of these substances in the field is scarce. However, the use of HSs to help overcome Fe and Zn deficiencies has been studied in field trials.[2,11] Naturally occurring organic materials rich in HSs, such as peat and decomposed manure, have been used as fertilizers after the employment of an enrichment processes with Fe or Zn.[15,19,20] Chen and Barak[18] used a peat preparation for the remedy of Fe deficiency in peanuts grown in the field. Yields increased from 4740 kg ha^{-1} in control plants to 5540 kg ha^{-1} and 5810 kg ha^{-1} when fertilized with Fe-enriched peat and FeEDDHA (at a level considered adequate to provide a complete remedy), respectively. Bar-Tal, et al.[23] have shown that FA maintains $10^{-3.5}$ mM Zn in solution in the presence of Ca-montmorillonite at pH 7.5, whereas Zn levels decreased to $10^{-5.5}$ mM in the absence of FA. These and many other studies show that HSs can play an important role in the micronutrition of plants under field conditions.

A straightforward test conducted in NS either with or without plants provided evidence that HSs stimulate plant growth by enhancing Fe and possibly Zn availability. It is unlikely that stimulation will also take place in acidic soils in which Fe availability is sufficiently high. It is also expected that application of HSs will not stimulate plant growth in soils exhibiting high levels of dissolved OM (above 100 mg L^{-1}).

5 CONCLUSIONS

Plant growth effects of HSs, often ascribed to hormone-like activity, have been widely reported. Plant parameters affected are root and shoot weight, root initiation, seedling emergence and growth, rhizosphere microbial population, nutrient uptake and flowering. These effects were found for both HAs and FAs as well as for compost-derived HSs. In most publications the activity of HAs was found to be similar to that of FAs although some researchers reported higher activity for low molecular weight products. Plant hormone-like activity attributed to HSs does not involve auxins, cytokinins or absisic acid, although if the appropriate precursors are present these plant growth regulators may be synthesized in the rhizosphere by microbes. Therefore it is concluded that growth enhancement of plants in NS and soil by HS should mostly be attributed to the maintenance of Fe and Zn in solution at sufficient levels. This effect is pH dependent and its importance decreases with decreasing pH.

ACKNOWLEDGEMENTS

We thank Dr. R. Liu, Milegua Layese and David Johnson for technical assistance. Dr. Donald White provided growth chamber facilities and useful discussions. We thank the Toro Company for financial support. This is contribution no. 991250070 of the Minnesota Agricultural Experiments Station Journal Series.

References

1.   Y. Chen and T. Aviad, in 'Humic Substances in Soil and Crop Sciences: Selected Readings', P. MacCarthy, C. E. Clapp, R. L. Malcolm and P. R. Bloom, (eds.), Soil Science Society of America, Madison, WI, 1990, p. 161.
2.   Y. Chen, in 'Humic Substances in Terrestrial Ecosystems', A. Piccolo, (ed.), Elsevier, Amsterdam, 1996, p. 507.
3.   S. Nardi, G. Concheri and G. Dell'Agnola, in 'Humic Substances in Terrestrial Ecosystems', A. Piccolo, (ed.), Elsevier, Amsterdam, 1996, p. 361.

4. D. Vaughan and R. E. Malcolm, in 'Soil Organic Matter and Biological Activity', D. Vaughan and R. E. Malcolm, (eds.), Martinus Nijhoff, Dordrecht, 1985, p. 37.

5. D. Vaughan, R. E. Malcolm and B. G. Ord, in 'Soil Organic Matter and Biological Activity', D. Vaughan and R. E. Malcolm, (eds.), Martinus Nijhoff, Dordrecht, 1985, p. 77.

6. R. Pinton, Z. Varanini, V. Vizzoto and A. Maggioni, *Plant Soil,* 1992, **142**, 203.

7. Y. Chen, H. Magen and J. Riov, in 'Humic Substances in the Global Environment and Implications on Human Health', N. Senesi and T. Miano, (eds.), Elsevier, Amsterdam, 1994, p. 427.

8. V. Romheld and H. Marschner, *Adv. Plant Nutr.*, 1986, **2**, 155.

9. Y. Chen and Y. Hadar, 'Iron Nutrition and Interactions in Plants', Kluwer, London, 1991.

10. C. E. Clapp, R. Liu, V. W. Cline, Y. Chen and M. H. B. Hayes, in 'Humic Substances: Structure, Properties and Uses', G. Davies and E. A. Ghabbour, (eds.), Royal Society of Chemistry, Cambridge, 1998, p. 227.

11. Y. Chen and F. J. Stevenson, in 'The Role of Organic Matter in Modern Agriculture', Y. Chen and Y. Avnimelech, (eds.), Martinus Nijhoff, Dordrecht, 1986, p. 73.

12. P. C. De Kock and E. L. Strmecki, *Physiol. Plant*, 1954, **7**, 503.

13. K. V. Dyakonova and A. E. Maksimova, *Trans. Jt. Meet. Comm. 2 and 4, Int. Soc. Soil Sci.*, Dokuchaev Soil Inst., Moscow, 1967, p. 79.

14. D. J. Linehan and H. Shepherd, *Plant Soil*, 1979, **52**, 281.

15. P. Barak and Y. Chen, *Soil Sci. Soc. Am. J.*, 1982, **46**,1019.

16. Y. Chen, J. Navrot and P. Barak, *J. Plant Nutr.*, 1982, **5**, 927.

17. Y. Chen, B. Steinitz, A. Cohen and Y. Elber, *Scientia Horticulturae,* 1982, **18**,169.

18. Y. Chen and P. Barak, in 'Proc. Int. Symp. Peat Agric. Hort., 2nd', K. M. Schallinger, (ed.), Volcani Center, Bet-Dagan, Israel, 1983, p. 195.

19. E. Bar-Ness and Y. Chen, *Plant Soil*, 1991, **130**, 35.

20. E. Bar-Ness and Y. Chen, *Plant Soil*, 1991, **130**, 45.

21. D. Vaughan and I. R. McDonald, *Soil Biol. Biochem.*, 1976, **8**, 415.

22. V. K. Jalali and P. N. Takkar, *Indian J. Agric. Sci.*, 1979, **49**, 622.

23. A. Bar-Tal, B. Bar-Yosef and Y. Chen, *Soil Sci.* 1988, **146**, 367.

USING ACTIVATED HUMIC ACIDS IN THE DETOXIFICATION OF SOILS CONTAMINATED WITH POLYCHLORINATED BIPHENYLS. TESTS CONDUCTED IN THE CITY OF SERPUKHOV, RUSSIA

Alexander Shulgin,[1] Alexander Shapovalov,[1] Yuriy Putsykin,[2] Cicilia Bobovnikov,[3] Galena Pleskachevski[4] and Andrew J. Eckles, III[5]

[1] Special Bio-Physical Technology and [2] Research Institute of Plant Protection Chemicals, Moscow, Russia, [3] Science-Industrial Complex "Typhoon," Obnisk, Russia, [4] Center of Epidemiology, State Health Department, Serpukhov, Russia; [5] Stable Earth Technology and Science Coordinators International, Louisville, KY 40299- 4470, USA

1 INTRODUCTION

The health of existing and future generations depends on the ecological solutions used in their countries. This is why governments, political parties, public organizations, mass media and most of the people of highly developed industrial countries pay considerable attention to the improvement of their environmental conditions.

It has been recognized during recent years that among the many kinds of xenobiotics there is a group with very high toxicity that qualify as supertoxins. This group consists of organic substances such as polychlorinated dioxins, dibenzofurans and biphenyls, pesticides containing chlorine or phosphorus and polyaromatic hydro-carbons.[1] The human body receives supertoxins while breathing, drinking water or eating animal or vegetable food. Supertoxins may have mutagenic, tetragenic and carcinogenic effects. They can induce porphyria, suppression of the immune system, cachexia of the body and otherwise affect the internal organs.[2]

The unique features of polychlorinated biphenyls (PCBs) make them a specific kind of toxin. PCBs have high chemical and thermal stability. They were widely used everywhere until 1980. The total amount of PCBs used by the world industry was probably over one million tons. Laws prohibiting the manufacture and use of PCBs are now in force throughout the world. Nevertheless, significant levels of PCB residues are still found in the environment and in living organisms, including humans.

Highly contaminated soils have been found in most areas where PCBs were recently produced or used. As an example, we can observe the environmental situation in the City of Serpukhov, Russia, where PCBs were used for 25 years as a filling for electrical transformers.

Even though the use of PCBs was halted 10 years ago, both the soil and the produce grown in and on that soil continues to be severely contaminated by PCBs. The most highly contaminated soils are in the southern part of the city, which has private gardens that supply the city markets with fresh vegetables, berries and greens. Therefore, it was a matter of specific interest to explore the use of a technology for *in-situ* soil detoxification by means of activated humic acids (AHAs). A. I. Shulgin, A. A. Shapovalov and U. G. Putsykin of Moscow, Russia developed this unique form of a humic compound.

It is well known that ecologically toxic substances have an ability to bind to a soil organic substance or to its clay mineral fraction. In the presence of a significant amount of soil organic matter, the adsorbability of a mineral fraction practically does not exist and the binding of the toxin is associated only with the organic matter.[3] The main components of soil organic matter are represented by humic substances, which contain up to 60-70% of humic acids.[4]

The influence of natural organic matter (NOM) on the toxicity and bio-availability of polychlorinated biphenyls has been considered in detail.[5] It has been shown that the main mechanism of binding of organic xenobiotics is a hydrophobic interaction of PCBs with humic acids, thus creating PCB-humic acid complexes. These complexes significantly reduce the bioavailability and toxicity of xenobiotics. It also has been concluded that in addition to the adsorption of PCBs on humic acids there is a possibility of the covalent binding of xenobiotics.[6] The main mechanism of creating the covalent links for organic matter containing phenol structures represents acidic binding to humic acids. This process induces the polymerization of xenobiotics with humic acids and the creation of polymers that are extremely stable with regard to acidic and alkaline hydrolysis, thermal degradation and microbial decomposition.

Until recently, an in-situ method of detoxifying large areas of soils contaminated with organic xenobiotics by the introduction of various humic acids was impractical and not used. One of the main reasons was a lack of an efficient and feasible industrial technology for consistently producing an effective humic acid in sufficient amounts. But now an efficient industrial technology for manufacturing a unique activated humic acid (AHA) using inexpensive and available materials has been developed and introduced. The active centers of reactions of the natural humic acids are blocked by different admixtures and this is an important distinction between natural humic acids and AHA. AHA has many unique features resulting from its being a complicated organic-mineral compound. The organic part of AHA contains many functional groups (mainly carboxyl and hydroxyl). The inorganic part of AHA contains micromicellar formations consisting of aluminum, silicon and iron oxides. Carboxyl groups have a high reactivity and affinity for water. Hydrophobic areas of the humic macromolecule have an ability to sorb organic xenobiotics. All these features determine the possibility of using AHA for detoxification of soils contaminated with PCBs.

2 RESULTS AND DISCUSSION

Table 1 demonstrates the ability of AHA to interact with different PCB congeners. The AHA was added to the contaminated soil as a water solution in which the AHA concentration is 100 mg/L. The total amount of AHA added to the soil was between 0.3% and 1% by weight, as estimated by acre/foot calculations.

As can be seen from Table 1, after 120 hours of contact between the PCBs and AHA the congeners' concentration was reduced by approximately 65%.

The direct influence of AHA on PCBs as regards soil reclamation and detoxification has been demonstrated by introducing AHA into the contaminated soil. The AHA was diluted with water in the ratio of 1 part AHA to 10 parts water. This solution was then mixed into the contaminated soil.

All the measurements were made in 1995 in an area adjacent to the transformer factory in the City of Serpukhov. The content of PCBs in the soil before and after treatment was determined. The macroflora of the soil also were studied. The determination of PCB content in soil before and after the treatment with activated humic acids was performed by gas chromatography using a Hewlett-Packard chromatograph with a capillary tube 25 mm in length and 0.25 micron in diameter. The packing was the phase HP-1. A mixture of 62 percent synthetic PCBs in isooctane was used as a calibrating standard.

Table 1 *Changes in content of particular groups of PCB congeners after AHA soil treatment*

PCB congenors	Initial PCB congenors content, mg/kg	PCB congenors content after 120 hr contact with AHA, mg/kg	PCB congenors content reduction, %
Mono	0.444	0.15	65
Di-	4.54	2.27	50
Tri-	11.5	3.44	70
Tetra-	3.74	1.31	65
Total	20.2	7.18	64.5

Table 2 shows the PCB content before and after the AHA soil treatment in soils with different initial PCB contents. It is evident that there is a sizeable reduction of PCB concentration. In highly contaminated soil, the reduction reaches 20-35% after 60 days. A 52-56% reduction of PCB levels was found in soils having 0.5-3.7 mg/kg of PCB contaminants.

Table 2 *Reduction of PCB content in natural conditions after AHA soil treatment*

Initial PCB content , mg/kg	PCB content after 60 days contact with AHA, mg/kg	Reduction of PCB content, %
89.1	17.2	19.3
12.3	4.5	42.2
12.9	4.54	35.2
3.7	1.95	52.3
0.5	0.28	56.0

Dichlorinated biphenyls and several kinds of trichlorinated biphenyls were not found in soil after 60 days of contact with AHA. The concentration of several types of tri- and tetrachlorinated biphenyls were ten times lower than at the beginning of the experiment. For pentachlorinated and other groups of highly chlorinated biphenyls, the reduction of PCB concentration after AHA soil treatment was considerably less. It is well known that the less chlorinated biphenyls have a higher degree of sorption and higher aqueous solubility than highly chlorinated congenors.

Basically, up to 70% of the soils of the City of Serpukhov were contaminated with less chlorinated biphenyls because electrical transformers had been filled with Arochlorine 1242. The main constituents of this material were less chlorinated biphenyls and only 2.5% of highly chlorinated biphenyls.

The growth of cultured nitrogen fixing bacteria in contaminated soils is very sensitive to the influence of various kinds of toxin and provides a clear index of soil contaminant reduction. The results of this bio-test are shown in Table 3.

Table 3 *Changes in Azobacter after AHA soil treatment*[a]

Variant	Accumulation of Azotobacter on the surface of soil particles, %
Initial soil, untreated with AHA	58
Soil after 25 days of AHA treatment	80
Soil after 60 days of AHA treatment	91
Soil after 25 days of treatment with AHA and with microorganisms	99
Soil after 60 days of treatment with AHA and with microorganisms	99

[a] Every figure was obtained from the average of two separate tests

Table 3 shows that addition of AHA to a soil provides efficient soil detoxification in a short period of time. Adding native microorganisms increases this effect. It has been found that native microorganisms, represented by the pseudomonades, develop very quickly in the presence of AHA. Their concentration in the soil treated only with AHA and in the soil treated with AHA and with microorganisms remains at 10^8 CCU/g after 25 days.

The testing area has been used for farming. That was good reason to conduct additional PCB concentration tests on samples of vegetables produced in this region. Three vegetables, white cabbage (*Slava*), the carrot type *Shantane* and the beet type *Bordo* were chosen for testing. Table 4 shows the averaged figures for the total PCB concentration in these vegetables at the end of a farming season.

Table 4 *Changes of PCB concentrations in vegetables after AHA soil treatment*

	PCB concentration, mg/L		
	Carrot Roots	Beet Roots	Cabbage Leaves
Untreated soil	9.4	1.53	6.7
Soil treated with AHA	1	0.07	3.6

Table 4 demonstrates that the decontamination of a soil with AHA reduces the PCB content in carrots by 89%, by more than 95% in beets and by approximately 46% in cabbage.

Related tests also showed a significant improvement in several agrochemical characteristics of soils treated with AHA. Among these improvements were an increase moisture holding capacity, an improvement of soil structure and an increase in the availability of plant nutrients. In addition, the productivity of the treated soils was increased by 30-40%.

4 CONCLUSIONS

High efficiency, low cost, simple technology, feasibility, the possibility of cultivating ecologically clean produce on previously contaminated lands, a significant improvement in soil agrochemical characteristics and increases in soil productivity are the specific benefits of soil detoxification with AHA. The results of this study recommend AHA technology for much wider use in the remediation of PCB contaminated soils.

This report describes a field test lasting only 60 days. However a follow-up study was conducted 12 months later. The same test procedures were used in order to detect any residual PCB in the treated soils. This follow-up test showed that the residual levels of PCBs were undetectable at the 1 part per million level.

The form of activated humic acid used in this work is called Stabilite™ in the U.S. Stable Earth Technology (SET) was formed by the Russian developer and four U.S. companies for the manufacture and distribution of Stabilite™. Stabilite™ has been found to be valuable in the remediation of heavy metal contamination of soil and water as well as in the remediation of hydrocarbon soil contaminants.

References

1.   A. D. Kuntzevitch, *Successes in chemistry*, 1991, **60**, 530.
2.   V. N. Maistrenko, R. Z. Khamitov and G. K. Budnikov, 'Ecological and analytical monitoring of supertoxicants', Moscow, 1996, p. 319.
3.   C. W. Bailey and J. L. White, *J. Agric. Food Chem.*, 1964, **12**, 324.
4.   D. S. Orlov, N. U. Lozanovskaya and P. D. Popov, 'Soil organic substances and organic fertilizers', Moscow University Publishing House, Moscow, 1989, p. 97.
5.   P. E. Landrum, *Environ. Toxicol. Chem.*, 1985, **4**, 459.
6.   Y. M. Bollag and C. Myers, *Sci. Total Environ.*, 1992, **117/118**, 357.

Subject Index

ab initio FEFF6 code, 195
Abscisic acid, 255
Absorption, 135, 147
  band, 153, 210
  behavior, 143
  coefficient, 192, 193,199
  edge, 191, 193
  K edge, 192
  maxima, 176
  measurements, 129, 132
  spectra, 130, 132-134, 140, 153, 155, 205
  spectroscopy, 147
  tails, 132, 134
Abundance, 29, 73, 123, 125, 174
Abundance pattern, 123, 125, 126
Acceleration voltage, 123
Acenaphthene, 228
Acetonitrile, 21, 22, 123, 256
Acid
  group structure, 33
  precipitation, 241
Acidic
  group, 70, 75
  hydrolysis, 152
  strengths, 107
Activated
  carbon, 241, 253
  humic acid, 266-269
Adjustable parameters, 196
Adsorbent, 241-243
Adsorption, 21, 32, 72, 75, 76, 93, 182, 188, 216, 223, 231, 241, 247, 248, 266
Aggregates, 4, 71, 74, 76, 77, 83, 84, 89, 91, 93, 96, 108, 203
Aggregation, 31, 43, 44, 46, 70, 73, 74, 76, 79, 87, 89, 91, 93, 96, 166, 227
Agriculture, 9, 16, 252
Alga cell walls, 66

Aliphatic, 13, 20, 23, 24, 26, 30, 38, 42, 43, 55, 59, 64, 66, 69, 71, 73, 76, 165, 171, 208, 242
Aliphatic alcohol, 211
Alkoxyl, 19
Alkyl
  chains, 73
  halides, 251
Aluminum, 9, 75, 132, 138, 139, 205, 207, 209-212, 214, 215, 217-219, 266
  Al^{3+}, 87, 96, 205, 218, 219
  Al^{3+}-FA, 96
  Al-O-Si, 211
Ametryne, 234, 236
Amicon, 102, 158
Amides, 9, 23, 203, 207, 208, 210, 211, 219
Amino acids, 2, 19, 31, 152, 165
Amorphous, 63-66, 134, 223, 230
Amphiphilic, 69, 71-73, 76, 79, 82-84
Anaerobic, 169
Analyte losses, 224, 238
Anilines, 248
Anionic groups, 182
Anodic stripping voltammetry, 107, 203
Anthocyanin, 20
Anthracene, 130, 227, 228, 230, 233
Anthropogenic origin, 212, 226, 227, 230
Apoferritin, 205
Aquatic systems, 166
Aromatic
  bands, 26, 28, 30
  components, 69, 188
  compounds, 129, 130, 135, 142, 143, 165, 173
  constituents, 73
  moieties, 38, 42, 165
  ring, 13, 23, 29, 42, 74, 171, 208
  units, 28, 30, 165

Aromaticity, 28, 73, 75, 79, 151, 152, 155
Asphaltenes, 134
Atomic absorption spectrophotometry, 203, 242
Atomic emission detection, 226, 231
Atomic force microscopy, 5, 87, 88, 89, 91, 93, 96
Auxins, 262
Axial, 191, 195-198
Azobacter, 268

Background electrolyte (BGE), 108, 112, 114, 116, 117
Background function, 193
Barrier conditions, 252, 254
Barrier wall, 252
Bean-like structures, 94, 96
Benzene, 248
Benzo[a]pyrene, 140
Benzoic acids, 101
Benzophenone, 130
Best fit parameters, 197
Binding orders, 218
Binding sites, 2, 7, 35, 39, 41-43, 191, 199, 244
Bioavailability, 87, 99, 234, 266
Biogeopolymer, 32, 34, 35, 39, 44, 45, 46, 47
Biomass, 255, 260
Biopolymers, 5, 31, 32, 34, 35, 39, 46, 66
Biosynthetic, 2, 5
Biphenyls, 248, 265
Blue Dextran, 181
Bond lengths, 191, 194-196
Bonding geometry, 191
Boric acid, 108, 109
Bound metal forms, 258
Bovine serum albumin, 123
Breakthrough curves, 247, 254
Bromide quencher, 81
Bronchitis, 87
Brown water, 204
BTX aromatics, 233
Buffer, 91, 109, 205, 256
Building blocks, 2, 5, 49, 71, 76, 112, 118, 171
Bullet-proof vests, 66

C/N ratios, 12, 171, 180

C=C, 13, 171, 208
C=N, 208
C=O, 13, 50, 171, 207, 208, 210
Cachexia, 265
Caffeic acid, 108, 109
Calcium, 244, 245, 247
    $CaCO_3$, 242, 257, 258
    $Ca(NO_3)_2$, 256
    $CaSO_4$, 242, 257
Calibration, 35, 108, 109, 123, 140, 158, 207, 214, 218, 225, 226, 234, 236, 256, 257
Capillary electrophoresis, 102, 109, 118, 203
Capillary isotachophoresis, 107
Capillary zone electrophoresis, 101-103, 105, 107-109, 112, 114, 116
Carbohydrates, 20, 23, 28-30, 38, 39, 42, 43, 55, 172, 175
Carbonyl, 13, 23, 26, 28, 29, 55, 56, 64, 171, 242
Carbowax-divinylbenzene, 226
Carboxyl groups, 33, 42, 81, 236, 266
Carboxylate groups, 32, 42, 172, 173
Carboxylic, 10, 20, 23, 28, 29, 43, 55, 171-173, 175, 203, 211, 219, 244
    acids, 91, 93, 174, 242
    groups, 109, 173, 188, 203, 207, 210
Carcinogen, 265
Carotenoids, 20
Cartridge filters, 147
Catabolic breakdown, 19
Catalytic effects, 9, 10, 11, 13, 16
Cation, 22, 81, 82, 84, 218
Cation exchange capacity, 242, 244
Cations
    N-alkylammonium, 73
Cadmium, 87-89, 93-99, 107, 205, 207, 209, 211-215, 217-219, 244, 252
    adsorbed, 93, 94
    complexed, 93, 96
    concentration, 88, 93, 94, 96, 214
    distribution, 88, 89, 96
    ions, 88, 96
    salt, 93
    solution, 93
    treatment, 93, 96, 98
    $Cd(NO_3)_2$, 94, 96, 98, 205
    Cd^{2+}, 96, 205
Cd-humate, 88, 93-97, 99

Centrifugation, 10, 170, 205, 234
Cerium, 244, 248, 249, 252
Cesium, 244
C-H stretching, 171
CH₃CN, 180, 182
Chain mobility, 64
Characterization, 20, 54, 60, 63, 126, 129, 136, 143, 155, 188, 199, 218
Charge transfer, 46, 121, 250
Chelate, 242
Chelating material, 252
Chelex®, 122, 124
Chemical
    compositions, 73
    oxidation, 173
    structure, 97, 121, 194
Chlorinated
    biphenyls, 267
    hydrocarbons, 248, 250, 252
    solvents, 241
Chlorine, 6, 224, 226, 231
Chloroform, 6, 256
Chlorophyll, 19, 256, 257, 259-261
    meter, 257
Chloroplasts, 19, 20
5-chlorosalicylic acid, 108, 110
Chlorosis, 255, 259, 261
Chromatograms, 181-184, 186, 188, 205, 206, 213, 214, 218, 219
Chromium, 206, 207, 212, 219, 242, 244, 247, 250
    Cr(III), 242-248
    Cr(VI), 248, 249, 251, 252
Chrysene, 130, 227-229
Clay, 72, 75-80, 87, 266
Cleavage, 174, 176
Clouding phenomenon, 84
Cluster ions, 121, 123, 124
C-O stretching, 13, 172
Coaggregation, 75, 76
Coagulation, 70, 72, 75-77
Coals, 1, 33, 34, 63, 108, 118, 226, 228, 232, 233, 242
    brown coal, 72, 130
Coatings, 70, 224, 226, 231
Coiling, 79, 228
Coils, 71, 76
Colloid, 4, 69, 71, 76
    behaviour, 70

character, 69, 76
destabilization, 75
processes, 71, 72, 76, 77
properties, 69, 72
solutions, 72, 73, 74, 76
stability, 70, 72, 75
state, 69-72, 76, 77
system, 69, 74, 76
Column modes, 244
Columns, 22, 179-184, 226, 241, 242, 247, 250
Commercial, 6, 7, 10, 102, 234, 252
Commercial scale applications, 247
Complexation, 42, 46, 74, 75, 87, 107, 129, 192, 194, 197-199, 204, 233, 261
Composite, 102
    ET band, 151
    materials, 69
Compost, 19, 26, 29, 30, 72, 262
    rice straw, 180
Compression, 70, 93
Configuration, 84, 89, 93, 121, 166
Conformation, 43, 46, 47, 64, 73, 87, 89, 91, 93, 94, 96, 121, 134, 166, 214
    changes, 42-45, 70, 72, 76, 162
    data, 43
    effects, 43
    gauche, 64
    processes, 31
    relaxation, 135
    searches, 46
    states, 70
    trans, 64
Conjugation, 79, 160, 173
Contact time, 22, 35, 36, 51, 59, 242, 243, 247, 248, 250, 251, 254
Contactmode, 91
Contaminants, 49, 63, 129, 203, 242, 244, 245, 250-254, 267
    environmental samples, 234
    groundwater, 250, 253
    soils, 212, 265-269
    water, 199, 241
Contamination, 204, 205, 241
Continuous distribution method, 135
Contour maps, 158
COO⁻, 66, 171, 172, 203, 208
COOH, 12, 13, 42, 171-173, 208, 210
Coordination
    complexes, 203

Coordination, *continued*
  numbers, 191, 195
  shell, 191, 198, 199
  sites, 87
Copper, 35, 41-44, 107, 191-199, 205,
207, 209, 212, 214, 215, 217-219, 244,
252, 261
  Cu(II), 35, 41-44, 107, 192, 194, 197,
  198
  Cu/C molar ratios, 192-199
  Cu^{2+}, 87, 96, 191, 205, 210
  Cu^{2+}-FA, 96
  CuCl$_2$, 256
  Cu-HS, 191, 192, 194, 196, 199
  Cu-hydrophobic acids fraction, 193-
  198
  Cu-N, 194-199
  CuN$_x$O$_{6-x}$ octahedron, 196, 198
  Cu-O, 191, 194-199
Copper complexes with
  citric acid, 192, 198
  ethylenediamine, 192, 199
  glycine, 192, 198
  salicylic acid, 192, 198
  aqua-complex [Cu(H$_2$O)$_6$]$^{2+}$, 192, 198
Copper-nitrogen interactions, 198
Correlations, 74, 151, 152, 227, 230, 231,
234, 238
Covalent binding, 266
Critical micelle concentration, 69, 72, 73,
79, 83
Cross polarization, 23, 35, 49, 63
Cross-linking, 242, 254
Crystalline, 23, 49, 64- 67
Crystallinity, 32, 50
Cutin, 23
α-cyano-4-hydroxy cinnamic acid, 110
Cyclohexane, 131
Cytochrome C, 123, 205
Cytokinin, 255
Czech soil, 118

Data analysis, 126, 136, 194, 197-199
Decomposition status, 124-126
Decontamination, 241, 252
Deconvolution, 147
Degradation, 5, 7, 19, 20, 26, 29, 32, 33,
42, 49, 66, 121, 171, 251
  thermal, 266

Demethylation, 20
Department of Energy, 199, 241
Derivatives, 5, 153, 155, 170, 173, 196,
236, 250, 254
Derivatization, 108, 109, 236
Desorption hysteresis, 229
Detoxification, 7, 266, 268
Diagenesis, 66
Dialysis, 224, 237
Dibenzofurans, 265
9,10-dichloranthracene, 130
Dichlorinated biphenyls, 267
Differential scanning calorimetry, 32, 34
Diffusion coefficients, 137, 139, 140
Dinitrophenols, 234
Discrete component approach, 135
Disorder, 191, 195, 196
Dispersion, 70, 89, 223
Dissociation, 43, 70, 74, 75, 93, 107
  constants, 75
Dissolved
  humic organic matter, 223-239
  organic carbon, 19-21, 26, 149-151,
    157, 166, 192
  organic compounds, 162
  organic matter, 203
Distances, 151, 191, 195-199
Distribution function, 194
Disulfane, 214
Disulfide, 214
DNA, 101, 121
Domain sizes, 63
Dose-response relationships, 224
Drift distance, 132
DRIFT fourier transform infra red
spectroscopy, 205, 207-211, 219
Drift tube, 131, 137
Dynamic light scattering, 70, 73, 74
Dynamic scaling theory, 74

E_0, 133, 134, 153, 194, 196
E$_4$/E$_6$ ratios, 10, 12, 13, 132, 134, 204,
207
Electric double layer, 70
Electric field, 70, 74, 101, 105, 137
Electrical conductivity, 88, 89
Electrolyte, 43, 44, 70, 72, 74-77, 84, 88,
96, 114, 203
Electron
  density, 74

donor-acceptor, 233
impact MS, 231
microprobe, 88, 96
Electron spin resonance (ESR), 5, 11, 13,
14, 192, 194, 198
Electronic
states, 133-135, 140, 143
structure, 191
Electroosmotic flow (EOF), 102, 108
Electropherograms, 114, 116
Electrophoretic buffer, 101, 102
Electrospray ionization (ESI), 121, 227
Electrostatic stabilization, 75
Elemental
analysis, 55, 121, 227
composition, 12, 171
Emergent properties, 32, 43
Emission-excitation wavelength pairs,
162-166
Empirical formula, 1, 2
Empty Bed Contact Time, 243, 247, 248
Energy, 31, 46, 47, 108, 121, 133-135,
143, 153, 198
of light, 153
origin, 193, 196
reference, 194
Entangled polymer, 101
Enthalpies, 34
Environment, 7, 20, 31, 32, 34, 42, 43,
46, 66, 82, 93, 96, 99, 107, 135, 191,
194, 196, 203, 212, 223, 265
Enzymatic, 5, 9, 19
Enzyme activities, 255
Eosine yellowish, 130
Equatorial, 191, 195-198
Equilibrium, 45, 93, 224, 225, 227, 231,
232, 237, 256, 257
constant, 231
Erythrosine B, 130
Esters, 23, 108
Ether demethylation, 20
Ethers, 171, 172
Ethylene, 64, 251
EXORCYCLE, 54
Expansion, 70, 140, 184
Extended X-ray absorption fine-structure
spectroscopy, 191-199
Extraction, 4, 9, 26, 28, 50, 124, 158,
204, 214, 219, 224-226

Faraday plate, 132
Fast atom bombardment, 121
Fatty acids, 19, 34, 69
Femtomol analysis, 122
Ferguson plots, 105
Ferric phosphate, 261
Fertilizers, 7, 261
Fiber
lasers, 130
shaped particles, 89
fouling, 229
uptakes, 227
Fibrous structures, 96
Field-flow fractionation, 69, 183
Fingerprint identification, 122
First derivatives, 153-155, 193, 198
Flocculation, 70, 75, 224
Flory-Huggins theory, 230
Flow rate, 132, 180, 181, 205, 243
Fluoranthene, 228
Fluorene, 228
Fluorescence, 10, 13, 15, 32, 34, 73, 80-
82, 84, 130-136, 143, 147, 148, 150, 152,
157, 159, 160-162, 165, 166, 170, 174,
179, 181, 186
absorption, 130
chromophores, 147, 151, 153, 173
decay, 134-137, 143
emission, 10, 11, 13, 15, 80-82, 131-
133, 136, 137, 149-153, 155, 158,
159, 166, 170, 173-176, 180-182,
184, 186, 226
emission bandpass, 158
emission spectra, 159
excitation, 10, 11, 13, 15, 80, 81, 130-
134, 143, 147, 158-160, 170, 173,
175, 180-182, 184
excitation bandpass, 158
excitation lineshapes, 160
excitation spectra, 11, 158, 159, 166
intensity, 81, 136, 150, 159, 160, 162,
165, 166, 173, 179, 184
fluorophore structure, 184, 188
fluorophores, 134, 135, 137, 160, 162,
165
mode, 192
quenching, 224
signals, 160, 166
spectra, 13, 147, 150, 155, 158, 170,
173, 175, 180, 181

Fluorescence, *continued*
  spectrophotometer, 80, 147, 158, 170, 180
  spectroscopy, 158, 166, 173
  structures, 179
  synchronous scan, 10, 13, 15, 158
  total luminescence, 158, 170, 175
  total luminescence spectra, 162-164
Fluorescent substances, 179-182, 184, 186, 188
Focal diameters, 130
Folding, 4, 45, 46, 47, 79, 228
Fourier transform, 5, 194, 197
Fourier transform infra red spectroscopy, 5, 10, 13, 14, 19, 22, 23, 25, 28, 30, 34, 102, 124, 147, 169-172, 174, 175, 192, 208, 210
Fractal, 2, 4, 45, 69, 70, 72, 74, 134
Fractal dimensionality, 44, 45
Fraction, 4, 21, 22, 26-29, 32, 43-46, 64, 66, 73-75, 102, 148, 150-152, 158-160, 162, 165, 166, 217, 218, 224, 225, 231, 234, 258
  collector, 181
Fractionation, 4, 20, 21, 29, 32, 43, 70, 75, 77, 148, 166, 200
Free energy relationship, 229
Free radical, 5, 11, 16, 19, 20
Fullerene, 130, 131, 139, 140
Fulvic acid, 7, 32, 34, 36-38, 41-44, 47, 59, 71, 75, 76, 79, 87, 88, 91, 96, 122, 130, 132, 134, 137, 139-143, 157, 169, 173, 179, 226-230, 238, 258, 260-262
  Armadale, 32, 33, 44
  IHSS peat, 257
  Laurentian, 34, 36, 38, 41
  soil, 43
Function, 13, 31, 43, 45, 63, 110, 111, 113, 134-136, 153, 195, 197, 205, 207, 223, 251,
Functional groups, 20, 35, 42, 45, 55, 56, 69, 70, 73-76, 91, 93, 96, 121, 203, 209, 219, 242, 266
Functionality, 37, 38, 43, 153, 155, 203, 207
Fused silica fibres, 224

Gas chromatography with electron capture detection, 242, 243

Gas chromatography-mass spectrometry, 226, 227, 242, 251
Gas phase, 140
ion mobility, 137
Gauss formula, 149
Gaussian function, 153
GC injector, 225, 227, 236
Gels, 101, 103
  electrophoresis, 101, 103
  fibers, 102, 105, 106
  filtration, 203
Genesis, 63
Gentisinic acid, 108, 109, 123
Geochem-PC, 256
Germanate, 109
Glass transition, 32-34
Globular protein standards, 214
Golf courses, 259
Graphite atomic absorption spectrophotometry, 203
Gyration radii, 151

H_3BO_3, 256
HA hump, 114, 116
Halobenzenes, 248
Hanning window function, 197
Headspace partitioning, 224
Heavy metals, 7, 63, 191, 199, 241, 252
HEPES, 256
Herbicides, 234, 248
Heteroaggregates, 77
Heterocoagulation, 76
Heterocyclic, 188
Heteroflocculation, 75
Heterogeneity, 29, 32, 123, 126, 135, 175
Heterogeneous, 32, 34, 73, 74, 76, 129, 203
  structures, 107
High density polyethylene, 66
High performance size exclusion chromatography, 4, 179-186, 188, 203-206, 214, 216-219
High structural disorder, 134
Higher order derivatives, 153
H-N-H angle, 195
H-O-H angle, 195
Hole-filling mechanism, 229
Homogeneous, 33, 134, 173
Hormone-like activity, 255, 262
Human albumin, 205

Humans, 265
HUMASORB-CS™, 242-248, 250-252, 254
HUMASORB-L™, 242
HUMASORB-S™, 242-244, 246, 247, 254
Humate, 6, 7, 70, 72-74, 76, 252
Humic acid, 1-16, 32-39, 42, 49-51, 54-60, 64, 65, 70-84, 87-98, 107-118, 124, 130, 139, 141, 148, 157, 169, 170-176, 179-188, 203-219, 226-238, 242-244, 248, 251, 253-262, 266-269
    Aldrich, 102, 108, 123-125, 130, 232
    Andosol, 181, 183
    brown forest soil, 181, 184
    coal derived, 33, 34, 252
    compost, 179, 181, 183, 184
    Czech, 116
    Dando, 180, 181, 183, 184
    Fluka, 108, 114, 116
    fraction, 184, 186, 188, 206
    Fujisaki, 180, 181, 183, 184, 186
    molecules, 69, 70, 81, 102, 184
    IHSS, 50, 54-58, 108, 111, 112, 114, 116-118
    IHSS Leonardite, 50, 108, 114, 116, 256, 257
    IHSS peat, 110, 111, 114, 115, 117
    IHSS soil, 88, 108, 116
    IHSS soil Summit Hill, 108
    Inogashira, 180, 181, 183, 184
    Kawatabi, 180, 181, 183, 184
    Leonardite, 50, 79, 102, 109, 114, 173, 244, 256, 258
    Nishimeya, 180, 181, 183, 184, 186
    peat, 108, 170, 173, 175
    soil, 1, 2, 5, 9, 12-14, 16, 34, 110, 173, 174, 179-181, 184, 185, 188, 203, 204, 207-217, 219
    Takizawa, 180, 181, 183, 184, 186
    Tsukano, 180, 181, 183, 186-188
    water, 203
Humic
    colloids, 74
    fraction, 186
    material, 5, 10, 16, 33, 34, 42, 43, 46, 69, 70, 72, 77, 79, 107
    nanoparticles, 74, 75
    organic matter, 223, 224, 226, 229-231, 233-238
    polymer, 79, 81-84
    pseudomicelle, 82, 83
Humic substances, 1, 2, 4-9, 17, 19, 31, 32, 34-36, 38, 44, 46, 50, 51, 59, 63, 66, 69-72, 75-77, 87, 101-106, 108, 119, 121-126, 129-132, 134-144, 147, 151, 153, 155, 157, 169, 170, 173, 179, 182, 183-186, 188, 191, 192, 199, 204, 214, 220, 248, 255-257, 259-262, 266
Humification, 9-13, 16, 19, 123, 171, 175
Humus, 19, 45, 124
Hydrocarbons, 241
Hydrodynamic
    injection, 112, 114, 116, 117
    radii, 74, 213
Hydrogen bonding, 5, 43, 46, 84, 91, 93, 96, 101, 102, 122, 207, 250
Hydrophilic, 22, 69, 71, 76, 79, 82, 84, 101, 147, 148, 227, 229, 230
Hydrophobic, 28, 43, 46, 69, 70, 71, 73, 76, 79, 81, 147, 148, 182, 223, 224, 229, 230, 234, 238, 250, 266
    acids fraction, 148, 150, 151, 192
    domains, 228
    interactions, 233, 236
    organic compounds, 223, 224, 226, 229, 231, 237
    organic matter, 79, 233
    partitioning, 227, 238
    polymer, 230
Hydroquinone, 5
Hydroxyl, 10, 19, 42, 242, 266
Hydroxyphenylacetates, 165
Hypotheses, 121

Image intensifier, 131
Immune system suppression, 265
In situ, 4, 129, 236, 252, 265, 266
Indoleacetic acid, 255
Inductively coupled plasma, 88, 242
Inductively coupled plasma-mass spectrometry, 203-207, 212-219
Industrial polymers, 31, 32, 34
Industrial wastes, 241
Industry, 6, 16, 131, 158, 170, 205, 257, 259, 265
In-fibre derivatization, 234, 236, 237, 239
Infrared spectroscopy, 207
Inhomogeneity, 41, 74, 123

Inner filter effects, 80
Intensity variations, 209
Intermolecular
    aggregation, 228
    interactions, 69, 223
Internal standard, 67, 207
International Humic Substances Society,
6, 7, 50, 80, 108, 257
Ion chromatograms, 215-218
Ion exchange, 233, 241, 247, 250, 252,
253
Ion
    mobility spectrometry, 137, 138
    selective electrodes, 203
    trace ratio, 227, 238
Ionic strength, 70, 71, 74, 76, 79, 87, 89,
91, 93, 96, 182, 214, 237
Iron, 9, 10, 16, 34, 39, 43, 50, 205, 207,
209-212, 216-219, 252, 255-262
    availability, 254, 259, 262
    nutrition, 259
    oxides, 10-13, 16, 266
    translocation, 261
    $Fe(OH)_3$, 256
    Fe^{2+}-FA, 96
    Fe^{3+}, 87, 96, 205, 210, 219
    Fe^{3+}-FA, 96
    Fe-O-Si, 211
    $FePO_4$, 256
    $FeSO_4.7H_2O$, 256, 257
Irradiation energy, 166
Irreversible, 72
Isochratic, 205
Isochromats, 51
Isolated-atom smooth background
function, 193
Isolation, 1, 147, 148, 158, 226
Isotherms, 223, 229
Isotope dilution, 218
Itai-itai disease, 87

Jahn-Teller effect, 197
Japanese Humic Substances Society, 180

Kerogen, 66
Keto, 203
Kidney, 87
Kinetics, 34, 43, 130, 134, 137, 140, 143,
226, 238

Laser
    ablation, 122, 130
    desorption, 121, 123, 129, 137
    Desorption/Ionization (LDI), 110-113,
    118
    energy, 110-113, 118, 132
    spectroscopy, 129, 143
Laser-based ion mobility spectrometry
(LIMS), 130, 131, 137-140, 143
Laser-induced
    fluorescence (LIF), 129, 134-136, 143
    multiphoton ionization, 137
Leachate, 20-27, 29, 30, 256
Leaching, 19, 21
Lead, 20, 38, 40, 49, 50, 79, 107, 137,
204, 205, 207, 209-212, 214, 215, 217-
219, 244, 245, 247, 248, 252, 253
    arsenate, 212
Life cycle cost savings, 253
Ligand exchange, 250
Light scattering, 32, 43-45, 46, 129
Lignins, 2, 4, 5, 20, 23, 26-30, 171, 174
Linear polyelectrolytes, 70
Liophilic, 71, 76
Lipids, 9, 20, 174
Lipophilic materials, 230
Liquid-liquid extraction, 225
Local
    atomic structure, 191
    ordering, 63
    structure, 197
Low cost, 253, 268
Low divergence, 130
Lysozyme, 123

Macroions, 70, 72, 75
Macromolecular structures, 162, 166
Macromolecules, 63, 101, 103, 122, 138,
143, 162, 165, 203, 242
Magnesium
    $Mg(ClO_4)_2$, 205
    Mg^{2+}, 81, 82, 83, 84
    $MgBr_2$, 80, 81
    $MgCl_2$, 80, 82, 83
    $MgSO_4$, 256
Magnetic circular dichroism, 34
MALDI-TOF mass spectrometry, 108,
109, 118, 121-126, 127

Manganese, 9, 16, 203, 205, 207, 209, 210, 212, 214, 215, 217-219, 256, 258, 259, 261
  Mn^{2+}, 205, 210, 218
  $MnCl_2.4H_2O$, 256
  oxides, 9-16
Manure, 261
Many-body factor, 195, 196
Mass
  distribution, 122, 124, 139
  ranges, 123
Mechanism, 6, 7, 42, 66, 76, 80, 81, 101, 102, 121, 184, 188, 194, 251, 259, 266
Membrane
  filters, 158
  permeability, 255
  separation, 241
Mercury, 158, 170, 219, 244, 251, 253
MES, 256
Meshwork of fibers, 103
Mesityl oxide (MSO), 102
Meso-structure, 35, 38, 39, 44
Metals, 7, 63, 203, 206, 207, 212, 217-219, 241, 242, 244-247, 252-254
  ^{27}Al, 206, 212, 214
  As, 204, 212, 219, 244, 253
  ^{75}As, 206, 212, 214
  ^{9}Be, 207
  ^{209}Bi, 207
  ^{114}Cd, 206, 212, 214
  ^{63}Cu, 206, 212, 214
  ^{57}Fe, 206, 214
  ^{59}Fe, 256, 258
  ^{202}Hg, 206, 214
  ^{115}In, 207
  ^{54}Mn, 256, 258
  ^{55}Mn, 206, 212, 214
  ^{65}Mn, 258
  ^{208}Pb, 206, 214
  ^{45}Sc, 207
  ^{64}Zn, 206, 212, 214
  ^{65}Zn, 256, 258
  binding, 39, 42-44, 87, 203, 209, 212, 216-219, 244
  binding sites, 199, 205
  complexation, 203, 211, 219
  humate, 203
  ions, 35, 41-43, 71, 72, 76, 82, 84, 87, 88, 96, 107, 206, 214, 241, 243, 258
  loaded chromatograms, 217

spike recoveries, 219
FA, 96
HA, 5, 7, 191, 203, 204, 207, 209, 211, 216-219
Methanol, 130, 131
Method-related biases, 234
Methyl ether, 23
Methylene groups, 23
Micellar, 2, 69, 70, 73, 79
Micelles, 4, 71, 73, 76, 79, 83, 84
  formation, 72, 82, 83
  like, 79, 81
Micellization, 83, 227
Microbial
  decomposition, 171, 266
  digestion, 241
Micronutrients, 252, 261
Microorganisms, 171, 173, 268
Microstates, 135, 137
Microwave oven, 207
Minerals, 9, 63
  nutrition, 255
  surfaces, 72, 77, 233, 236
Mobility, 32, 64, 66, 102-105, 130, 131, 137-140, 223, 233
Mode-coupling, 130
Model structure, 196
Moisture holding, 268
Molecular
  configuration, 159, 165
  conformation, 88, 93
  fractions, 166, 203, 204, 206, 213-219
  fragmentation, 162
  ions, 121, 125, 126, 137, 228, 238
  radius, 101-103
  size, 73, 79, 101, 103, 105, 159-162, 165, 166, 179, 203, 204, 213
  species, 203, 214, 217-219
  structures, 1, 3, 160, 165
  weight, 13, 32, 47, 63, 69, 70, 79, 87, 89, 91, 101-108, 118, 121, 124, 151, 155, 158-160, 162, 165, 166, 179-181, 183, 184, 186-188, 205, 213-216, 227, 262
Molybdate, 109
Monomer, 31, 32, 34, 83
Monosaccharides, 19
Montmorillonite, 75, 76, 262
Morphological conformation, 88
Multiphoton ionization, 129, 130

Multiple scattering, 191, 197
Mutagen, 265

Nanospheres, 70, 71, 76
Naphthalene, 227-230, 233, 237, 238
Natural organic matter, 21, 147-153, 155, 165, 166, 172, 192, 266
Natural waters, 93, 129
Nearest neighbors, 40, 42, 191, 194
Negative chemical ionisation MS, 231, 232
Nickel, 43, 244
Nitric acid, 204, 207, 212, 219
Nitroanilines, 232, 234
Nitrobenzenes, 232
Nitrogen
  atoms, 199, 203
  contents, 43, 171, 180
  laser, 108, 123, 131
Nitrophenols, 232, 234
Nominal, 121, 123, 162, 180
Non spherical molecules, 101, 105
Non-linear
  isotherms, 34
  effects, 130, 197, 229
Non-uniformity, 191, 199
Normalized peak intensities, 217, 218
Nuclear magnetic resonance
spectroscopy, 5, 20, 22, 23, 28, 30, 32, 34-36, 39, 40, 42, 46, 49-51, 54, 58-60, 63, 64, 66, 67, 73, 121, 147, 155, 251
  ^{13}C, 4, 5, 19-22, 26, 31, 34-39, 41, 43, 49-52, 54, 56, 57, 59, 60, 64-66, 73, 147, 151, 155, 226
  ^{1}H, 5, 22, 34, 35, 37, 41, 49, 51, 56, 64, 66
  ^{15}N, 4, 35
  acquisition parameters, 22
  chemical shift, 35, 55, 56, 59, 64
  CP/T$_1$-TOSS, 50, 51, 55, 57
  cross-polarization magic-angle-spinning, 22-25, 29, 30, 35-37, 39, 41, 49-51, 54-56, 58-60, 73, 147
  dead time, 49, 51, 59
  differential signal loss, 59
  dipolar dephasing, 22, 38, 39, 42, 66
  direct polarization (DP), 23, 55, 63-65
  direct polarization magic angle spinning, 22-24, 29, 49-51, 54-56, 58-60

distorted baselines, 49, 51, 55, 59
echo time, 54
free induction decay (FID), 22
Hahn echo, 51, 53, 55
Hartman-Hahn matching, 35-37, 50, 59
high field instruments, 36
inverse gated-decoupling, 22
magic angle spinning (MAS), 34, 36, 50, 51, 53, 54, 59, 64
paramagnetic relaxation, 39, 40
proton decoupling, 64
ramp-CPMAS, 35-37, 39, 50, 51, 56, 58-60
sidebands, 49, 50, 55, 56, 59, 64
spinning speed, 22, 35-37, 50, 51, 54, 55, 59, 60, 64
T_1^C, 49-51, 55-57
total sideband suppression (TOSS), 49-51, 55-60
Nucleic acids, 4, 32
Nutrient solutions, 255-262
Nutrients, 93, 96, 203, 255

OCH$_3$ groups, 174
Octahedron, 191, 195, 197, 199
Octanol-water coefficients, 226-229, 231-235
Ogston model, 103, 105
OH
  deformation, 13, 172, 208, 210
  stretching, 23, 171
Okefenokee Swamp, 157
Oligosaccharides, 121
Olive-oil, 9
Organic
  analytes, 224
  carbon, 223, 234, 236
  compounds, 7, 21, 70, 73, 140, 142, 143, 173, 238, 241, 250, 251
  contaminants, 242, 250
  materials, 32, 50, 76, 87, 255, 261
  matter, 9-13, 16, 125, 159, 162, 166, 169- 172, 255, 266
  molecules, 49, 140, 152, 159, 166
  removal capacity, 250
  xenobiotics, 266
Organisms, 63, 66, 265
Organophilic, 223
Organotin cation species, 236

Oriented bilayers, 70
Origin, 2, 6, 9, 10, 63, 102-104, 108, 118, 123, 124, 180, 197
Outer layer potential, 74
Ovalbumin, 123
Oxidation, 5, 19, 20, 29, 175, 251
Oxide/water interfaces, 72
Oxidizers, 63
Oxoanions, 241, 254

PAH, 223, 227-230, 238
Paper chromatography, 2
Paramagnetic metal effects, 31
Partial specific volume, 106
Particulate humic organic matter, 223, 224, 229
Partition coefficients, 102, 103, 223-225, 227-231, 234
Partitioning, 79, 82, 223, 225, 230, 234
   behaviour, 229
   mechanism, 229, 230
   process, 231
PCB
   congeners, 224, 231, 232, 266, 267
   humic acid complexes, 266
PCE (trichloroethylene), 243, 248, 250-252
PCNI index, 217-219
Peat, 7, 64, 72, 73, 105, 108, 110, 111, 114-117, 169, 170-172, 174, 175, 261
   Baltic, 102, 105
   Finnish, 102
PEEK™ (polyetheretherketone), 205
Peptide hydrolases, 19
Peptides, 2, 123
Peroxyl, 19
Pesticides, 204, 212, 224, 226, 233, 234, 241, 248
   containing chlorine, 265
   containing phosphorus, 265
pH, 5, 10, 11, 20-22, 43-45, 70- 72, 74-76, 79-82, 87-96, 98, 102, 103, 107, 109, 112, 114, 116-118, 130, 131, 134, 137, 141, 142, 158, 170-181, 192, 193, 204, 205, 214, 227, 234, 236, 237, 242-244, 247, 256-259, 262
Phase separation, 70, 75, 84, 224
Phenanthrene, 228, 230

Phenols, 9, 12, 13, 20, 42, 55, 91, 93, 160, 163, 165, 171, 172, 174, 188, 203, 208, 210, 211, 219, 236, 242, 244, 266
Phenoxy radical, 5
Phenylalanine, 152
Phenylmethylpolysiloxane, 227
Phosphate buffer, 103, 118, 180
Photo-
   alteration, 63
   chemical, 63, 140
   degradation, 162
   excitation, 132
   oxidation, 157-163, 165, 166, 174
   reaction, 173-176
Photon
   absorption, 137, 143
   energy, 132
Photox, 158, 159, 162
Physisorption, 237
Phytosiderophores, 257
Plant, 2, 4, 9, 19, 28, 67, 125, 169, 170, 204, 226, 227, 257, 259-261
   Aspen (*Populus tremuloides*), 19-21, 23-26, 29, 30
   beet, 261, 268
   berries, 265
   cabbage, 268
   carrot, 268
   cell, 2, 4, 19, 20
   cell membranes, 20
   creeping bentgrass (*Agrostis palustris*), 256, 257, 259, 260
   dicotyledonous plants, 256, 259
   greens, 265
   growth, 6, 63, 254-257, 259-262, 268
   growth chamber, 257
   growth enhancement, 255, 260, 262
   growth regulators, 255, 260, 262
   growth stimulation, 255, 256, 259, 261
   leaves, 2, 4, 19-27, 29, 30, 157, 212, 257, 259, 260, 268
   leaves leachates, 2, 26
   leaves size, 259
   melon (*Cucumis melo*), 256, 257, 259
   monocotyledonous, 256
   nutrients, 268
   root, 105, 106, 255, 257, 260-262
   root growth, 257
   root tissue, 261
   senescence, 4, 19, 20, 21, 26

Plant, *continued*
    shoot, 255, 257, 260-262
    species, 256
    terrestrial, 66, 87
    tissue, 19, 23, 257, 261
    turfgrasses, 257, 259, 260
    vegetables, 265, 268
Plutonium, 248, 251
Poisson equation, 93
Pollutants, 93, 96, 107, 224, 230
Poly(methylene) crystallites, 63, 66
Poly(ε-caprolactone), 66
Polyacrylamide gel electrophoresis, 101
Polyacrylate, 80, 224, 226, 233, 234
Polyaromatic hydrocarbons, 265
Polycarboxylates, 261
Polychlorinated
    biphenyls, 226, 227, 230, 234, 238,
        239, 248, 265-267, 269
    dioxins, 265
Polycondensation, 173
Polycyclic aromatic compounds (PACS),
129-131, 140
Polydimethylsiloxane, 224-233
Polydispersity, 70, 72, 74, 75, 77, 79,
103, 184, 203, 213
Polyelectrolyte, 1, 70, 72, 76, 107
Polyethylene glycols, 101-105, 181
Polymer, 31, 32, 34, 47, 50, 66, 69, 79,
81, 84, 89, 106, 224, 243
Polymerization, 9, 16, 19, 69, 266
Polysaccharide, 4, 19, 172, 211
Polystyrene-sulfonates, 214
Porphyria, 265
Potassium
    KCl, 73, 256
    KH$_2$PO$_4$, 256
    K$_2$SO$_4$, 256
Potentiometry, 107
Power flux densities, 130, 139
Precipitation, 70, 72, 76, 77, 242, 261
Primary structures, 31
Productivity, 268, 269
Prolate ellipsoid, 105
Prometryne, 234, 236
Proteins, 2, 4, 19, 31, 32, 47, 121, 123,
135, 152, 203
    folding, 46
Pseudo-chelation, 42-44

Pseudomicellar structures, 79, 82-84, 227
Pullulans, 180
δ-pulse, 134
Pyrene, 80-82, 130, 227, 228, 234, 237,
238
Pyrolysis GC-MS, 147
Pyrolysis mass spectrometry, 32, 121

Qualitative, 39, 41, 47, 60, 138, 139, 143,
194, 199, 217, 219
Quality, 39, 107, 153, 172
Quantitative, 35, 36, 39, 41, 49, 54, 55,
59, 60, 197, 218
Quaternary structures, 31, 32, 46
Quenching process, 81, 143
Quinine sulfate, 170
Quinoids, 5, 160, 165
Quinolines, 237

Radial pair distribution, 194
Radiationless transitions, 152
Radionuclides, 241, 242, 252, 254
Raman band, 151
Random coil, 105, 214
Rate constants, 140-142
R-COOH, 107
Reaction mechanisms, 29, 203
Reactions, 1, 5, 9, 19, 20, 49, 67, 87, 135,
233, 241, 266
Reactivity, 63, 147, 266
Redox potential, 10, 11
Red-shifts, 151
Reducing agents, 63, 251
Reduction, 5, 28, 123, 124, 165, 194,
242, 248, 251, 267, 268
Reductive dehalogenation, 251
Regeneration, 254
Relative fluorescence intensity, 15, 159,
162, 165, 166, 173
Relaxation rate, 73
Renal cortex, 87
Reptation, 101, 103, 105
Reptation mechanism, 101
Resin chromatography
    GPC, 4
    XAD, 4, 21, 22, 32, 148, 188, 226
Resolution, 30, 35, 42, 87, 96, 123-125,
131, 132, 140, 170, 173, 205, 214, 231
Reversible, 72, 76

Ring-fission, 20
River waters, 93
RNA, 121
R-OH, 107
Rose bengal, 130
Rotor frequency, 59

Saccharides, 29
Salting-out, 73
Sampling depth, 171
Sampling time, 227, 228, 232, 236
Saturation capacity, 247, 248
Scanning electron microscopy, 45, 87, 91, 93
Scattering angle dependence, 73
SDS-protein complexes, 101, 105
Second derivatives, 174, 175, 176
Secondary
    effluent, 130
    synthesis, 87
Sediments, 50, 76, 77, 226, 231-234, 236, 237
Segmental dynamics, 63
Self-assembly, 72, 76
Self-organizing system, 31
Semiquinone, 19
Separation, 4, 20, 21, 32, 70, 75, 77, 101-103, 107, 108, 116, 118, 131, 137, 147, 183, 184, 186, 188, 203, 204, 214, 219, 234, 252
    mechanism, 105
Sephadex gel chromatography, 179, 181, 186-188
Sequestering, 212, 228
Serpukhov, Russia, 265, 267
Sewage, 204
Shapes, 4, 13, 89, 96
Signal/background, 214, 215
Silicon, 89, 266
Simulations, 198, 199
Sinapinic acid, 108, 109, 123
Single-photon counting, 134
Site-binding model, 70
Site-specific competition, 229
Size, 10, 21, 43, 56, 69, 70, 73-77, 79, 87, 89, 91, 101-103, 105, 152, 159, 166, 169, 203
Size exclusion
    chromatogram, 213, 215-217

chromatography, 101-103, 203, 205, 206, 214, 216, 219
Size fractions, 75, 157, 158
Small angle X-ray scattering, 70, 73, 74
Sodium
    Na^+, 72, 89, 93
    Na_2CO_3, 108, 116, 117, 242
    Na_2SO_4, 242
    $NaNO_3$, 88, 89, 91-93, 95, 98
    acetate, 214
    pyrophosphate, 102, 122, 204
    tetraethylborate, 234, 236
    tetraphenylborate, 236
Soil, 1, 2, 4, 6, 7, 9, 12, 13, 15, 16, 34, 50, 63, 72, 77, 87, 99, 103, 108, 122, 124, 129, 130, 179, 203, 204, 207, 212, 214, 219, 232, 234-237, 255, 258, 261, 262, 265-269
    acidic, 262
    amendment, 169, 252
    application, 9
    artificial horticultural, 169
    brown forest, 180
    brown lowland, 180
    clay, 204
    clay loam, 204
    clays, 9
    components, 87
    conditioners, 6
    contaminants, 266, 268, 269
    Dando, 179
    decomposition, 66
    decontamination, 268
    detoxification, 265, 269
    forest, 34, 43, 66, 123, 124, 126, 204, 208
    formation, 169
    Fujisaki, 179
    gray lowland paddy, 180
    humus, 19
    Inogashira, 179
    Kawatabi, 179
    leachate, 4
    loam, 80
    macroflora, 267
    mineral, 9
    mineral surfaces, 70
    Nishimeya, 179
    organic, 122-124, 126
    organic amendments, 16

Soil, *continued*
  organic forest, 123
  organic matter, 19, 42, 63, 66, 67, 107, 204, 219, 266
  pH, 107
  porewater, 234
  productivity, 107
  profiles, 66, 123
  properties, 63, 66
  quality, 107
  reclamation, 266
  samples, 123, 18
  sandy, 169
  sandy loam, 204
  silt, 80, 204
  solution, 93, 261
  structure, 268
  surface, 19, 21, 87
  systems, 19
  Takizawa, 179
  treatment, 267, 268
  Tsukano, 179
  water, 21
Solar energy, 157
Solid
  interfaces, 72, 76
  phase microextraction, 224-234, 236-239
Solubility, 66, 73, 80, 87, 107, 109, 122, 123, 131, 207, 224, 229, 230, 231, 233, 242-244, 254, 261, 267
  enhancement, 224
  parameters, 229-231, 233
Solubilization, 73
Solution
  conditions, 70, 72, 73, 76, 77, 79, 82, 84
  viscosity, 32
Solvated electrons, 133
Sorption, 34, 66, 108, 223-231, 233, 234, 236-238, 267
  coefficients, 225, 231, 233, 234, 236, 238
  equilibrium, 231, 236, 238
  kinetics, 227, 237
  mechanisms, 233, 234
Specific polar interactions, 233
Spectral shift, 166
Spherical particles, 89, 91, 93, 96
SPME-HPLC coupling, 239

Spontaneous processes, 71
St. Mary's River, 157
Stability tests, 242, 243
Standard calomel electrode, 109
Stationary phases, 182, 184
Statistical methods, 126
Stern-Volmer constant, 141
Strontium, 244
Structure, 1, 2, 5, 6, 7, 30-34, 42, 43, 46, 49, 63, 64, 66, 69, 70, 74, 79, 81, 87, 89, 91, 93, 107, 121, 129, 136, 147, 155, 162, 165, 170, 171, 195, 223, 242, 266
Sub-picosecond, 130
Sugar, 2, 9, 29
Sulfur, 192, 214, 244
  containing triazines, 234
  contents, 43
  functional groups, 214
Superoxide, 19
Supertoxins, 265
Surface
  area, 241
  features, 88, 89, 91, 93, 96
  tension, 69, 72, 73, 79, 80, 82, 83, 84
Surfactant, 4, 70, 72, 73, 76, 79, 83, 84
$SUVA_{254}$, 149-152, 155
Suwannee River, 32, 33, 43, 147, 149-153, 155, 157, 159-164, 166, 192, 237
Synchronous
  scan, 170, 173-176
  scanning, 166, 170
  spectra, 158, 160, 161, 166
Synthetic chelates, 259
Syringic acid, 174

Tacticity, 32
Tannins, 20, 28
Terbutryne, 234, 236
Terrestrial carbon reserve, 223
Tertiary, 31, 32, 43, 46
Tetraalkylammonium, 237
Tetrabutyltin, 234
Tetracene, 130
Tetrachloroethylene (TCE), 243, 248, 250-252
Tetragen, 265
Tetragonally distorted octahedral structure, 194
Theoretical
  code FEFF8, 192, 197, 198

fits, 197
Thermal energy, 134
Thermodynamics, 226
  equilibrium, 31
Thiol, 214
Third generation X-ray synchrotron, 199
Three-dimensional
  profiles, 87
  structures, 121
Time-resolved
  absorption spectroscopy, 140
  emission (TRE), 131, 136, 137
  fluorescence, 131, 134-136
Toluene, 131, 139, 140, 233
Total acidity, 10, 12, 13, 75
Total luminescence spectra, 162, 165, 166
Total organic carbon, 170
Toxic metal removal, 245
Trace metals, 204, 205, 207, 214, 219, 257
Transformation, 21, 31, 132, 232
Transient absorption spectroscopy, 132
Transition metal, 203, 255, 261
Transmission electron microscopy, 87, 91
Transphilic, 147
Treatment, 4, 6, 9-14, 16, 170, 252-254, 257, 259, 267, 268
Trialkylorganotin, 237
Triazine sorption, 234
Triphenyltin, 234
Triplet
  energy range, 143
  excited species, 140
  quantum yields, 142
  quenching, 131, 140-143
  state energy, 142, 143
  triplet absorption (TTA), 129, 130, 140, 141
Tris-hydroxymethylaminomethane (TRIS), 102, 103
Tryptophane, 135, 152
Tunability, 130
Tungstate, 109
Two-photon excitation, 139
Tyrosine, 152

Ultrafiltered HSs, 73-75, 102, 140
Ultrafiltration, 101, 102, 157

Ultrahydrophilic, 147, 148
Ultrasonification, 88
Uncoated capillaries, 101, 102
Uncoiling phenomena, 166
Uranium, 244
Urea, 109-115, 118, 248
Uronic acids, 28
UV-visible, 2, 10, 73, 102, 103, 108, 130, 135, 143, 147-150, 155, 179, 180, 184, 186, 203-207
  absorbance, 213, 215-217
  λ_{max}, 150, 151, 153, 155, 159
UV-vis-NIR, 130

Vanadium, 251
Van der Waals, 46, 223, 250
Vanillic acid, 174
Vibrational spectroscopy, 121
Vibronic peaks, 82
Volatile organic compounds, 241

Waste streams, 242-245, 247, 248, 250, 252
Waste waters, 9-16, 130, 226-228, 232, 233, 241, 252
Water, 6, 10, 20-22, 34, 35, 42, 63, 70-74, 76, 77, 80-82, 84, 88, 91, 93, 129-131, 133, 136, 151, 157-159, 162, 165, 169, 171, 173, 204, 205, 207, 214, 223, 227-234, 236, 241-243, 245, 247, 250, 252, 253, 257, 265, 266, 269
  black, 157
  brown lake, 130
  holding capacity, 257
  molecules, 72, 195, 198
  municipal, 6
  regimes, 169
  resources, 241
  river, 165
  Suwannee River, 147, 157, 158, 160, 162, 165, 166
Wavelength shifts, 209
Waxes, 23, 29

Xenobiotics, 63, 129, 265, 266
X-ray, 5, 32, 44, 45, 96
  absorption coefficient, 191, 193
  absorption fine structure (XAFS), 5, 199

X-ray, *continued*
  absorption measurements, 192
  absorption near-edge structure
    spectroscopy, 191-194, 196-199
  absorption spectra, 193
  absorption spectroscopy, 191
  diffraction, 34, 46

  energy, 192
  mapping, 89, 96, 98

Zero-valent iron, 243, 250, 251, 254
Zinc, 207, 210, 212, 214, 215, 218, 219, 244, 255- 262
  $ZnSO_4.7H_2O$, 256, 257